Robert S. De Santo

Concepts of
Applied Ecology

Springer-Verlag
New York
Heidelberg
Berlin

301.31

D 441

Robert S. De Santo
Chief Ecologist
De Leuw, Cather & Company
1201 Connecticut Avenue, N.W.
Washington, D.C. 20036
USA

Cover photo: Shallow coastal marsh at high tide, North Atlantic. Barrier beach background, saltmarsh grass *(Spartina alterniflora)* foreground and saltmeadow grass *(Spartina patens)* near foreground. August. Nikon f = 55 mm Micro-NIKOR-P Auto. Ectachrome, 35 mm R.S.De Santo.

Library of Congress Cataloging in Publication Data

De Santo, Robert S
  Concepts of applied ecology.

  (Heidelberg science library)
  Includes bibliographies and index.
  1. Ecology. I. Title. II. Series.
QH541.D47   301.31   77-16051

© 1978 by Springer-Verlag, New York Inc.

9 8 7 6 5 4 3 2 1

ISBN 0-387-90301-1 Springer-Verlag New York

ISBN 3-540-90301-1 Springer-Verlag Berlin Heidelberg

To Polly, Pamela, Amanda, Pumpkin,
Blacktoe, Dusty, and Gummy,
my family for whom I write.

Preface

This book represents the interests and attitudes, the information, and the philosophy that define my work and career as it has evolved over the years. Not written as a substitute for any of the many textbooks on ecology, it is meant to present the simplest and most direct approach to a complex field as distilled out of my work as an applied ecologist, who deals with concrete daily problems in the real-world context of economics, politics, and logistics. I hope that it is useful to the reader who seeks an overview of applied ecology, including sufficient specific detail to make that reader more comfortable with the field and more conversant with the capabilities and limits of ecologists and their tools.

Each chapter is followed by a bibliography which has two functions. The first is to represent the main sources or reviews of information upon which the associated chapter is partly based. The second is to give sources for some of the examples utilized in the chapter and some of the illustrations summarizing and clarifying the text, which have been adapted, cited, or derived, from those references. In that sense, I must most sincerely thank all those fellow ecologists who have preceeded me and who have made my work far more diverse and interesting to me than might otherwise have been the case. I have tried to build upon those who have come before, just as others will build beyond what I can see from our present vantage point.

As references I have listed most of those works that I believe are most clearly written, inclusive, and accessible. Seeking them would create a library of ecology which I believe could serve any serious student well by offering a firm beginning in understanding ecology, environmental management, and the art of applied ecology. To aid that student's self-confidence, an extensive glossary has been prepared that I believe to represent a good, if not fully complete, compendium of ecological argot.

All of the various facets of this book, including the bibliography, conversion tables, and glossary, were derived from my experiences and from those sources of special significance to me in my work and in my own understanding of ecology. Therefore, my appreciation to the authors whom I have cited includes not only a measure for the information and encouragement that their works have given me toward writing this book, but an equal measure for the support that their works have given in my practice of ecology as a consultant.

I also owe a considerable debt to my past teachers, my students (from whom I believe I have learned a great deal), and those of my colleagues with whom I have worked. I hesitate mentioning one because I must necessarily leave out others who are also important to me but are too numerous to list. The late Dr. Paul Warren, my mentor at Tufts University, was the first teacher I had who successfully invited me to think, and I shall forever be in his debt for his encouragement. Next to Dr. Warren, I must also thank those who have questioned, tested, and opposed my ideas. Without the struggle they have caused me to suffer through from time to time, I would not have written this book. Such testing is as clearly unpleasant as it is necessary and I am strangely grateful for those devil's advocates.

I also gratefully acknowledge the encouragement and unflagging support that has been given me by my editor, Mr. Victor H. Borsodi, without whose moral support I doubt this book would have been undertaken and completed. I am singularly flattered to recognize his interest in my work, and I am most grateful to him.

I am pleased to acknowledge the skill and speed of my friend Wayne Morgan, whose drafting and graphic skills I hope have clarified the text. His suggestions and work are much appreciated.

East Lyme, Connecticut   Robert S. De Santo

Contents

1 Introduction

Grant us the courage, the will, and the wisdom to be good stewards of Creation. To neither selfishly waste nor to wantonly destroy the handiwork of nature.

There are as many passions—past, present, and to come—associated with the art and science of ecology as there are with any other human activity. Indeed, our intense emotions are so human that we easily grow blind to that truth which tells us that our thoughts and study, our hopes and our problems, are human fabrications that would simply melt away without us. Of course, such a humiliating consideration should not at all suggest that the natural laws of science would be absent in a world without humans; however, we should see that it is the human alone who thrives on discovering principles, constructing "progress," and conquering those problems which the uniquely ambitious human species seems fated to create and struggle with, while all other creatures seem content with their lots.

The human struggle to conquer nature is a fabled characteristic for which we continue to award great respect. Yet, the real power which we humans have manufactured through the Industrial Revolution makes it vital that we serve as good stewards and become better ecologists than have been our progenitors.

This book seeks to explore the perspective and the hope experienced by one ecologist interested in education and applied ecology. It is offered as a guide to the philosophy of ecology while defining some specific concepts and tools of this science.

There is little dispute that the most effective method of teaching is for one experienced person to guide a student. Tutelage is especially useful when very complex and interrelated studies are required and where the basis of knowledge is either so sketchy as to be hard for a novice to collect without help, or where the knowledge is so voluminous as to be overwhelming to that novice. The art and science of ecology seems to be a most curious dichotomy in which both too little specific knowledge and too much general knowledge exist only to confuse and frustrate the student. It is, therefore, an adolescent science.

Since the early 1960s, many educational institutions have attempted to answer the needs of their students by providing a plethora of curricular options allowing the student to choose, and often insisting that the student choose, a card of courses from different academic departments. The objective has been to make that student an interdisciplinary ecology major. This cafeteria approach to "customized" education has not been successful in efficiently teaching us to be better ecologists and has, in fact, retarded the growth of applied problem solving once those students find themselves stranded in the real world.

Seen from the real-world side of the looking glass, the academically pure curriculum does not reflect dispassionate and constructive teaching. It seems to misfit the earnest and naive student for the shock of confronting a real and unanswered, interdisciplinary (that is, ecological) problem. This sad failure of most educational systems cannot be brought to the altar of any particular priesthood. It comes from the weakness we have for trying to mass produce a product that requires a wisdom not easily transferable and certainly not capable of mass production. The skill needed by a student to make better judgements than can his or her tutor comes with experience in interdisciplinary problem solving.

This book does not pretend to guide the student through a course in ecology. Instead, it has the primary objective of bringing that student to a better understanding of ecologic thought, a better understanding of an approach to the fundamentals of ecology, and a mental set useful to a fuller comprehension of the art and science of ecology.

This book is divided into seven major divisions, including the appendices. Each section is introduced by an essay that is meant to tune the student to a philosophical understanding of the appropriateness of the material that follows relative to applied ecology.

This book means to show that the art and science of ecology is a human construction that has a very substantial philosophic structure. The student of this intricate field must certainly seek support from an interdisciplinary understanding not commonly reinforced by the fragmented courses of study offered in a formal and classical educational system. The alternative has come to be an unusually painful "on the job training"—the only significant classroom for flexing and conditioning all the prerequisite intellectual muscles of an ecologist.

Ecology was defined in 1929 by V. E. Shelford as the science of communities, and today this definition remains as good as any and better than most. It says that ecology is concerned with life and interrelationships. This book explores ecology and those concepts and tools which are used to define this science. It is not meant to be used only as a text or glossary; it has been written to help the student find a path to an understanding of a useful philosophy and regimen for ecologic studies. The book means to accomplish this goal by inviting the student to use the three major tools it provides. The first is a text that sets the stage with descriptions of a philosophy used to manage the main elements of ecology. The text also provides the essential approaches for analysis and ecologic management. The second tool is represented by the literature cited and the bibliography, which have been purposefully selected in order to supplement the text and, more importantly, to support the student in his or her efforts to go further than is possible through these relatively few pages. The third tool consists of the glossary, which reflects the accepted usage of special words in ecology as derived from the literature of the field. It has been drawn exclusively from the literature of the twentieth century and is as comprehensive as time and space permit. It is hoped that it will assist the student to better understand the literature and to more accurately and clearly communicate with other students and professionals in the field.

The definition of an ecologist becomes increasingly illusive and confusing the more concentrated our effort grows in defining one. Because ecology is the science that studies organisms and their relationship to the environment, the ecologist is the person who studies that science. As there are many varieties of chemists and engineers, of psychologists and lawyers, however, so there are many varieties of ecologists. Furthermore, there are not only such specialists in the field as terrestial or aquatic ecologists (marine or fresh water); paleoecologists, plant ecologists, or animal ecologists; and many more, but there are also mathematical ecologists and systems ecologists, chemical ecologists and organismic ecologists. If we look closely enough, we find that no two ecologists are identical, just as no two humans are identical. Yet, many who are interested in this field do not realize that important fact. There is no one course of study that leads an individual to understanding and success. In fact, there are as many right courses as there are wrong or incomplete courses. Perhaps Figure 1.1 can help illustrate the fabric of ecology and the general areas of specialization that may be identified.

Such a maze of choices as shown in Figure 1.1 confront any student interested in ecology. Choices are necessary in order to construct a career and the clearest choices can only be confidently made through experience.

There is no question that some individuals involved have the same kind of gifts of talent in science as do some of those involved in other successful careers. Some people have a talent for music, or sculpture. In science, some have a talent for laboratory work, some for field work. Some have a natural aptitude for theoretical concepts, some for the practical solution of real-world problems. These aptitudes are present in us all, and the trick to becoming a fulfilled ecologist is critically dependent upon seeking to identify and utilize those natural aptitudes which you have to develop within the pyramid of interests outlined in Figure 1.1.

Talents and opportunities to develop those talents are not usually given out. They must be consciously sought by the student.

Each course, each project, each learning experience, must be seen as contributing to some facet of an evolving career focused on ecology and on

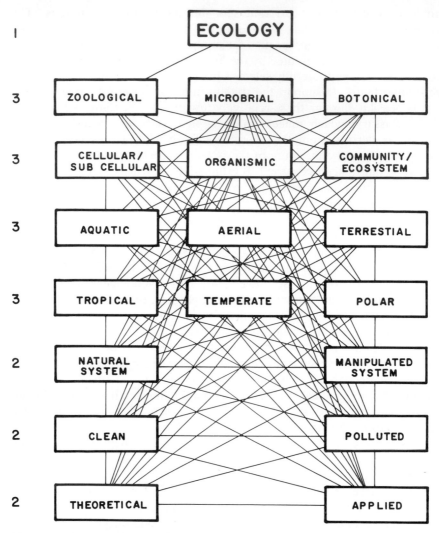

Figure 1.1. In attempting to define the career objectives of an ecologist, many schemes may be used. One hierarchy of career choices is represented here on seven levels. Each choice may be made in combination or independently. Boxes represent special interests and curricular choices and lines represent possible links between these interests within the career of an individual ecologist. There are 64,827 possible combinations of career interest within this one matrix. Ideally, an ecologist should know something of all these choices while specializing in some limited area of particular personal interest.

the budding ecologist's growing propensity for seeking to understand the relationships between living and nonliving elements of the environment. Relatively few of these ecologic relationships are fully understood, and such full understanding is really not the key to being an ecologist. Instead, the true master ecologist is one whose professional career lies balanced within the matrix of Figure 1.1 and whose objectives are to extend the art and science of ecology beyond the existing frontier. Such a person recognizes the need for the research and developments of the theoretical side of ecology, but that person also addresses the need for applied ecology—the frontier of investigating and attempting to solve real-world problems.

Therefore, although this idealized ecologist must be well founded in the sciences in order to master the field, he or she must also be conversant in other fields as well. In particular, he or she must be interested in people, in their behavior, their sociology, their government, their law, and even their religion, because the significance of an ecologist's career should be largely measured by the ability the ecologist has to communicate and understand.

Bibliography

Boughey, A. S. 1971. *Fundamental Ecology*. Intext Educational Publishing, Scranton, Penn., 222 pp.

Chadwick, M. J., and G. T. Goodman, 1975. *The Ecology of Resource Degradation and Renewal*. John Wiley & Sons, New York, 480 pp.

Ehrenfeld, D. W. 1976. The Conservation of Nonresources. *Am. Scientist,* **64**, 648–656.

Graber, L. H. 1976. *Wilderness as Sacred Space*. Association of American Geographers, Washington, D.C., 124 pp.

Hutchinson, G. E. 1970. The Biosphere. *Scientific American,* **223**(3): 44–53.

Karmondy, E. J., 1976. *Concepts of Ecology,* 2nd ed. Prentice-Hall Inc., Englewood Cliffs, N.J., 238 pp.

Odum, E. P., 1971. *Fundamental of Ecology*. W. B. Saunders, Philadelphia, 574 pp.

Odum, E. P. 1977. The emergence of ecology as a new integrative discipline. *Science,* **195**, 1289–1293.

2 Ecosystem Organization and Human Ecology

Each of us perceives our relationship with nature from our own private vantage point, constructed both consciously and subconsciously as we live. Our perception, therefore, is not only quite personal but very often changeable as our experience expands or narrows our lives.

The precepts that humans use in order to direct our view of nature are largely based on our social history. Looking from the vantage point of the society of the western world, as this book does, a strong bias must be dealt with that would otherwise distort an inspection of ecology.

That bias is nurtured by the very fiber of western civilization, and others, which devoutly seeks a false logic that sets us apart from nature. People have always struggled to live in peace with nature, but on their terms and resulting from their conquest of nature. Yet we all feel some uncanny presence of mind that makes us question the wisdom of "man the conqueror." Herman Melville's Ahab would strike out at the sun if it frustrated him, just as he struck out at the great white whale. Ahab was swept away as you might swat a fly. Shelley wrote to capture the same sort of human pride which, frustrated by nature, describes the crumbling remains of a past civilization in some unknown desert. The windworn inscription on a toppled statue reads "Look on my works, ye Mighty, and despair!" The irony of those chiseled words of Ozymandias should serve an ecologist very well as a caution in helping to search for a perception of humans and their true place in nature.

Walt Kelly's Pogo said, "We have met the enemy and he is us!" That is often the case when "man the conqueror" strikes out to manipulate nature in some grand and often misguided way. Yet the ecologist must not preach a return to our primitive beginnings where our human spirit and presence was certainly more digestible by nature than it is today. We are not there now, nor can we ever return unless we are forced to return by catastrophe. Instead, we must all have something of the ecologist's temper in order that we may serve as stewards to our environment. The ecologist strives to insure that we, as human "conquerors," suffer no wanton waste nor make the world ecologically strained and damaged for future generations. The ecologist sees our species as a tiny, albeit decisively important, part of nature.

In order to understand the role the ecologist can play in guiding our relationship with nature, we must explore the main aspects of the language, the science, and the art of ecology. These aspects are not at all mysterious and, in addition, there is no need to look into all the nooks and crannies of the broad array of specific topics present in the field. Samples will serve to guide us through enough detail to gain confidence in understanding what ecology is without becoming intimidated by the science involved. Our objective is to evolve and describe the philosophy of an ecologist as it is reflected in the use of scientific concepts, tools, and problem solving in the field. In this sense, the first perception that any novice ecologist seeks is of the place of humans in nature and the organization of that nature into definable entities, such as the organizational concept of the *ecosystem.*

The ecosystem is a remarkable, and often confusing, concept which may refer to either a geographically large or small area, depending on the context in which it is used and the objectives of the ecologist using the term. An ecosystem is generally defined as including all the parts of a particular environment under consideration. Those parts include both living and nonliving aspects, all contributing in some direct or indirect way to the functions—the complex interacting subsystems—within the space under study. For example, a marsh, a forest, an ocean, or a small aquarium is each an ecosystem. Each one functions with fewer or more subsystems than the others, but all contain both living and nonliving aspects that are interacting and interdependent, and all are relatively self-sufficient as a whole.

Within the somewhat arbitrary boundaries that we usually set for a particular ecosystem, we can naturally expect to include populations of plants, animals, bacteria, and viruses. These aspects of the ecosystem comprise a *community,* or several communities, of interacting organisms of different species which inhabit that ecosystem. A small ecosystem can be made by enclosing a small clump of soil and moss in a sealed jar. Placed in a reasonably warm and lighted area, this microcosm will live and grow for several weeks, if not months. Even as it "dies," the main ecologic principles of ecosystem functioning are still demonstrated, although the particular system may be rapidly changing. In contrast, the largest ecosystem that we can similarly consider is the planet Earth, with its vast numbers of communities of organisms. The living layer of our planet is generally referred to by ecologists as the *ecosphere* or the *biosphere,* but even on so grand a scale it still can be defined by the same fundamental principles of ecology seen in the miniaturized ecosystem described above.

One of the most interesting and quite remarkable aspects of a study of ecosystems is the characteristic they all possess of adjusting to those changes which moderate their overall reactions. This characteristic may be

called *homeostasis*—a maintenance of constancy and uniformity of functions within an organism, a community of organisms, or an ecosystem.
Such a steady state is not as much steady as it is a constant, usually mild,
fluctuation about some average level of functioning. It appears that a
healthy ecosystem, large or small, is remarkably self-regulating based upon
a feedback of information, of populations, of limiting forces, which control
the living and nonliving aspects of that ecosystem. This condition provides
a greater or lesser degree of homeostasis (i.e., equilibrium) within the
ecosystem and between the main constituents of that ecosystem.

Those constituents include all the distinct pieces within the "body" of
any ecosystem and they consist of the *biotic,* or living, and the *abiotic,* or
nonliving, parts of the puzzle. The biotic components include all the
organisms that inhabit the ecosystem and the abiotic components include
all the rest—the physical environment and the chemical elements and
compounds that are not a part of the living world but upon which that life
depends.

A basic concept necessary to successfully negotiate a course toward
understanding ecology is that although our most immediate tendency is to
study the living aspects of a system, we must be disciplined into contem-
plating the nonliving factors with equal weight and respect. Both components are clearly interrelated. Indeed, the abiotic environments of air,
water, and land serve as the broadest division of the biosphere into three
levels which can exist as stable entities in the absence of life. The converse
is obviously not the case. Biotic environments, life as we know it, cannot
persist in the absence of land, sea, and air.

In summary of these basic precepts, we can formulate two laws to help
guide us.

1. There can be an infinite division of the biosphere into smaller and
   smaller ecosystems, down to the least single living unit within its living
   space.
2. The finite separation of one ecosystem from another is a convenience
   used by ecologists in order to simplify our study but it is artificial and
   does not necessarily reflect the true interdependence of real ecosystems.

Obeying both laws, we can define an ecosystem as a part of the biosphere
that is more or less distinct from contiguous parts of that biosphere. We can
further define that ecosystem as possessing four basic elements.

1. Abiotic constituents, such as oxygen, water, nutrients, and light
2. Biotic constituents, such as plants, animals, and microbes
3. A source of energy and its utilization to power the ecosystem
4. Renewable nutrient input to supply food to the biotic constituents

Within most ecosystems the biotic constituents may be divided into those
species which can manufacture their own food and those which cannot.
The *autotrophs* are self-sufficient because they can carry out the physiolog-

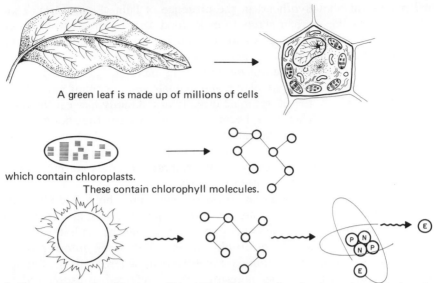

A green leaf is made up of millions of cells

which contain chloroplasts.
These contain chlorophyll molecules.

A photon of light from the sun collides with a chlorophyll molecule causing an electron to be freed from one of its atoms.

The electron is transferred within the chloroplast interacting with ADP molecules.

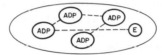

The process gives each such ADP molecule sufficient additional energy to transform it into an ATP molecule.

$$\boxed{ATP} + H_2O + CO_2 \longrightarrow$$

This energized molecule of ATP, when added to a living, cellular system in the plant containing water and carbon dioxide, will react to yield sugar and oxygen. The ATP is reduced in its energy level back to ADP.

Figure 2.1. The process of photosynthesis is the basis for all life. Directly or indirectly, green plants provide the food which all other forms of life rely upon for their existence. Only green plants can convert the energy of light into the stored chemical energy of organic chemicals (ie. food).

ical process of photosynthesis in the presence of light (Figure 2.1). That process allows these green plants to make food, that is, sugar, from water and carbon dioxide. As far as is known, there are no other means in nature whereby the stored chemical energy of food is created from basic nonorganic chemicals through the utilization of light energy.

Heterotrophs are all those plants and animals which cannot make their own food and which, therefore, must directly or indirectly harvest the food made by autotrophs. Therefore, because the sun provides the energy in the form of light, which permits the autotrophs to grow and make food, it is the sun that allows most ecosystems to continue. Like the hand that winds the spring of a clock, directly or indirectly the sun provides the external energy supporting the life activities within all ecosystems.

Capturing the sun's energy and transforming it into the stored chemical energy of food, the autotrophs are the *primary producers* of an ecosystem and they are usually consumed (eaten) by at least some of the heterotrophs in that ecosystem. Those heterotrophs feeding solely upon autotrophs are referred to as *primary consumers* (vegetarians), whereas those heterotrophs which feed upon other heterotrophs are referred to as *secondary consumers* (Figure 2.2). Heterotrophs, such as fungi, which live on dead and decaying organic material are referred to as *decomposers* and may be plants (saprophytes) or animals (saprobes). These decomposers are secondary consumers because they rely upon autotrophs or upon other heterotrophs for food. In extracting materials and energy from that food, decomposers tend to break down the constituents of that food into its simpler elements. Such disassembly of dead material tends to remobilize and recycle the basic constituents of ecosystems.

At each generalized phase in the transfer of the sun's converted energy through an ecosystem, some of that energy is lost as heat, ultimately to be radiated back into space. However, the sun provides much more energy than can be utilized within these systems, and provided that no other factor limits its life, the system will tend to function in an ecologically orderly fashion. Energy passed from the sun to the autotrophs, and from the heterotrophs to the decomposers, represents the accumulation of a mass of new organic material. This may be thought of as the *biomass sequence* within an ecosystem. The disassembly or breakdown of this accumulating biomass within an ecosystem by the resident decomposers of that ecosystem, may be thought of as the *recycling sequence*. In it, biomass is broken up into its components, which become the raw material, the *nutrient pool* necessary for the sustenance of the autotrophs; autotrophs must draw from that nutrient pool in order to indirectly utilize the sun's energy. The biomass sequence is *anabolic,* because material is built up. The recycling sequence is *catabolic,* because materials are broken down. Both are necessary to establish a system that is ecologically healthy and stable.

Recognizing the existence of interdependent cycles of anabolism and catabolism in a healthy and stable ecosystem is very fundamental to

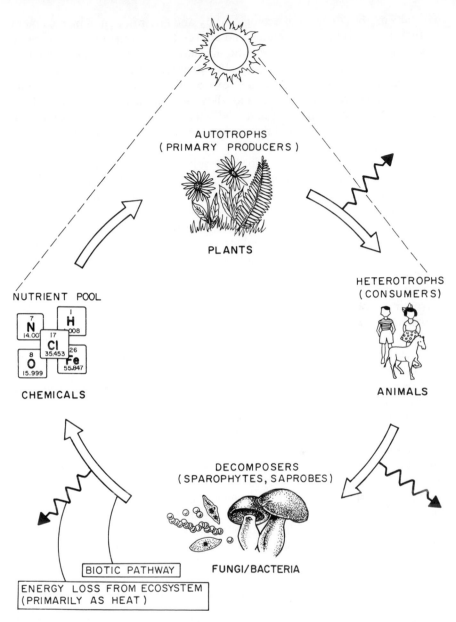

Figure 2.2. Within any ecosystem, there exists a food and energy cycle, which must be interdependent with numerous other ecosystems as illustrated here. Energy which is external to a particular system must provide food or energy to keep that ecosystem functioning. Light energy from the sun is the root source which ultimately supports (powers) all ecosystems.

virtually all the concepts of ecologic study and environmental management. In fact, the recycling of elements and energy through the biotic and abiotic constituents of an ecosystem can be divided into quite discrete sequences in order to simplify an examination. For example, the biomass sequence in a typical ecosystem may include several distinct levels of interactions. These can be readily defined and they lead to a consideration of the dynamic balance normally maintained between the energy extracted from the sun by an ecosystem and the consequent multilevel productivity of that ecosystem.

Charles Elton, an English biologist born in 1900, is generally credited with forming the basic principles of modern ecology. In particular, his investigations into population numbers and their relationship to food led to his elucidation of certain key concepts that stand today as useful and clear as they did in 1927, when he first published them in a textbook.

In considering the interrelated populations of autotrophs and hetero-trophs in an ecosystem, Elton defined a pyramid of numbers (now often called an Eltonian pyramid) which essentially states that those organisms which represent the bulk of the food in an ecosystem are at the base of a pyramid of relative mass (Figure 2.3). Organisms that feed on this base represent fewer in number and a smaller gross mass but a larger individual size. Other levels on the ascending pyramid represent other organisms of the ecosystem, each successively fewer in number and larger in size, but smaller in the aggregated mass.

For example, if we examine an ecologically sound lake and identify all the autotrophs and heterotrophs and their respective total biomasses, we can construct the Eltonian pyramid in Figure 2.3.

At the base of the pyramid we would usually find a variety of autotrophs converting a small fraction of the sun's light energy, and some available inorganic nutrient chemicals in the lake, into the organic material the excess of which provides food for the next level on this food pyramid. If we assign an arbitrary value of 100 biomass units to this autotroph level, the base level, we find that the next level, the heterotrophs, which require the autotrophs for their food, have a collectively smaller biomass of about 10. These heterotrophs, the primary consumers, are usually larger than the autotrophs (the producers) in individual size and usually feed directly upon the producers. They are, therefore, vegetarians or *herbivores,* whether they are cows in a field or microscopic animals feeding on microscopic plants (algae) in an experimental lake.

If we then examine the next level of this food pyramid, we are likely to find a consumer that lives by depending upon the biomass produced by the primary consumer level. In that case, these heterotrophs would be labeled secondary consumers and carnivores because they would feed upon the primary consumers. Furthermore, we would find that the total biomass of this third level of the food pyramid would be 100 times smaller than the first level, the producers, and 10 times smaller than the second level, the primary consumers.

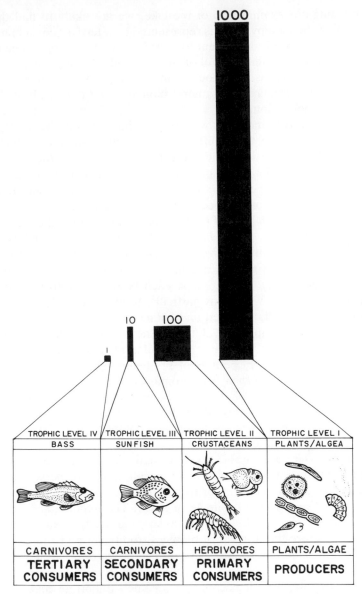

Figure 2.3. The Eltonian (Food) Pyramid is a concept which relates to consumption efficiency. This diagram reflects that concept relative to the fact that the biomass of one trophic level requires ten times its equivalent mass in food provided by the next lower trophic level. Therefore, one unit of bass biomass requires ten units of sunfish biomass as food which, in turn, requires one hundred units of crustacean biomass, and so on.

Continuing our examination of the lake, we are likely to find that one more level of the food pyramid is represented by a tertiary consumer or top carnivore. That level is dependent for survival upon the biomass generated on the secondary consumer level of the pyramid. When present, the total biomass of this level is one-thousandth of the producers' biomass, one-hundredth of the primary consumers' biomass, and one-tenth of the secondary consumers' biomass.

Based upon this framework, we can examine not only the interrelationships present between these different levels but also those which exist within any one of the levels. Obviously, one level is either directly or indirectly very dependent upon other levels. A particular species of plant or animal is also interdependent within the family of organisms that may occupy the same level of the food pyramid. This interdependence, and the definable nature of each level of the pyramid, allows us to label each step in the pyramid a nutrition or *trophic level*. That is, each step or level in the food pyramid may contain several species which all secure their food (nutrition) in the same general manner. All producers, all herbivores, all carnivores, and all decomposers may each be lumped into particular and different trophic levels within any particular food pyramid. Furthermore, as a general rule in a fully matured and balanced ecosystem, the transfer of biomass from one trophic level to the next higher one is about 10% efficient. This essentially means that under ideal natural conditions, if primary consumers eat 10 pounds of producers, they might gain one net pound in the increase of their own biomass.

In most instances, the relationship of one trophic level to others is not as clear and definable as it may be in this hypothetical example. The food pyramid is fabricated to help us understand complex interrelationships, not to duplicate real world events.

Certainly, there are trophic levels of activity within every ecosystem, but the organisms at one trophic level are physically dispersed and inter-mixed with organisms from other trophic levels. As we will see, no stable and "healthy" ecosystem is simple or small. The hypothetical food pyra-mid exists only as an initial aid to understanding. It is somewhat more realistic to picture the transfer of biomass (food energy) from one trophic level to another by developing a diagram of predator/prey interdependen-cies to better understand a particular ecosystem. Even when greatly simpli-fied, for example, by not accounting for such important and ubiquitous facets as insects, microbes, and the decomposer trophic level, a *food web* remains a formidable condensed expression of the dynamic functioning of an ecosystem (Figure 2.4). At the least, it leads us to the realization that it would be most unusual for any one species of organism to be limited in its interdependence to only one other species. That web of interdependence may center upon food but, whatever the specific circumstances, interde-pendence usually is an expression of some form of *ecological competition* between the organisms of an ecosystem. Whether active or passive, that

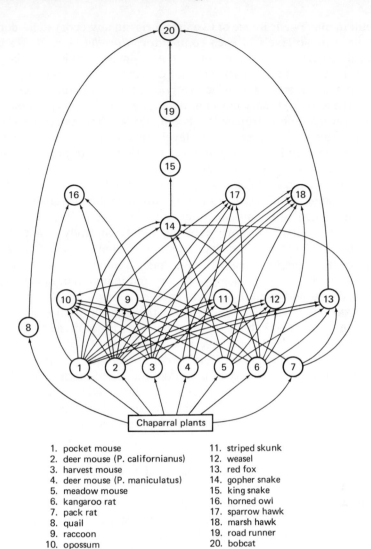

1. pocket mouse
2. deer mouse (P. californianus)
3. harvest mouse
4. deer mouse (P. maniculatus)
5. meadow mouse
6. kangaroo rat
7. pack rat
8. quail
9. raccoon
10. opossum
11. striped skunk
12. weasel
13. red fox
14. gopher snake
15. king snake
16. horned owl
17. sparrow hawk
18. marsh hawk
19. road runner
20. bobcat

Figure 2.4. The basic food web of the Chaparral Ecosystem (from Boughey, 1971b) deliberately omits in this diagram the shrub-mule deer-cougar food chain. It is sometimes absent in the field because of human disturbance of this ecosystem resulting from the hunting of both animals. Hunters from the middle Pleistocene are believed to have disrupted yet other food chains of this ecosystem by exterminating such animals as sloths, camels, and horses and thus causing their predators such as the saber-tooth "lion" to pass into extinction. The "undisturbed" chaparral ecosystem of today has been vastly and irreversibly changed by approximately 20,000 years of human occupation. (Note that insects have been omitted from the diagram. Each of the listed animals probably eats one or more insect species at some stage in its life cycle.)

competition may occur in one of two categories in any ecosystem, depending on the trophic levels of each competitor. *Parallel competition* is the ecological struggle that may occur within a single trophic level between different species seeking the same food. For example, if cows compete with deer for the same grasses on which to graze, there is parallel competition between these two primary consumers. *Perpendicular competition* exists when organisms on one trophic level compete for their existence with the organisms from some other trophic level. In a balanced ecosystem, this form of competition is usually expressed by the predator/prey relationship of higher trophic level organisms preying on lower trophic level organisms.

Many of an applied ecologist's most trying problems deal with unraveling and managing the aberrations of competition created within ecosystems by humans. Such problems are often caused by our intervention with technology into otherwise slowly evolving and dynamically balanced biotic and abiotic recycling within ecosystems. This seems a very fundamental problem because the technology that we have created is entirely foreign to the natural world. That technology of tools and innovation, of science and art, of medicine and even of war has evolved along with our species and represents a time span of only about one thousandth of all biologic time. The past three and one-half billion years has seen life evolved to its present state on the earth. The human race is a latecomer.

The simplest logic for an ecologic perception of our present technologic problems rests on a sensitivity to a foundation in evolution upon which all life rests. Life has evolved slowly relative to humankind's inherent ability to invent and thus "evolve" our miraculous technology from Coca Cola to DDT and from the wheel to the transistor. The repercussions that often result when our technology manipulates the otherwise natural world, stem from the fact that many of the products of that technology are as alien to the natural cycles of our world as if they had been delivered here from a dissimilar universe.

Whatever the change, an ecosystem can accommodate the change provided the integrity of its trophic levels is maintained and energy continues to flow through the ecosystem in an orderly, balanced manner.

Furthermore, the rate of that energy transfer and its utilization should be examined in order to better understand some more of the stability and evolution of an ecosystem. Such consideration relates to ecologic *productivity* and the associated energy flow within ecosystems.

When we examine productivity, we are confronted by two clear categories that we must consider. One relates to the productivity of the autotrophic elements of an ecosystem and the other relates to the heterotrophs and their productivity. In either case, a measure of *calories* present (the amount of heat energy needed to raise 1 g of pure water under 1 atm of pressure $1°C$) may be used in order to summarize the biomass produced and/or the heat energy expended. That is, any ecosystem contains some amount of biomass which, through its collective life processes, generates some heat that is lost to the environment. If all this lost heat were trapped and if all the biomass were used to fuel a fire to heat water, the number of calories represented by that particular ecosystem could be recorded. That record would provide some information about the amount of energy potential represented by that ecosystem and it would also provide a means of making comparisons between similar and dissimilar ecosystems relative to their productivity.

With this concept of caloric values in mind, ecologists can report three basic measures of productivity with respect to an ecosystem or its subsystems. For an ecosystem of autotrophs, we can calculate the total energy expended by that system in living as well as the total energy stored in the biomass of the system. Because autotrophs are primary producers, as previously defined, this present measure of the total energy of this system is called *gross primary productivity*. Estimates of this measure for the major ecosystems found on Earth have been made which clearly indicate, as one should logically expect, that not all ecosystems are equally productive. In fact, some ecosystems are immensely more productive, relative to their total caloric content, than are others (Figure 2.5). Furthermore, those ecosystems which are most productive are usually those relatively more resistant to irreversible ecologic disruption by man. We will consider this aspect of ecologic balance and stability further on in this book. In any event, gross primary productivity is reported as a rate at which an ecosystem, occupied by a single or a group of primary producers (i.e., autotrophs), generates energy as heat or food measured as calories per unit area per unit of time (e.g., cal/m²/day).

Another measure of the same or some different ecosystem may be reported in which the calories lost from the system as heat are not reported. This results in a figure lower than the gross primary productivity (which includes heat) that is called the *net primary productivity*. Therefore, if we were to collect all the green plants grown in a field at the end of one growing season, including their roots, and if we oxidized (i.e., burned) all of that harvested biomass, we could calculate the net primary productivity which that field of plants provided per unit of time.

Another useful measure of productivity is *secondary productivity*. It may be calculated as gross or net values and refers to trophic levels other than the producer level of green plants. Obviously, gross and net secondary productivity must be evaluated with as much care as is gross and net primary productivity if one wishes to evaluate the flow of energy through any ecosystem. However, this is often a very difficult task because ecosystems are usually complex mixtures of producers and consumers. Any one ecosystem will contain all the concomitant integrations of primary and secondary consumers, which makes a calculation of productivity very difficult.

Some simplification of the ecosystem concept has been sought by defining a group of subunits that can be carved out of an ecosystem, thus simplifying our examination of it. These subunits are labeled *communities*. Each community represents a different and definable functional unit of

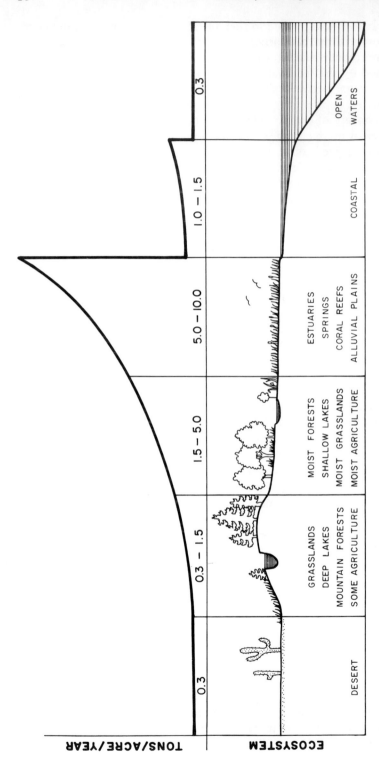

Figure 2.5. The world distribution of primary productivity is illustrated here based upon data reported in the literature. The ranges reported reflect instances of manipulating ecosystems naturally and artificially causing productivity to be greater or lesser depending upon the particular circumstances within a particular ecosystem. (Adapted from Odum, 1963.)

autotrophs and/or heterotrophs that may be found within a particular ecosystem and each is usually named for the dominant or most conspicuous biotic or abiotic aspect of that community. For example, the intertidal ecosystem, which is located between the extreme high-water level and the extreme low-water level along any marine coastline, may be divided into a number of ecologic communities. These communities may be present singly or in any combination within the ecosystem as a whole. Intertidal communities may include the marsh community, the tidepool community, the sandy and/or rocky beach community, and the mud flat community. These are identified by the name of abiotic characteristics which each has. Others are labeled by conspicuous biotic characteristics, such as the sagebrush community, the oyster bed community, or the chaparral community, for example.

Although the division of ecosystems into community subdivisions is useful for calculating specific ecologic productivity or for predicting specific environmental impacts that may result from some particular activity, it is important to recognize that no community, in fact, no one ecosystem, is wholly separate from its contiguous communities or ecosystems. Energy flows from one to the others in various forms and the overall ecology we study is something of an algebraic summation of all productive and consumptive processes within the whole system. Therefore, in many instances the actual difference between ecosystem and community is a relative one. Ecosystems are usually larger and contain more than one community system, each being a major or a minor community depending upon the dependence of a minor community for a major community and for the provision of energy.

By examining various communities, it is possible to investigate the relative productivity of these systems, a measure that summarizes all of the ecologic relationships within each community when taken out of the context of a larger ecosystem. Knowing whether a community is contributing to the productivity of an ecosystem, is an important facet of environmental management and such basic information aids greatly in our effective practice of sound conservation.

One means of measuring this overall productivity of a community is reported by comparing the amount of energy required by that community to maintain its existence with the new biomass that is generated by the autotrophs of the community. *Respiration,* defined in the next chapter, is a measure of the energy needed to run the community, whereas biomass increase is a measure of net production.

Both these measures may be reported as grams of biomass produced (P) and grams of food consumed through the respiration associated with the community (R). A community in which production is greater than respiration ($P/R > 1$) is autotrophic and will be growing in mass (Figure 2.6). One in which respiration is greater than production ($P/R < 1$) is heterotrophic and will be declining in mass unless some external source of energy is provided in order to subsidize these energy requirements. A community in

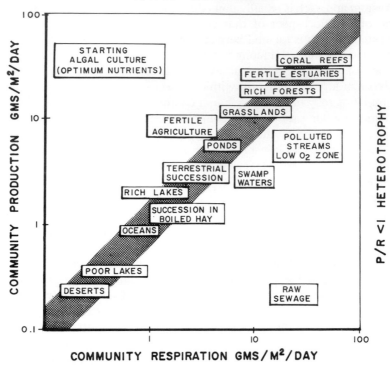

Figure 2.6. The relationship between the productivity of a community and its overall consumption can be based on an analysis of the photosynthetic production (P) as compared with the respiratory consumption (R) of well-defined communities such as those plotted above. If this sort of production is greater than this sort of consumption, the community is autotrophic and more potential food is generated than is consumed. The communities which lie along the diagonal line average an equilibrium between production and consumption over a yearly cycle. In general, as a young and growing community tends to reach a climax in ecological succession, it tends toward a balance between P and R. (Redrawn from Odum, 1956.)

which production and respiration are equivalent ($P/R = 1$) has a net biomass that is neither losing nor gaining. This represents a stable and balanced community, examples of which are fully matured communities developed over a long period of time. Changes in such a stable community tend to ecologically encourage the establishment of a more mature community and discourage establishment of a less mature community. The progression of orderly sequences, seen within the context of a natural, ecologic community, relates to both the evolution and relative productivity of the community through a conceptualized process known as *ecologic succession*.

We can safely assume that any observed ecosystem with its contained communities has not existed unchanging from its inception. From examining numerous ecosystems, we can theorize that each one has evolved through a series of overlapping communities. One has developed into the next with imperceptible boundary lines between adjoining stages. Each stage is labeled a *sere* or *seral stage* and each represents a community evolving into some other seral stage. However, it appears that this pattern of ecologic succession is not endless and that there is a relatively stable and final seral stage, known as the *climax sere*.

It is often difficult for us to confidently conclude that a particular sere is a climax because ecologic stability is so relative a concept. Therefore, it is more important for us to understand the concept of ecologic succession and its climax sere than it is to search for the perfect example in the real world. If obtainable, that climax would be a community with productivity and respiration balanced, able to perpetuate itself forever provided that all physical and biologic conditions also were perpetuating.

Both large and long studies, as well as small and short studies, can be used to illustrate the principles of ecologic succession. Each such study is an important exercise as a means of relating the ecologic principles of succession within a forest over a period of 150 years to identical principles within a protozoan culture over a period of 90 days. In the first instance, we can quite reliably predict the seral stages through which a cleared plot of land will evolve, such as may result after a cultivated field of corn, for example, is abandoned in the Piedmont region of the southeastern United States (Figure 2.7). Beginning with the bare soil of the fields, grasses become established from seeds carried on the wind. This early seral stage contains both plants and animals that are able to encroach and establish themselves in this new and barren area. They are called *pioneer species* and eventually are replaced by other species defining the next seral stage. In our example, we can expect shrubs to appear with the grasses within the first 20 years after abandonment by the farmer. Apparently because of changed soil conditions and the natural dispersion of seeds of all types from contiguous areas, we are likely to see the appearance of pine trees in the grass–shrub field under consideration. In that case, a pine forest will probably develop within about 100 years. This community is not the seral climax, however. Deciduous trees are generally fast growing, and because they usually produce dark shade their presence near a pine forest usually signals the eventual extinction of the pine trees. Pine trees are generally shade sensitive and in competition between pine trees and deciduous trees

Figure 2.7. The idealized process of secondary succession is diagramed above. It leads from a bare field, such as that of a farm abandoned just after it has been tilled, to the ecological climax forest. It is based upon a diagram and the description of these ecosystems of the piedmont region of the southeastern United States. (From Odum, 1971.)

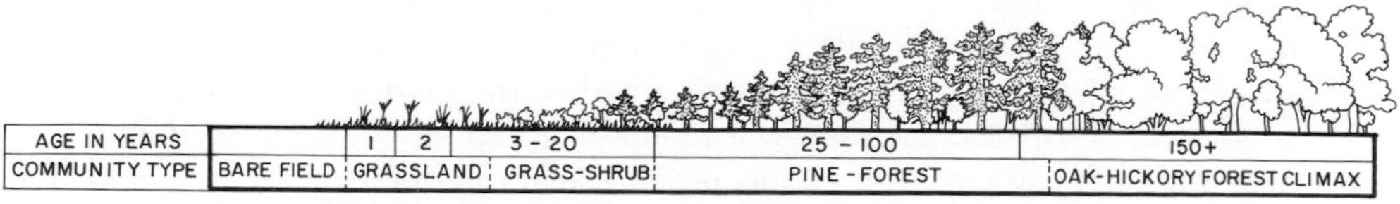

AGE IN YEARS		1	2	3 – 20	25 – 100	150+
COMMUNITY TYPE	BARE FIELD	GRASSLAND	GRASS-SHRUB		PINE - FOREST	OAK- HICKORY FOREST CLIMAX

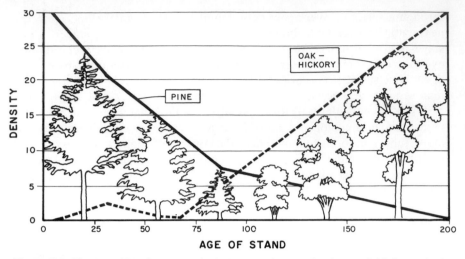

Figure 2.8. The transition from one seral stage to the next is often an initially gradual process which soon becomes geometric as the transition accelerates. The slope of the graphs here show this principle. The relative size of each tree represents the tree's density not its height. Age is given in years and density is given in individuals per hectare. (Redrawn from data presented in Oosting, 1942.)

for light, all other things being equal, the deciduous trees will win and shade out the pine trees (Figure 2.8).

When the ecologic succession is taking place in an area where a preexisting community has been removed and where contiguous areas contain native biologic material, we call the progression *secondary succession*. The abandoned field study is an example. In contrast to that, *primary succession* is used to describe the ecologic events that create a succession of seral stages within areas not previously occupied by any community. Present day volcanic activity is perhaps the best example of large-scale primary succession.

For example, in the summer of 1883, Krakatoa, a volcanic island 25 miles west of Java, exploded, engulfing the land in lava and gas. Once cooled, Krakatoa existed as a sterile land upon which ecologic studies would record the pathways of primary succession. Within about 50 years after the explosion, more than 1200 species of animals were found to have colonized the new land, having arrived there by air and sea.

Somewhat the same principles of primary succession can be observed in simple laboratory experiments, which record the successive dominance of different organisms in cultures of protozoan communities over relatively short periods of time. If a sterile broth is inoculated (i.e., seeded) with a few drops of raw water, an ecologic succession usually results that illustrates the nature of seral stages within a limited environment (Figure 2.9). Those species of protozoans which are best suited to the existing environmental conditions at a particular time will increase in biomass and, in so doing, they will alter various factors and prepare that environment for the next seral stage. The propensity of one group or species to succeed means that

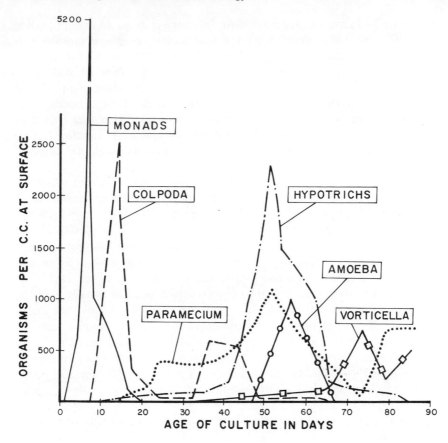

Figure 2.9. Ecological succession in a protozoan culture illustrates the dominance of essentially one species at a time. As ecological conditions age and change, different niches are created which allow the dominance of one species over another. (After Woodruff, 1912.)

they will dominate the other populations that are present but less suited to the particular environment now present. In many respects, the *pioneer species* which arrived on Krakatoa prepared the way for many of those which followed.

The essential difference between primary and secondary succession is that the abiotic environment in primary succession has received no prior biotic contributions, whereas in secondary succession, the abiotic environment has received prior biotic contributions. Therefore, secondary succession usually proceeds at a much more rapid pace than does primary succession.

However, regardless of how well we understand the specific nature of the succession or lack of it within any ecosystem or within any community that may be a part of that ecosystem, we must also seek an understanding of the balance between the spaces and functions that comprise a community. Space and function are other concepts necessary to an understanding of the philosophy of ecology. We know that ecology guides us in a study of living organisms and their relationship to the environment. So far we have looked at the ecosystem and the community concept. We must also examine the ecologic concept of the individual and its relationship to the community and to the ecosystem. This examination leads to an understanding of where an organism lives (its living space) and what that organism must do in order to succeed at living within its space (its function). Two concepts have been developed that can assist in the ecologic definition of these two avenues of interest.

In the first and more simple instance, the ecologic space in which an organism lives within a community is called the *habitat* of that organism. It is actually that physical part of the community structure in which the organism finds its home. The habitat is, therefore, the sum total of all the environmental conditions present in the specific place occupied by an organism. This concept is often referred to as "the address" of an organism, the physical place to which you would go were you to search for the organism in question. The concept may be expanded, and usually is in the literature, to include all the organisms of a particular species, not just a single individual. Furthermore, in those instances where a community is not very diversified, it is likely that the habitat of an organism within it will encompass the whole community. The habitat of a particular species of clam may be the whole strand of intertidal sand along a beach. The same beach may simultaneously be the habitat for all the other organisms of that sandy intertidal community as well. The habitat of a fox, in contrast, may be a den within a much more diversified upland forest community. The fox may be relatively alone in its habitat.

It must be well understood that habitat is basically a physical concept and that the total existence of any organism requires an understanding that extends far beyond the habitat concept of community structure. That ultimate in ecologic species-specific concepts is the *niche*. It is defined as the total expression of all ecologic conditions, reactions, and capacities that describe the total functioning of an organism within its habitat. The first suggestion of the niche as an ecologic concept seems to have been presented in 1917 by Joseph Grinnell, who investigated "The Niche Relationships of the California Thrasher." He concluded that the range of activities and distribution of this bird were limited by a narrow phase of conditions associated with the western United States chaparral ecosystem of low dense scrub vegetation characterized by shrubs and dwarf trees, mostly evergreen and often with hard leaves, such as oak and buckbrush.

The niche concept was further developed by Charles Elton in 1927, when he considered what an organism was "doing" in a community, rather than what it looked like, to be that organism's niche. In 1934, G. F. Gouse cited Elton's earlier work and further explored the definition of the ecologic niche of a species within a community. The most inclusive discussion of the niche concept, however, appeared in a 1957 paper by G. Evelyn Hutchinson. He emphasized the concept as a multifaceted definition of the unique ecologic facets of any one specific species. Taken together, then, an ecologic niche is a three-dimensional figure with a virtually limitless number of facets, each representing one ecologic feature of the single species that is uniquely described by that niche. Therefore, if we seek to understand the full meaning of ecology, we must seek to understand all its parts. Understanding the niche concept is central to that understanding. Yet in truth, we deal with the total niche concept only as an abstraction. We are limited because we do not understand many of the details of any organism's life and because we cannot effectively depict all the multivarient factors that comprise the life and reactions of any one species. Nevertheless, the ecologic niche concept may be imagined as a graph of axes, each showing the narrow or broad range of conditions and capacities that define the total life of an organism.

If we could plot all the facets of an organism's life, we would have a picture of the organism's niche. How do we draw multiaxial figures when there are at least many hundreds, if not thousands, of information categories that must be gathered and presented if we are to define the total ecologic processes and interrelationships of one organism? Most ecologists do seem to agree that the niche concept, however abstracted because of our present-day, incomplete knowledge, is envisioned as a three-dimensional figure with an internal focal point. That point is the organism living within the biotic, abiotic, and behavioral limits of the figure. Without trying to set finite axes to define the limits of each niche facet, we should still envision any niche as a unique three-dimensional solid. As such, it can serve a very real need which exists for a more concrete and clear concept than seems possible given the technical and intricate examples present in much of the current literature.

However, bearing in mind the need to avoid oversimplifying, yet seeking to more clearly depict the ecologic niche concept, we can consider the following views. The total existence of any living organism can be sorted into three categories. The first category contains all those biotic and abiotic facets that completely define an organism's *life of internal interactions*. The second category contains all those biotic and abiotic facets that completely define an organism's *life of external interactions*. The third category is represented by a pure dimension of *time*. Taken together, these three categories can be used to define the unique living space and activities of any species. If each category is pictured as a two-dimensional plane, each category contains an infinite number of points and each point represents a specific bit of information.

An organism's life of internal interactions includes such information as the physiology and internal regulation of that organism's life. Its external interactions include all biotic and abiotic facets of its existence external to its physical body. The climatic limits within which this organism can live represent one type of external information requiring identification on this plane of external interactions (Figure 2.10). The third plane, time, is

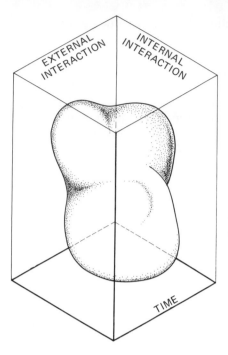

Figure 2.10
The Ecological Niche Concept can be
defined as an n-dimensional hypervolume
because the niche of any species is
unique. The niche is a three-dimensional,
irregular figure whose limits are defined by
the parameters of the internal and external
environments within which an organism
exists. Logically then, members of the
same species occupy very similar, though
not absolutely identical, niches.

necessary because time progresses at a steady pace and, in ecologic
systems, is often an extremely critical facet, determining the nature and/or
the extent of ecologic interactions. For example, both the internal and
external interactions of organisms may be markedly different at different
times. One needs no ecologic training to know that all life changes with the
relentless change of time.

If we define these three planes of life as intersecting at right angles, we
can use them as a framework from which to project the unique three-
dimensional figure that defines the ecologic niche of any particular organism. That figure may be subdivided, relative to the changes within the niche
through which the organism passes as it ages. All details must be accommodated; for example, our figure of a niche may be somewhat altered if we
consider the figure that defines the niche of a male versus that of a female
organism of a particular species.

Based on this concept, we can better examine the statement that
because the niche of one species is unique, no two species can occupy the
same niche at the same time. This is generally called the *competitive
exclusion principle.* It was well summarized by Garrett Hardin, in 1960,
when he framed his review upon the acknowledged ambiguous statement
that "complete competitors cannot coexist." In brief, the principle directs
that if two different species occupy the same ecologic niche, if they are
found in the same geographic area, and if one species multiplies even the
very slightest bit faster than does the other species, the first species will
completely displace the second species, which will become extinct in that

geographic area. Furthermore, we must assume that the principle is applied to ecologically stable communities and that coexisting species are not limited in their interactions by the same factors.

This is another way of expressing competitive exclusion. When different species interact differently with factors in their mutual environment, those species can coexist. When they interact with the same response, their ecologic niches overlap and they cannot coexist in a stable environment.

Whether stable or unstable, environmental factors are often the main external factors impacting the niche of an organism. They may determine the ultimate success or failure of that organism in its specific existence. For example, the niche of an organism includes appropriate internal and external interactions with the food web in which that organism participates. The web is important to the particular ecosystem involved and may be manipulated, often by man, and often to the real or potential long-range detriment of ecosystem stability. For example, if we reduce the number of trophic levels in a food pyramid, we insure a more efficient conversion of the biomass energy from the lowest level to the highest level. That is, we may grow alfalfa for feeding beef cattle, which are then harvested by man for food. The ecologic instability that threatens in this plan concerns the pure culture of an organism and the lack of diversity of niches within the food web, which, if present, better provides *ecologic stability*.

The evolution of life has included a logically parallel evolution of ecologic niches, associated ecosystems, and food webs. This diversity of niches provides ecologic stability, which usually resists violent fluctuation within a subject ecosystem. The simplification of an ecosystem, such as results from most farming operations, tends to *stress* the ecologic balance that would otherwise exist. In simpler terms, any tendency toward managing a level of artificiality in any ecosystem, or in any community within an ecosystem, tends to unstabilize that ecosystem and all its associated communities, habitats, and niches.

Such artificiality usually reduces the number of different niches present in the managed communities. Such a reduction in diversity may cause the overall productivity of the community to change violently should any internal or external interaction of the artificially favored niches initiate ecologic change. Diversity of niches represented within a community allows the failure of one niche to be cancelled by the success of some other niche. One niche's poison, indeed, may very literally be another niche's food. However, if niche diversity is low, this tendency for *averaged stability* is also low or absent and the community may collapse and disappear. As indicated earlier, the instability of ecosystems and communities is often a result of human intervention in some artificial and stressful way. Specifics of this sort of problem will be discussed later in Chapters 5 and 6. However, before we proceed to such considerations, it is appropriate to examine some of the fundamental social and cultural aspects of our own species niche through what we have come to label *human ecology*.

Although human ecology has been restricted by some to a consideration of internal interactions relative to the internal physiology of the human body and its response and accommodation to environmental alterations, the field should be all inclusive. Human ecology should be, and generally is, the investigation of the human niche and consideration of those unique qualities for both good and bad that the human species possesses. Furthermore, because humans are members of the biosphere and because all the concepts and principles of ecology, except perhaps those found in the Bible, were framed by humans, it is important to have some concept of how humans have evolved into the species we see today occupying an extremely important and influential niche.

From my point of view, the most significant aspect of human ecology is this simple fact. The *cultural evolution* of our species was largely derived from progenitors whose environment selected for a *concentration characteristic*. Our early ancestors were gregarious, finding relative power and security in forming groups, which grew in size and density. We still usually find individual strength in a group and individual insecurity in solitude.

A major part of the success of such a concentration of humanoids was attributable to the inheritance and transmission of experience and information. This human trait reached its optimum level, for our purposes here, with the appearance of spoken and written language. The evolution of recorded information about humans and our existences made at least the external interactions of new generations less and less surprising and new than they were for earlier generations. The information explosion of the last few decades is the natural outcome of the recognition by our progenitors that knowledge and experience can be transferred and that this transference is helpful in making existence easier and more logical. This certainly must have been true of the early hunting–gathering bands, when information about the habits and distribution of game, for example, would serve the group well if passed from older members to younger. Our beginnings were successful and were governed in large part by the interest that must have been shown in our relationship to the environment.

There are, now as then, factors within our niche that limit our existence and regulate the density and distribution of our species, as is the case for other species in other niches as well. Some human cultures still reflect a direct nontechnologic response to the human niche, such as the Australian aborigine tribes that remain hunters–gatherers in a very harsh natural environment. In their existence, we can see very clearly how species success is regulated by harsh and limiting environmental factors (Figure 2.11). In principle, they have essentially the same position in the human niche for the Australian bushman as they do for those of our species living in modern cities.

Indeed, we all occupy essentially the same human niche. Of course, as individuals we may have differences but these are slight. Relatively slight physical or metabolic differences have evolved as an enhancement of adaptation to an existence within a particular ecologic community. An example is the differences between the diet of the bushman and the diet of an eskimo. Perhaps the divergent habitats of these two different human

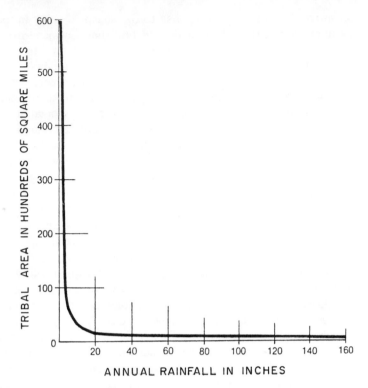

Figure 2.11. There are very logical limits imposed upon a species by the environment. Pushed to the outer limits of a niche, the pressured species must spend all its time and energy struggling to sustain life at that limit. Yet, the boundary of any niche is seldom absolute. Here, the range of aboriginal Australian tribes is shown to be inversely related to rainfall. Where there is little rain, and therefore, little food, the tribe must increase its range in order to survive. (Redrawn from Boughey, 1971a.)

races is a good example of a fundamental characteristic of our species, which allows humans to artificially extend their niche. By use of technologic devices to manage our environments, we can exist in habitats that would not otherwise fall within the niche capacities of the species. Clothing is one such technologic advance, as are all our industrial activities, which allow us to survive in hostile environments through fabrication of "unnatural" support systems. As we shall see later in this book, the artificiality of these support means often lead humans to create ecologic problems that would otherwise not exist and that are quite difficult to correct once they exist. Victor Frankenstein had a problem which ultimately took more to undo than to do. The *ripple impact* of one ecologic action may cause other expected or unexpected actions to occur in other parts of the involved ecosystem. Predicting and managing such real or possible consequences of an action is the basic substance of applied ecology and environmental management.

The earliest applied ecologists were the humanoids who gathered food and either moved as nomads in response to the seasonal shifts in the food resources upon which they depended or established permanent occupations on sites that had sufficient year-round supplies of food to support a population. That culture seems to have been followed by the first cultivated plant species, or *cultigens,* ushering in the primeval farmer, who could select and support the culturing of food plants and make humans less dependent on hunting for food. Of the 3000 of cultigens associated with the evolution of the humanoid species, we now have about 12 favored species, which feed about 90% of the world population.

Our next step in the technologic expansion of our niche was the domestication of animals. That leads us to a consideration of the ecologic relationships which are fundamentally changed when any agent, such as ourselves, is able to effectively manage ecological principles and expand the niche in which that agent exists. One obvious result of our abilities, and perhaps our lack of grace, is our present biosphere, upon which we have had much ecologic influence. The pathways that we have followed leading to our present world seem quite logical and, if given another opportunity to grow food instead of hunting it thousands of years ago, we undoubtedly would repeat our choice. It was a most significant choice because it meant that a group could grow more than enough food to feed itself. In the "spare time" that the hunter could not expect to have when more hunters required him to hunt longer and less successfully, the farmer turned some human effort to the quality of existence, and new human communities (i.e., cities) appeared in our history. They were a result of the increased productivity which farming and irrigation brought. They were necessary because people had to live close together for them to more efficiently regulate water and food and capitalize upon this new found strength. Our embryonic culture derived this strength from environmental management, however sound or unsound our ecologic instinct.

In fact, many of the principles of ecology can be redefined as expressions of human culture in much the same way as they are used to describe the so-called natural world. Our progress through various societal levels includes at least the following such principles.

First, *diversity,* in this human context, refers to the ever increasing division of labor in our communities. Our growing technologic sophistication has an ecologic parallel in the principle of *niche diversification,* through which a living space becomes more and more crowded with an ever evolving biotic community of different species. Each one occupies a somewhat different niche, each therefore having a somewhat different "occupation."

Second, *competition* is an expression of ecologic pressures between two organisms in overlapping niches, which conflict with each other in pursuit of the finite natural resources they each seek. In human society, this competition is often between the specialist of one generation or technologic level and another seeking a place (i.e., job) in the next advancement of that society. One group of specialists seeks to do a better job than is offered by a competing group of specialists. One group tends to become obsolete and is replaced by the other more "efficient" group, just as organisms that are more efficient in a niche which overlaps another one often eliminate the less efficient organisms.

Third, while both diversity and competition grow as the biosphere evolves and as human society evolves, *dominance* decreases. In society, kings are becoming more and more obsolete and the power of kings is relentlessly divided among groups of people, such as between politicians and/or celebrities of all sorts. Ecologically, a community will not be dominated by one species provided the diversity of life in the community is not limited and provided competition is not unevenly balanced with respect to the niches present in that community. A natural and healthy tropical rainforest or coral reef does not contain one niche dominating the others. There are many different species with few representatives of any one particular species—the ultimate in socialistic philosophy, where no one element has enough dominance to do anything but coexist within a low profile in community functioning.

Fourth, *succession* is perhaps the most fascinating aspect of the art and science of either ecology or sociology. It appears that a community which produces an excess of food, that is a community for which autotrophic net productivity is greater than net respiration ($P/R > 1$), can expand. Its excess store of energy can be used to grow or to power other communities. In the same way, a society that has an excess store of time or energy (i.e., food) can evolve to a higher societal grade just as our ancestral irrigators/farmers evolved cities while their hunter–gather coevals would expend all their energy and time simply trying to find enough to eat to stay alive.

In thinking on the principles of ecology and on our large or small interest in understanding what makes an ecologist tick, we may have to use many devices aimed at disarming some of the complexities that shroud the field. Relating the principles of ecology to our own human lives is one way of helping because there is no fundamental difference between the manner in which we each fill our individual ecologic niche and the way in which any other living organism fills its niche. Because an ecologist is interested in understanding one or more aspects of the ecologic niche concept, nonecologists can understand the same philosophy and methods of approach by following and expanding upon the role playing outlined above.

Even though the following sections of this book are often technical in their approach to defining the tools and processes of ecologic study, it should be understood that the technology an ecologist uses is the same, philosophically, as the tools a diplomat or a mathematician may use to make her- or himself feel comfortable in his or her field of interest. Some of us are quite powerfully motivated to study a field in order to feel at home and secure in at least that one field. Knowledge makes us feel secure and at peace. Few of us find any rest with insecurity as a traveling companion. Perhaps ecologists are driven into their field because by learning the facets of ecology and the many interlocking principles of that art and science, they never find themselves in totally unfamiliar surroundings, either figuratively or literally.

Bibliography

Baker, H. G. 1965. *Plants and Civilization*. Wadsworth Publ., Belmont, Calif.

Berry, B. J. L., and J. D. Kasarda. 1977. *Contemporary Urban Ecology*. Macmillan, New York, 497 pp.

Birdsell, J. B. 1953. Some environmental and cultural factors influencing the structuring of Australian aboriginal populations. *Am. Naturalist, 87*, 171–207.

Bormann, F. H., and G. E. Likens. 1967. Nutrient cycling. *Science, 155*, 424–429.

Bormann, F. H., and G. E. Likens. 1970. The nutrient cycles of an ecosystem. *Sci. Am., 223*(4), 92–101.

Boughey, A. S. 1971a. *Man and the Environment*. Macmillan, New York, 472 pp.

Boughey, A. S. 1971b. *Fundamental Ecology*. Intext Educational Publishers, San Francisco, 222 pp.

Burkhill, I. H. 1953. Habits of man and the origins of cultivated plants of the Old World. *Proc. Linnean Soc.* (London), *164*, 12–42.

Cloud, P. 1974. Evolution of ecosystems. *Am. Scientist, 62*(1), 54–66.

Cole, L. C. 1966. Man's ecosystem. *Bioscience, 16*, 243–238.

Combel, H., and R. J. Braidwood. 1970. An early farming village in Turkey. *Sci. Am., 222*(3), 50–56.

Doxiadis, C. A. 1968. Man's movement and his city. *Science, 162*, 326–334.

Elton, C. S. 1927. *Animal Ecology*. Macmillan, New York.

Gottman, J. 1961. *Megaloplis*. MIT Press, Cambridge, Mass., 810 pp.

Grimmell, J. 1917. The niche relationship of the California Thrasher. *AUK, 34*, 427–433.

Hardin, G. 1960. The competitive exclusion principle. *Science, 131*, 1292–1297.

Helback, H. 1960. Ecological effects of irrigation in ancient Mesopotamia. *Iraq, 22*, 186–196.

Hutchinson, G. E. 1957. Concluding remarks. *Cold Spring Harbor Symp. Quant. Biol., 22*, 415–427.

Hutchinson, G. E. 1959. Homage to Santa Rosalia, or why are there so many kinds of animals. *Am. Naturalist, 93*, 145–159.

Kormondy, E. J. 1976. *Concepts of Ecology*. Prentice-Hall, Englewood Cliffs, N.J., 238 pp.

Kozlowsky, D. G. 1968. A critical evaluation of the trophic level concept 1, ecological efficiencies. *Ecology, 49*, 48–60.

Linderman, R. 1942. The trophic dynamic aspect of ecology. *Ecology, 23*, 399–418.

Lsoac, E. 1962. On the domestication of cattle. *Science, 137*, 195–204.

MacArthur, R. H., and J. W. MacArthur. 1961. On bird species diversity. *Ecology, 42*, 594–598.

Odum, E. D. 1963. Limits of remote ecosystems containing man. *Am. Biol. Teacher, 25*, 429–443.

Odum, E. P. 1969. The strategy of ecosystem development. *Science, 164*, 262–270.

Odum, E. P. 1971. *Fundamentals of ecology*. W. B. Saunders, Philadelphia, 574 pp.

Odum, H. T. 1956. Primary production in flowing waters. *Limnology and Oceanography*. Vol. 1:102–117.

Oosting, H. J. 1942. An ecological analysis of the plant communities of Piedmont, North Carolina. *Am. Midland Naturalist,* **28**, 1–126.

Oosting, H. J. 1950. *The Study of Plant Communities.* W. H. Freeman and Co., San Francisco.

Paine, R. T. 1966. Food web complexity and species diversity. *Am. Naturalist,* **100**, 65–75.

Perkins, D. Jr., and P. Doly. 1968. A hunter's village in Neolithic Turkey. *Sci. Am.,* **219**(5), 98–106.

Rickleys, R. E. 1973. *Ecology.* Chiron Press, Newton, Mass., 861 pp.

Sauer, C. O. 1962. Seashore–primitive home of man? *Am. Philosoph. Soc. Proc.,* **106**, 41–47.

Shelford, V. E. 1929. *Laboratory and Field Ecology.* Williams and Williams, Baltimore.

Shepard, P. 1967. Whatever happened to human ecology? *Bioscience,* **17**, 901–911.

Smith, P. E. L. 1976. Stone-age man on the Nile. *Sci. Am.,* **235**(2), 30–38.

Turner, F. B. (ed). 1968. Energy flow and ecological systems. *Am. Zoologist,* **8**, 10–69.

von Eckardt, W. 1964. *The Challenge of Megalopolis.* Macmillan Co., New York, 126 pp.

Whittaker, R. H. 1962. Classification of natural communities. *Bot. Rev.,* **28**, 1–239.

Whittaker, R. H. 1970. *Communities and Ecosystems.* Macmillan, New York.

Whittaker, R. H., and S. A. Levin (eds.). 1975. *Niche. Theory and Application.* Penn, Dowden, Hutchinson, & Ross, Inc., Stroudsburg, Penn., 448 pp.

Woodruff, L. L. 1912. Observations on the origin and sequence of the protozoan fauna of hay infusions. *J. Exp. Zool.* **12**, 205–264.

Woodwell, G. M., and R. H. Whittaker. 1968. Primary production in terrestrial communities. *Am. Zoologist,* **8**, 18–30.

Woodwell, G. M. 1970. The energy cycle of the biosphere. *Sci. Am.,* **223**(3), 64–74.

3 Units of Study

The conventional pattern for teaching ecology is usually to envelop the student in a shroud of detail and fundamentals and facts that seem calculated to confuse and shake that novice. Supposedly, science is founded on fact so perhaps rote fact is a logical place to start. In fact, however, the best scientists are people of broad interests. They are interesting and not at all shallow or draped in facts and are not especially interested in conversation limited to the circle of their special field of scientific fact. They are also people not generally impressed by their own originality and not especially concerned with the significance of their work or the recognitions given them or withheld from them. They are, however, obsessed by a need to understand how their work relates to their surrounding environment and world. This need and curiosity is particularly well developed in the ecologist even before he or she consciously decides to call her- or himself by that label.

Usually beginning as a science student, the budding ecologist seeks to take systems apart and to understand the components and how they fit together to ultimately make the whole organism, community, or ecosystem, work. There is, therefore, not much difference between the engineer and the ecologist. Many times these are careers with interchangeable beginnings and they should have converging ends. Nonetheless, we are here concerned with the fundamental biology study prerequisites of becoming an ecologist and the units of organized study which lead in that direction. These units may be artificially segregated into at least five topics, making a pathway that includes studying (1) the internal functioning of living organisms, which is physiology; (2) the study of life on land, which is terrestrial ecology; (3) the study of life in water, which is fresh-water ecology; (4) marine ecology; and (5) the study of ecologic communities of special concern, particularly those which are transitional and overlap the four units listed above. Understanding how the principles of ecology weave through these units of study will clarify the decision a student may make in order to better define his or her career objectives.

It is difficult to plunge into a long and tedious exercise of rote learning when the relevance of the lesson is not clear to the student at the outset. A consideration of animal and plant physiology, as a fundamental key to ultimately understanding ecologic systems and the philosophy of ecology, is offered here as an introduction to these units in an attempt to clarify the interrelationships between systems.

Animal and Plant Physiology

Physiology is generally defined as a study of the functions of the organs and parts of living organisms. Therefore, it relates to individuals as opposed to groups (i.e., populations). Because ecology is concerned with living organisms and their relationship to the environment, physiology is an important prerequisite to any comprehensive understanding of ecology. Although we cannot deal here with a detailed study of fundamental physiology, it is possible to allude to the basic areas in which any ecologist should have a good understanding. Furthermore, a few of the most basic units of physiologic study will be used in order to illustrate the interrelationship between physiologic and ecologic functions.

As a beginning, physiology may be subdivided into five major areas, each of which contributes, in a substantial way, to defining parts of the ecologic niche of any organism under study. These areas, whether in plants or animals, include (1) feeding and digestion (the chemical modification of ingested food substances); (2) respiration, photosynthesis, and internal circulation; (3) coordination and effector systems; or those which carry out such actions (eg., muscle systems); (4) receptors of external stimuli; and (5) reproduction.

Relative to a philosophic approach to understanding ecology, these five areas can be viewed as they define or alter the niche of any particular organism. They are of particular interest to the ecologist who seeks to predict the impact of some environmental alteration upon a community of definable organisms and abiotic factors.

In that sense, the feeding and digestion of an organism involves the internal incorporation of substances that are initially external to that organism. In ecologic terms, any living organism feeds on its environment, drawing in the required essentials while also drawing in unessential, and sometimes deleterious, materials. The sorting out of the good and the bad is partly a function of physiology, where internal filters function to screen out the undesirable and concentrate the desirable elements.

Respiration, photosynthesis, and circulation are physiologically much the same when viewed from an ecologic perspective. Respiration involves a complex series of biochemical reactions that convert the stored chemical energy of digested food into forms useful to supporting the life processes of an organism, utilizing oxygen and sugar and producing carbon dioxide, water, and energy. Photosynthesis, in contrast, involves a complex series of biochemical reactions that use carbon dioxide and water to convert light energy into sugar, useful in supporting the life processes of an organism, and byproducts of oxygen. The crucial difference is that the photosynthetic process often *produces* more stored energy (sugar) than is required to support the plant in question. The resulting excess sugar (energy) may be utilized as food by some other organism which feeds on the source organism.

Circulation may be considered as generally an adjunct to respiration and photosynthesis. In organisms larger than several microns in diameter, the internal distribution of food and other materials is necessary and is accomplished through some forms of circulation. Whether by blood or some other fluid, or by simple diffusion, circulation is an essential physiologic process. From an ecologic point of view, whether internal or external to an organism, specific types of circulation and mixing cause the environment to be either homogeneous or heterogeneous and thus determine the efficiency and nature of the physiologic and ecologic response an organism may make in its internal or external niche.

Coordination and effector systems, from a physiologic point of view, involve both internal and external factors, including nervous systems in animals and any process in plants or animals that distributes information from one point in the organism to another. The generation and distribution of growth hormones in plants is an example. In fact, the physiologic differences between internal coordination by nerves and by chemicals are not great.

The means by which nerve messages move involve the production and controlled release and movement of chemicals within the cells of the organism. The fundamental difference between these two forms of coordination is that chemical responses to the environment are usually much slower, such as growth movements, than are nerve-regulated responses, such as animal movement.

Coordination is usually discussed physiologically in conjunction with a consideration of effector systems, those systems which interpret information into actions, such as muscle systems or secretory systems. The basis of coordination requires the conveyance of bits of information from one point to another point, whether internal or external to an organism. In order for that level of coordination to effect some change, a system must be interposed that can *intercept, translate,* and *interpret* the bits of coordination information into movement or some other physiologic response. Effector systems are those which react to coordination information.

From an ecologic point of view, coordination and effector systems may be examined in the same basic manner as they are through physiologic study. However, they may also be conceived relative to the coordination of the ecologic factors that exist between organisms, between communities, or between ecosystems. Although this somewhat philosophic concept obviously involves understanding the physiology of individual living organisms, it also requires an understanding of those ecologic principles which govern the response of a community of organisms to the ecologic coordination of existing conditions and changing events.

Very often, the events causing ecologic reaction or change are physical alterations of an environment. Such alterations may or may not be instituted by biologic factors. Climatic or geologic changes are examples of this sort of abiotic change. Invasion of an environment by some dominating population or the appearance of anthropogenic substances are examples of biotic caused change. Both types of environmental change always effect some ecologic change through the response of the impacted community or ecosystem.

The philosophic perception of ecologic coordination focuses on environmental change as the direct means of coordination. It recognizes that the physiology of those organisms living within the changed environment determines specific responses, depending upon the species-specific interception, translation, and interpretation of environmental change as defined within the niche of the species. The expressions of individual response may vary greatly from one organism to another, being most nearly identical in individuals of the same species.

With the environment as the coordinating system in ecology, the effector systems are represented by the biotic factors within that environment. As with the traditional view of physiology, in ecologic systems all effector systems possess feedback regulation. The expression of a response to environmental change causes change of these responding biotic factors (organisms) (Figure 3.1). That change generates a somewhat different set of bits of information and effector systems usually respond to the new environment in a slightly different way than in the first instance. As response continues, a final and relatively stable level of ecologic readjustment is reached as a compensation for or reaction to the original environmental change.

Receptors are physiologically defined as organs or cells within an organism specialized to receive particular types of stimuli. The ear for sound, the eye for sight, and certain cells in the skin for touch and temperature are examples of receptors. They serve to convey information about the internal and external environment of an organism to the coordination and effector systems of that organism. This same terminology can be used for ecologic systems, because the interpretation of environmental parameters accrues at the level of individuals in a community. At a less literal and more philosophic level, the ecologic receptors of a community may be collectively defined as the whole physiologic range of species-specific receptors present in the community. Therefore, the community, comprised of many different species and viewed as a biotic–abiotic entity, receives and responds to information based on the cumulative operation and balance of all the receptors present.

If a particular species dominates the community, its receptors will

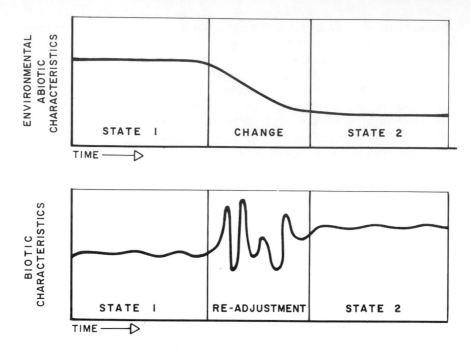

Figure 3.1. There is virtually always an ecological response to an abiotic change in the biosphere. An ecological response is essentially a biotic change which results from the physiological readjustment of the living system to an altered environment. The rate and extension of these physiological and ecological readjustments often fluctuate irregularly until some relatively stable state is achieved when a stable abiotic environment is present.

similarly dominate in conveying information to the coordination and effector systems within that community. Very often, we and our technology dominate an ecological system. In that circumstance, our receptors include our own senses and those we had created as technologic extensions of our senses. The interception, translation, and interpretation (expression) of the collected information may easily be one sided and therefore may not reflect the sort of natural system response diagrammed in Figure 3.1.

At least in the short run, using our experience learned from earlier generations coupled with our technology, we often have the ability to anticipate and manipulate ecologic change to our advantage. Other creatures do not seem to have this innovative ability. However, the larger the manipulations, the greater appears the risk of suffering unanticipated short-term or long-term consequences that we have not anticipated. Our bank of receptors still does not seem capable of sufficiently complete and comprehensive data collection to allow our coordination and effector systems to successfully manipulate the environment. Some examples of the receptors we use will be dealt with in this book in our consideration of the tools of analysis on which an ecologist may rely.

Reproduction as a physiologic process relates to the control and transmission of genetic information from one to the next generation of a particular species. With respect to ecologic processes, reproduction certainly includes these traditional physiologic processes. It also includes the community processes wherein an ecologic system grows in size or is translocated from one place to another. The reproduction of ecologic communities is never perfect because the multiple factors that go to make up any one community are never exactly the same in the reproduction of that community elsewhere.

Taken together, the continuum of ecology is philosophically identical to the continuum of physiology. Both studies involve complex, living systems. Both rely upon our technical skills of extending our senses in order to discover and understand the functioning and interrelationship of system components. Both require much more fact finding than we have presently amassed on either subject and both are interrelated themselves, so that understanding parts of one helps us to understand parts of the other.

As an emphasis of the fact that physiology and ecology are interdependent and that the differences between other life sciences are not nearly as clear as is often suggested, it may be helpful to seek a sweeping generality—for example, "Water is the most ubiquitous and important facet and basis for life." Offered as a focal point for the student interested in pursuing an understanding of ecology, we can believe that whatever its earthly form, life is basically a water solution. With a very few peculiar exceptions, most living organisms consist of 80 to 90% water. The blood plasma of our own bodies is about 93% water and our muscles are about 80% water; a jellyfish, in contrast, is about 98% water.

Furthermore, in all living systems, water is not only the "universal" solvent, it also transports food and waste products internally and often externally to an organism. It serves as a medium in which the physiologic activities of the living cells of an organism take place. In that sense, water provides the physical matrix of life. In addition, the contribution to life support by water is not at all entirely passive. The chemical nature of life requires water at many specific points. In particular, hydrolysis is a general chemical process fundamental to all life; it requires the alteration and/or decomposition of a chemical compound by the insertion of the elements of water.

As a medium, water can carry us through the interrelationships of science between disciplines, on the one hand, and between individuals, communities, and ecosystems, on the other. In fact, perhaps the most coherent scheme we can use to attack the problem of dissecting the continuum of ecology into more or less separate entities at the ecosystem and community level is to base classifications upon the character and quantity of water present in the system.

At a more fundamental level, and still dealing with the physiologic nature of ecology, we should seek to understand some of the characteristics of single cells. In some instances, a single cell is a complete single organism. In other instances, several cells are interrelated, forming a multicellular organism, such as a human. From an ecologic point of view, when multicellular organisms are examined on their cellular level, a parallel can be drawn between the cells of that body and the individual organisms of a community. Each cell participates with others, contributing to the definition of a physiologic niche in which the organism, as a whole, can survive. Even with cells that are specialized in their primary function, the basic characteristics possessed by all cells remain basically the same. Seeing how these characteristics interrelate with conditions internal or external to an organism is an important facet of sound ecologic learning. These basic characteristics include the following main considerations.

In 1665, Robert Hooke published a brief paper in which he observed a thin slice of cork with the aid of his microscope. He noted a structure consisting of " . . . a great many little boxes . . .," which he called cells. What he actually saw was the cellulose cell walls which, in the living tissue, surrounded the individual living cells from which the cork was derived. Extrapolating upon Hookes' observations, we can postulate that the cell is the fundamental building block of organisms. Some single cells are very large, such as a 14-cm ostrich egg, whereas others are small, such as blood cells (corpuscles) 10 microns in diameter. However, most cells have strikingly similar structures regardless of size or functions. These include an enclosing *membrane,* a *nucleus,* and a matrix of semifluid *cytoplasm* which fills the cell surrounding the nucleus.

Although the membrane is quite thin (about 75 Å thick) it is a remarkably dynamic boundary between internal and external cellular environments. It is therefore a crucial element as an intermediary between the coordination of external environmental factors and the ultimate level of organismic response at a cellular level. The cell membrane serves as a selective sieve allowing certain molecules to move freely or not so freely from one side to the other. Regardless of the specific structure of the cell membrane and the presence of pores or other means of regulating the movement of materials across the cell boundary, it functions to vary the cell's permeability. That characteristic of life is largely determined by the chemical organization of the living membrane.

Effected by the environmental milieu surrounding the cell, water passes through the membrane largely by *osmosis.* Osmosis refers uniquely to the movement of water across a membrane from the side of highest concentration of water molecules to the side of lower concentration. However, this sort of diffusion can be demonstrated with a nonliving membrane as well as in a living cell and does not explain the fact that many other molecules,

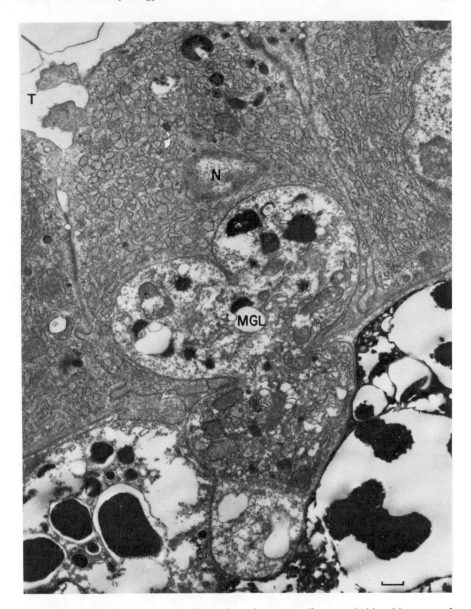

Figure 3.2. The cell membrane is dynamic and can engulf a remarkably wide range of variety and sizes of objects. Here a blood cell (MGL) in the blood stream of a tunicate (*Botryllus schlosseri*) is moving through a cell in the wall of the blood vessel whose nucleus is labeled N. This blood cell (MGL), which like an amoeba can move by forming pseudopods, will ultimately disintegrate once it leaves the circulatory system and enters the enveloping test (T) or tunic, of the tunicate. Scale = 0.25μm.

much larger than water molecules, pass through the membrane at variable rates. Some molecules penetrate the living membrane only at particular times in the life of a cell. Therefore, that membrane is apparently *selectively permeable*.

All membranes are composed in part of fats called *lipids*. Certain molecules, which dissolve such fats, penetrate cell membranes with relative ease. Larger and less solventlike materials may cross the cell membrane by being engulfed in the membrane and carried into or through the cell (Figure 3.2). The reverse also may occur in those cells which release materials as waste products or as secretions. Such materials are released to the outside of the cell (Figure 3.3). Some secretions may be released to act at some distance from the cell in the coordination of internal and external events in a community of cells. Other physiological cell membrane events may involve changes at the cell surface itself (Figure 3.4) and the capacity of the membrane to alter its structure or participate in events that are associated with living systems.

This dynamic nature of the cell membrane is all remarkable enough, but perhaps most remarkable and characteristic of life is the phenomenon in which ions and molecules are moved from one side of a cell membrane to

Figure 3.3. A high concentration of secretory droplets (SD) are seen here packed in several cells at the tip of a blind blood vessel, called an ampulla (AM), in the tunicate *Botryllus schlosseri*. The nucleus (N) can be seen in a number of these cells which separate the test [or tunic (T)] of the animal from the blood corpuscles (RBC) within the circulatory system. Two ampullae are labeled here while the third one, containing the secretory droplets, is not labeled. Scale = 5 μm. (Nomarski Optics.)

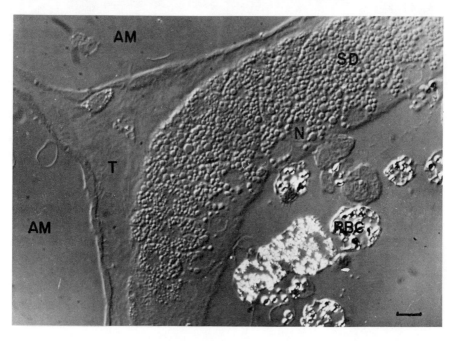

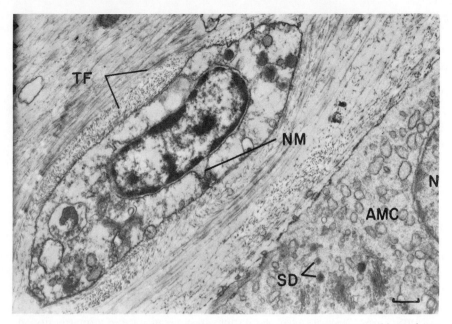

Figure 3.4. Some amoeboid cells in the test of the tunicate *Botryllus schlosseri* can apparently secrete fibers (TF) which appear to be composed of cellulose produced by the cell membrane of those cells. This particular cell is shown lying next to a cell of an ampulla (AMC) which contains some disintegrating secretory droplets (SD). A portion of its nucleus (N) is also shown as well as the abnormally appearing nuclear membrane (NM) of the cell within the test substance. Scale = 0.5 μm.

the opposite side by an *active transport*. Unlike diffusion, active transport requires an expenditure of energy. Active transport makes possible another fundamental condition of ecology expressed at the cellular level. A living cell, by the expenditure of energy through active transport, can sustain an internal composition that is different from that of the surrounding environment. *The cell, therefore, is to a lesser or greater degree independent of the surrounding environment.* Such internal homeostasis is vital to life and to the formulation of an ecologic philosophy. Biotic systems affect a control of their internal environment based on physiologic capabilities expressed by the limits of the ecologic niches of participating species. From the physiologic point of view, this concept distills down to the following essence. A particular molecule may penetrate the living cell membrane and thus enter a cell at one time but not another, depending upon the physiologic state of the cell in question.

Typically, each cell contains a more or less spherical nucleus which contains some very characteristic biochemicals, including deoxyribonucleic acid (DNA). Most of life as we know it has evolved under the biochemical direction of DNA or closely related compounds. Constructed from a sugar, a base, and an acid, each DNA molecule may contain a relatively long linear sequence of chemical base pairs whose coded order convey genetic information. That is, DNA directs a complex of biochemical reactions which, taken as a whole, define the life of the organism in question and the capacity of the internal and the external niche of that particular cell and organism.

Without knowing much more about the role which the nucleus plays in the physiologic expression of ecology, we can still draw some conclusions useful as guides for fitting the cell into the organism and the organism into the community or ecosystem. For example, as indicated the genetic information within a cell is usually carried on the ribbonlike molecules of DNA. Under favorable physiologic conditions, the cell and all its contents may divide. Such replication includes the DNA and that process is *nearly always perfect,* without any flaw in the faithful reproduction of the coded DNA molecules. The molecules are thus passed on to the next generation of cells. The fact that the duplication process is not always perfect is crucial. The rare flaw, a change or *mutation,* causes this fundamental life process to be changeable, that is, *mutable. Mutability is the basis of evolution.* Evolution is a very important aspect of ecology.

Ecology is comprised of biotic and abiotic changes, some seasonal, some expressed in terms of geologic time, some encompassed within fractions of a second. Some such changes may require rapid adjustments on the part of organisms living in that changed environment. Some may require slow adjustment. In either case, an overriding physiologic aspect of ecology provides for the homeostatic tendency for living systems to adjust to environmental fluctuations. That tendency can be compared to the shock absorber on a vehicle which smoothes the ride over a rough road.

When the ecologic roughness of life falls within the capacity of the niche of an organism, that organism adjusts, usually rapidly, to accommodate and to smooth out physiologically the environmental change. When the roughness of the environment goes beyond the capacity of the niche of a community of organism, genetic mutability may allow for the gradual change of the niche capacity through *genetic adaptation.* Darwin's concept of "the selection of the fittest" is a philosophic expression calling upon ecologic factors that allow evolution to alter niches and, therefore, to alter and adapt species.

For example, assume that the ecological niche of a species is described by the genetic information "ABCD" in the nuclear content of that species. Furthermore, assume that as a natural result of the slightly imperfect processes of replication of that information, there is a change, or mutation, in the genetic information to "ABCd" in the reproductive, or germ cells (e.g., egg or sperm, ova or pollen) of an individual of the species. If that altered information is (1) carried into the development of an individual of the next generation, and (2) if the information is physiologically expressed in that individual, and (3) if that expression alters the niche of that individual, making it more compatible with the environmental change imposed on the species, then that mutated species, occupying a shifted niche, will be better equipped to accommodate the environmental change than were its progenitors. *In this case, the egg comes before the chicken.* The mutated chicken will ultimately tend to replace the "old" chicken because the "new" chicken is better fitted to the new environment and the new constraints upon its niche.

This process is normally very slow because the physiology of genetics is conservative. The continuity of genetics tends to be stable. Species tend not to change. Niche capacity tends to be conserved, unchanging from one generation to the next. Mutation is rare and the ultimate expression of mutation is even rarer under normal circumstances. However, certain external agents, such as some chemicals and some radiation, may accelerate the natural rate of genetic mutations, making the effected organisms better or worse adapted to their internal or external environment. Usually, such *mutagenic factors* effect genetics in a negative way, causing *misadaptation.* The seat of these genetic events is mainly in the nucleus, whereas the expression of these events takes place by transmission and translation through the cytoplasm of the cell.

Surrounding the nucleus, and confined within the cell membrane, is the semigelatinous cytoplasm. It is a complex of distinct and characteristic substructures suspended in an aqueous mixture of colloids and solutions. About 90% of the cytoplasm consists of water. From an ecologic perspective, the cytoplasm of a cell serves as the physical matrix through which genetic information is transcribed and expressed physiologically. The cytoplasm thus responds to the internal and external environment of the cell, which results in the physiologic coordination of a community of cells. That coordination results in the ecologic coordination of a community of individuals within an ecosystem.

Serving as a simple example of the interrelationships between physiologic and ecologic factors within a single organism, we can examine a piece of the life cycle of the fruit fly, *Drosophila* sp. (Figure 3.5). During a portion of that cycle, the larva of the fruit fly feeds continuously, requiring relatively large amounts of saliva (Figure 3.6). The saliva is produced by the paired salivary glands, each comprised of several relatively giant cells that secrete the saliva into a collecting duct. Stained to show DNA (Figure 3.7),

Figure 3.5. The life cycle of an organism often involves physiological changes. The cells of the fruit fly, *Drosophila* pass through a number of physiological stages which require cells to undertake different functions with time. One dramatic specialization is undertaken by those cells forming the paired salivary glands found during larval development. This diagram shows a side view of a third and last instar (larva) and the relative size of the salivary glands. The length of the whole larva is about four millimeters.

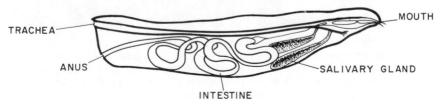

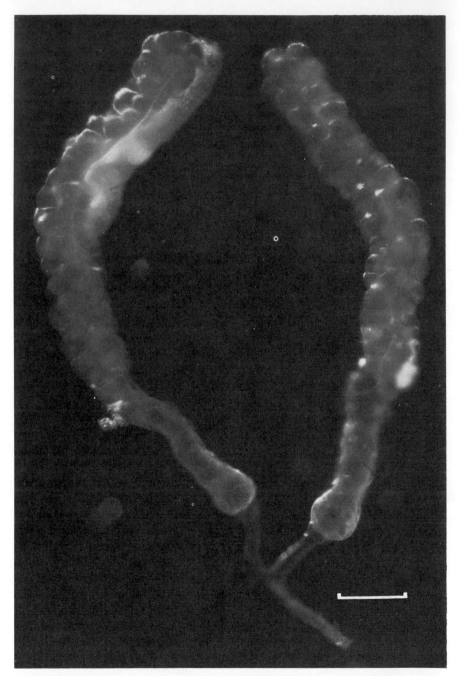

Figure 3.6. Each salivary gland from the third instar larva of the fruit fly *Drosophila* sp. contains about 150 individual cells which are seen here as a grape-like cluster along the salivary ducts. Each cell secretes saliva which is collected in the duct and delivered to the mouth region of the larva. The division of labor at the cellular level within the fruit fly exemplifies the specialization of cell function which is genetically controlled and which is expressed within the niche of a particular organism. Scale = 0.25 mm.

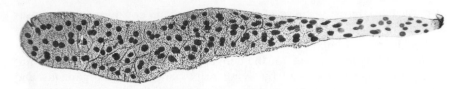

Figure 3.7. A stained salivary gland shows the nuclei of a single *Drosophila* gland. The stain colors only DNA molecules. Each cell contains one nucleus. These cells specialize in one predominant function, that of secreting saliva, while other cells in the same fruit fly, containing identical genetic information, perform other very different and coordinated functions.

each cell shows a dark nucleus that depicts the physical relationship of one cell to the next much more vividly than is otherwise possible. The specialization of these particular cells lends to their dominant physiologic function of secreting saliva *during a particular* stage in the life cycle of the parent organism. What directs that call to specialization is one of the many questions that we must answer. However, we can acknowledge that the specialization (i.e., differentiation) of certain cells to perform certain special functions is a physiologic circumstance. It is regulated directly by the DNA of those cells and indirectly by the masking or enhancing relationships which internal and external factors, including the environment, play on the genetic capacity (i.e., niche capacity) of those cells. In that sense, no single cell is inextricably bound to the destiny of performing one dominant function, such as secretion of saliva, as opposed to being a cell in the wing or brain of our subject fruit fly. *The capacity of one cell to be any cell in the parent organism is a principle resident in the nature of DNA and the physiologic definition of life.* Associating that concept with ecology and the relationships between biotic and abiotic factors in a community requires a hybrid logic. The pure logic of physics and math says that combining 2 + 2 always equals 4. The mutable logic of ecology says that combining two living functions may not always result in the expected summation. Ecologically, the whole is often greater than the sum of the parts of an ecosystem.

In part, our inability to anneal both logics is because the laws of physics are constant and unchangeable. The biotic laws of ecology are changing constantly as life and ecosystems mutate and evolve. Living organisms are only relatively predictable, not absolutely predictable, unlike the signs of a mathematical formulation. So the ecologist must seek a temperment that can accept things at face value, a temperment that is not so firm and unyielding that the unpredictable nature of life causes it sustained upset.

As a means of making the transition between biotic and abiotic interrelationships, between biologic systems and the environment, we can consider a few biologic constituents of some human systems. As examples of the interrelationships between our own physiologic/ecologic systems, two areas, (1) blood circulation and respiration and (2) the auditory system, are

of particular interest because of their importance in considering the consequences of pollution.

The human circulatory system permeates every living tissue in our
bodies, serving as a means for the delivery and pickup of all the prerequisites and products required by living cells. Single-celled organisms and
microscopic multicellular organisms are in such intimate contact with the
surrounding environment that an internal circulatory system is not necessary. However, with all larger organisms, some sort of conveyor must be
present at the cellular level of organization that can sustain the life processes of those cells which are not in direct contact with the environment or
which do not have direct access to food and to a "garbage disposal" system
(Figure 3.8). In humans, these requirements are met by our blood system.

One of the many important physiologic functions of this circulatory
system is to carry oxygen to cells, where it is needed in the metabolic
process of respiration. The exchange of oxygen and carbon dioxide
between an organism, its cells, and the environment is termed respiration.
Oxygen is used within a cell to oxidize organic molecules (food), which
contain stored energy, in such a way that some of that energy is transformed and used to support the life processes of the host cell. The controlled "burning" requires food (sugar) and oxygen, both carried to the cell
by the circulatory system. These are acted upon by the cellular processes of
respiration to produce carbon dioxide and water as waste products and
energy, primarily as ATP (Adenosine Tri Phosphate), as the useful product.
Looking only at the movement of oxygen from one point in the body to
another point involves the moving blood and the respiratory pigment,
hemoglobin.

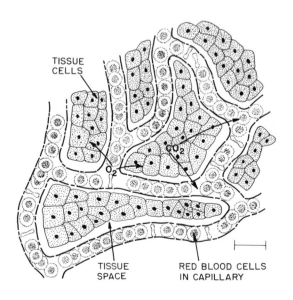

TISSUE
CELLS

CO_2

O_2

TISSUE
SPACE

RED BLOOD CELLS
IN CAPILLARY

Figure 3.8
Circulation at the cellular
level is diagramed here,
illustrating internal
respiration. Capillaries filled
with disk-shaped red blood
corpuscles pass adjacent to
tissue cells. Each cell of a
complex, multicellular
organism, such as man, must
be served by such a system
which carries materials to
and from the cells. One
crucial function of this
system is to allow for internal
respiration in which oxygen
(O_2) is carried to the cells
while carbon dioxide (CO_2) is
carried away from the cells.
Scale = 10 μm.

Hemoglobin is a relatively small organic molecule containing (1) iron; (2) a pigment, or color group, called porphyrin; and (3) a protein *(globin)*. Associated in this molecule, the iron and porphyrin are called *heme*. Each hemoglobin molecule can form a relatively unstable chemical association with oxygen, which is called *oxyhemoglobin*. This characteristic allows a fluid containing hemoglobin to absorb about 60 times more oxygen than could possibly be dissolved in a comparable volume of water. The ecologic relationship of this aspect of the respiratory and circulatory system concerns, at one level, the environmental circumstances that may alter the concentration of oxygen available to the blood, and, therefore, to the cells of an organism.

For example, if hemoglobin is exposed to a mixture of oxygen and carbon monoxide, two compounds will form, one combining oxygen with hemoglobin and one combining carbon monoxide with hemoglobin. However, the carbon monoxide compounds will be about 300 times more stable than the oxygen compound and, under those conditions, the blood may not be able to carry sufficient oxygen to the cells for them to function normally. In effect, the cells of a body will suffocate unless the carbon monoxide is removed by exposing the blood and its contained hemoglobin to "unpolluted" air.

In man, the site of gas exchanges between blood and the polluted or unpolluted atmosphere takes place in the lungs. A simple understanding of lung physiology is helpful in understanding the relatively intimate physical contact that always exists between the internal environmental aspects of an organism and the external environmental aspects of the niche surrounding that organism.

In man, the lungs have evolved to form a spongy, moist complex of more than 750,000,000 exceedingly small blind sacs called *pulmonary alveoli*. Blood flowing through the circulatory vessels of the lungs is separated from atmospheric air by not more than two cell thicknesses—the wall of the single-cell-thick capillary blood vessel and the single-cell-thick pulmonary alveoli. The exchange of gases, including oxygen, carbon dioxide, water vapor, and others, is therefore a surface-related phenomenon.

The combined surface area of all the pulmonary alveoli in the lungs of one human is greater than 1000 ft^2. Communication between many of the internal and external physiologic and ecologic events of an organism take place across this moist membrane separating internal and external environments. The interrelationships across this delicate and permeable barrier are chemical. Physical, rather than chemical, communication between internal and external events also takes place across a delicate membrane in the ear.

At least with respect to humans, one important parameter associated with our fullest utilization of our niche capacity relates to hearing. In its most practical sense, if an individual does not hear an oncoming train and fails to move from its path, the outcome is decidedly one sided. Less dramatic events are commonly encountered in which hearing is important in the ecology of a community, such as one organism finding a mate, establishing territorial limits, or receiving other information. They all may rely upon hearing and its physiologic processes, which translate vibrations in the external environment into nerve impulses that are interpreted internally by specific cells and structures.

In man, the vibrations, or sound waves, that reach an ear may, if powerful enough, cause the *ear drum,* or *tympanic membrane,* to vibrate at approximately the same frequency as the vibration of the air. Attached to the ear drum are three interconnected small bones (Figure 3.9) which translate the vibration of the ear drum into a piston motion. This transmits the sound vibrations from the ear drum to an inner chamber filled with fluid. The oscillations of this fluid excite auditory nerve cells, which carry information from this receiving or intercepting station, the ear, to the translation center, the brain. Interpretation of the information is largely based upon learning through experience, unlike the purely physiologic functioning of respiration discussed earlier. Hearing is a physiologic phenomenon made possible through the interrelationships of physics and the biologic interpretations of sound based on generated nerve impulses in the brain. *Therefore, unless there is an ear to hear the felling of a tree, that event makes no sound.* It makes only vibrations in the air.

A consideration of circulation and respiration illustrates the intimate physical contact that exists between even complex, large, organisms and the enveloping environment. All the cells of our body are either directly or indirectly bathed by that environment. Physical events in the environment,

Figure 3.9. As a mechanical device, the ear intercepts sound vibrations in the air and translates them into oscillations of fluid in the inner ear. There they cause nerve impulses to carry information to the brain. In the brain, the impulses are interpreted providing the organism with data about the external environment.

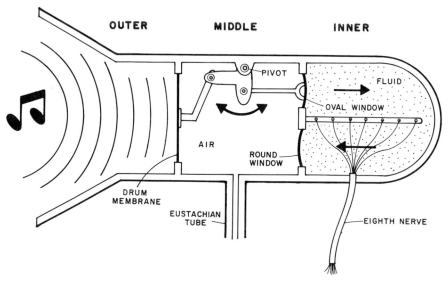

such as sound waves, are translated by a whole array of receiving devices (receptors, such as ears) and interpreted by cells in the brain or elsewhere. The individual organism is exposed to infinite variations of significant and insignificant internal and external environmental events. Intercepted, translated, and interpreted through the physiologic processes of a living organism, these events define the base data upon which the individual organism exists at any one instant of its life.

The physiologic "success" of an organism or of a whole species is inseparable from the ecologic success of that organism. Survival of the fittest refers not to genetic fitness, physiologic fitness, or any other suborder of fitness. It refers to ecologic fitness, which is success measured by survival. Survival is by luck, or through the capacity of an organism's niche, or through the ingenuity of an organism's strategy to hold dominion over its community or the world.

I suspect that the preeminent success that humans have fabricated will continue only so long as our ingenuity and technology can continue to outdistance *the ecologic laws of penance* summarized in the preceding sentence. These are laws that should be neither feared nor flouted. They are respected by the ecologist. In that sense, the applied ecologist seeks to artificially expand the human niche. He or she wishes to hold reasonably stable sway over those ecologic problems which might otherwise cause serious damage or collapse of those ecosystems essential to the survival of humans as a dominating species in our world. In order to seek that goal, the ecologist must understand enough about the various ecosystems of our world so that correct predictions, corrections, or guesses can be made in solving the problems she or he may encounter. In broadest terms, those concerns encompass terrestrial, marine, and fresh-water ecosystems, and we will examine them with respect to some fundamental ecologic characteristic of each.

Terrestrial Ecology

Recognizing the interdependence of cells within an organism through a study of physiology is the first step toward understanding the ecology of fundamentally different environments. The major physical differences between environments, including light, temperature, and rainfall, are reflected in the physiologic adaptations of organisms whose niches extend into these differing environments. Furthermore, the exclusiveness of an environment, accommodating a relatively limited number of niches, is often an expression of the relative physiologic "strain" which resident organisms must withstand. Expressed as a measure of primary productivity, especially in terrestrial systems, one can see remarkable increases in niches per community as one moves from polar communities to tropical communities which, when not limited by lack of water, may have a productivity ten times greater than arctic communities.

Variations in temperature, water availability, and other abiotic facets of the terrestrial environment seem to be far more heterogeneous than are the facets of the aquatic environments. Philosphically, it appears that the evolution of niches in the terrestrial environment has required the appearance of the "higher" life forms, those which can cope with harsher conditions. Seed plants, insects, and warm-blooded animals all have evolved out of the relatively hospitable aquatic environment to invade and ecologically dominate the terrestrial environment, along with viruses, bacteria, fungi, and protozoans. However, terrestrial and aquatic species of the ubiquitous microorganisms do not especially differ in their physiologic functioning. They live on a different scale of interactions than do larger organisms. They obey the same principles of ecology but they have a greater capacity to survive harsh environmental stresses. This is often accomplished through physiologic gyrations that effectively result in a condition of suspended animation, allowing microbes to survive hard times when other life forms die.

Those other life forms are apparently far more sensitive to environmental change. Philosophically this suggests that living organisms cope with a changeable environment in two basic ways. The "higher" and more complex forms of life have evolved a physiology that allows them to insulate their internal environment from the external factors that would otherwise disrupt them. The "lower" and most simple forms of life have evolved a physiology that allows them to passively exist, perhaps without functioning, in a hostile environment, waiting out the change. There are exceptions, of sorts, in the higher forms, such as those warm-blooded animals that hibernate or estivate as a physiologic means of waiting out adverse environmental circumstances.

Whatever the nature of their environment and the physiologic means by which a species survives in its niche, one of the most interesting aspects of ecology relates to the large-scale distribution of life on the earth. *Biogeography* may be divided into *phytogeography* and *zoogeography*. The result of biologic evolution and the geology of our world has led to a natural distribution and isolation of species. Six main regions, realms, or kingdoms around the world can be recognized by the presence of distinct vegetation types, and six very similar realms can be independently identified relative to the natural distribution of terrestrial animal life around the world (Figure 3.10).

Dependent for their evolution upon biotic and abiotic factors complete with the physical geologic insulation of one kingdom from another, *the endemic* (i.e., native) species of one kingdom may be perfectly able to survive and prosper in another kingdom whose abiotic factors are within the niche capacity of the species in question. In fact, the intentional or unintentional transplantation of a species between realms may result in the invasion of that new realm by that species if its niche is more efficient than that of the native species. Such success may cause us to label the invader a *"weed"* if it thrives in places where it is not wanted. For example the water

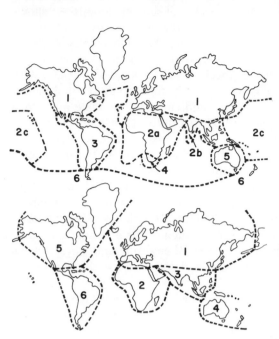

Figure 3.10
The biogeographic regions of the world for **flowering plants** are shown in the upper diagram based upon information presented by Good, 1953. Those regions include: (1) Boreal, (2) Paleotropical with subdivisions (2a) African, (2b) Indo-Malayan, and (2c) Polynesian, (3) Neotropical, (4) South African, (5) Australian, and (6) Antarctic. The biogeographic regions of the world for **animals** are shown in the lower diagram based upon information presented by deBeaufort, 1951. Those regions include: (1) Palearctic, (2) Ethiopian (African), (3) Oriental, (4) Australian, (5) Neoarctic, and (6) Neotropical. Because there is considerable exchange between the fauna of regions 1 and 5, these regions are often grouped together as one called the Holarctic region.

hyacinth *(Eichornia crassipes),* a floating water plant native to South America, was apparently introduced into the Southern United States as an ornamental whose clippings were sold at the 1884 New Orleans Cotton Exposition. Escaping some garden pond, the plant flourished to such an extent that it now effectively clogs canals and rivers in the southern United States, where a favorable climate allows it to survive with great success.

Rabbits, transplanted to Australia from Europe a hundred years ago, found themselves in an alien but favorable environment and multiplied at such a rate that their population spread as much as 70 miles per year devastating the dry landscape of vegetation. As with the water hyacinth, the invading European wild rabbits found their transplanted niche unlimited by some of those community factors, such as predators and disease, which stressed their existence in their native realms. The absence of these checks and balances allowed their numbers to grow geometrically, until natural stresses were reinstituted by natural or artificial means.

Myxomatosis is a respiratory viral disease of rabbits that was introduced into Australia after World War II. The disease killed most of the rabbits but natural mutation has led to the appearance of resistant individuals, and surviving rabbits are appearing now that escape this management practice of biologic control. No such effective control has been found for the water hyacinth, but one or more may be expected to evolve, certainly over geologic time. Human management incentives will not usually wait for such relatively slow evolution.

Nevertheless, the philosophy of ecology points out that a niche transplanted into a realm where ecologic stress is absent or diminished allows the species of that niche to rapidly dominate the new environment. It also tells us that the absence of ecologic stress on a niche invites the evolution of competition that leads to new stress between the original dominant and new, invading, and opportunistic species. For example, in the 1800s, European St. John's wort, or goatweed, was brought to the United States, where it flourished until 1940 as a weed and came to dominate more than 250 million acres of prime pasture land in California. Unhindered in its exploitation of a virgin realm, goatweed was constricted in its niche when an insect parasite appeared on the scene in 1940 to cause sufficient ecologic stress on the weed to prevent its continued success.

From the philosophic point of view, mature, that is stable, ecologic systems, tend to be socialistic—no individual species within a community does very well and dominates and no individual species does very poorly! Although also present in the aquatic environment, this principle is particularly accessible to observation in the terrestrial environment, where we can more conveniently dissect our own niche and compare it with the principal components of other niches.

In the face of an increasing population, the human species has flourished like a weed without indeed being a weed. It has ecologically dominated almost every realm that it has sought to conquer and it has usually done so by technologically expanding the niche beyond its natural limits. For example, human populations stressed and therefore limited by disease have mitigated the stress by medicine. Access to the natural resources needed to sustain our ecological aggression has been made possible by our industrial achievements. The philosophic conclusion drawn by an ecologist reflecting upon our ascension to global ecologic dominance points to the fabricated and artificial nature of that ascension. This does not at all mean that we are evil, that we should relinquish our dominion, or that we should revert to the primitive (i.e., natural) existence of our more ecologically balanced ancestors of the Middle Ages, or before. It does mean that we are technologically juggling many fragile and some potentially lethal aspects of our ingenious, artificially expanded niche. It also implies that for us to sustain this dominance, we will have to expand our juggling performance in order to accommodate the new ecologic stress that such dominance generates. Dealing with that performance requires an investigation of the main elements associated with the natural system in the absence of artificial manipulation. The terrestrial system is most directly involved in our niche.

The interrelationships found within the food pyramids of ecologic systems, as defined in Chapter 1, may be used as a guide in grouping the great diversity of life and associated niches in terrestrial ecosystems. This scheme divides all terrestrial life into two main categories, autotrophs and heterotrophs. It further groups heterotrophs into *phagotrophs* and *saprotrophs*. It is a division based upon the basic physiologic differences found in organisms that either make food (from water, minerals, and light) or do not make food.

Autotrophs, commonly called green plants, refer only to those species of plants that contain chlorophyll. In virtually all terrestrial ecosystems, these autotrophs dominate the landscape. They are relatively stable and in the sense of the physical structure of a community, they grow to form a relatively thin or thick layer between the earth and the air. That layer, if dense, protects the earth from the errosive powers of the atmosphere, including wind and water. It also provides a matrix in which organisms are mutually better able to exist because of the shelter, food, and substrate provided. All the nonautotrophs within that matrix are heterotrophs, dependent directly or indirectly upon the autotrophs' excess food to sustain their lives.

Phagotrophs are all the larger consumers, including such primary consumers as most insects and all other herbivores. The secondary consumers that prey upon herbivores are also phagotrophs. They are simply once removed from the source of the net food production that is provided by the autotrophs.

Saprotrophs are all the smaller consumers, mostly fungi, bacteria, and protozoans, that feed upon organic material just as do the phagotrophs. One major difference is that the saprotrophs are usually passive, feeding upon the dead (i.e., detrital) material produced by some other heterotroph or from some autotroph population. The saprotrophs digest the organic material upon which they feed to produce the mineral (i.e., inorganic) constituents from which that material has been originally synthesized by an autotroph. Therefore, saprotrophs are decomposer microorganisms that convert the organic residue produced by a community into a pool of basic mineral elements; these must be recycled in order to maintain the long-term ecologic stability of that community.

The habitat of most saprotrophs corresponds to the location of the highest concentration of organic residue from the community in question. In terrestrial systems, this normally corresponds to the uppermost layer of the surface of the earth, including the uppermost few inches of soil and the overlying litter that has collected there from the resident community.

One of the major contributors to ecologic instability is the long-term or permanent removal of materials from the recycling process normally found in nearly all natural and healthy (i.e., stable) ecosystems. The stress that such removal causes leads to the total depletion of the fundamental raw materials upon which that community depends for its ecologic balance. In a terrestrial system of autotrophs and heterotrophs, such as a forest or a garden, the removal of the prerequisite abiotic minerals from the system ultimately causes the biotic elements to starve and not repopulate the system until those minerals are returned to the cycle.

If the soil of the forest or garden is sterilized chemically or physically, removing the saprotrophs, the litter will not be recycled and any preexisting ecologic balance will be upset and then lost. The normal cycling of the organic litter of a terrestrial system dominated by vegetation begins with initial digestion by certain fungi and bacteria. They work on the larger molecules of the organic residue, including sugars, simple proteins, and the amino acids that make up those proteins. Other bacteria are able to decompose cellulose, a major constituent of terrestrial vegetation litter, whereas still other bacteria are associated with the final product, *humus*, resulting from these above interactions.

Humus is dark, moist in its natural state, and rich in organic material because it is not yet fully reduced to its most basic constituents by the participating saprotrophs. The *mineralization* of humus apparently requires a series of different physiologic manipulations performed upon humus by another series of saprotrophs, with the final result being the total decomposition of the humus into its constituent mineral forms. These include, for example, the release of nitrates and phosphates from the contained proteins of the litter/humus.

Under ideal circumstances of moisture and warmth, such as in tropical rainforests, the decomposition of litter occurs relatively rapidly. As much as a kilogram of saprotrophs (10^2 to 10^5 individuals) may be found in 1 m^2 of ground area. There, the depth of the humus is quite shallow. In less "favorable" climates, such as the temperate rainforest, the humus layer may be much thicker because lower temperatures reduce the metabolic activity of the saprotrophs, allowing the humus to accumulate.

Historically, the cultivation of virgin land usually proceeded with spectacularly productive results until the humus and natural mineralization residues of the land were farmed out. If crops were grown and removed, if only one species of plant was cultivated, if the fertility of the soil was not protected through application of organic litter, then the land would eventually become effectively infertile with no ecologic balance of autotrophs and saprotrophs. Such was the case when western colonists farmed New England until it was relatively infertile. Many Yankees then moved on to literally greener pastures in the Ohio Valley and further west.

With the natural ecologic succession of deciduous hardwood forests returning to the abandoned farm land of New England, the natural balance autotroph and saprotroph recycling of elements is also returning. However, the usefulness of our knowledge of ecology is limited because the pressures on the land are much different from what they were in the 1800s. If the human population is to be expanded, if intensive agriculture is a requirement of our society, then technologic solutions must augment our ecologic knowledge of how we can both use the land and save the land. An important part of that need rests on a better understanding of the soil as a central component in any terrestrial ecosystem. We must appreciate the fact that the biosphere is composed of the atmosphere, the hydrosphere (or water bodies of the earth), and the *pedosphere*, or land bodies.

The soils of the pedosphere, more than any other ecologic factor, are least amenable to a clear-cut separation of functions and constituents between biotic and abiotic factors. Soil is a combination of mineral and organic factors which, as suggested above, are normally in a dynamic balance. Soil is therefore neither mineral nor organic, but both; inseparable, but variable in the dominance or distribution of each factor, depending upon the ecologic state of the soil in question.

For example, soil is not a homogeneous mixture but usually occurs in relatively distinct layers, or *horizons,* parallel to the surface of the ground and each of different individual thicknesses (Figure 3.11). The uppermost layer of "topsoil" is the *A horizon* and contains the newly accumulating litter broken down into humus and being progressively decomposed until the organic content of the original litter is entirely converted to mineral constituents. Moving progressively downward from the ground surface, the A horizon is divided into an *A-0 horizon* of litter (the litter is sometimes subdivided into whole litter, duff, and leafmold); an *A-1 horizon,* which is the dark humus layer; and an *A-2 horizon,* lighter in color than A-1 because the humic acid and other dark colored constituents of the overlying layers have been leached out from, or altered in, the A-2 horizon.

The underlying *B horizon* is light colored and consists entirely of mineral soil, whose soluble constituents have derived and percolated down from the A horizon. Below the B horizon is the *C horizon,* which represents the native material from which the overlying horizons have derived and which

Figure 3.11. The idealized soil profiles of a virgin forested area compared with an eroded area are shown here. They represent conditions which are often found within a deciduous forest biome. The B_1 horizon contains leached materials including humic acids removed for the overlying layers. See the text for a further definition of layers.

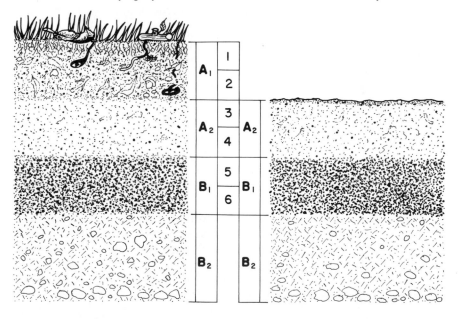

Virgin   Eroded

normally does not enter into the organic mineral recycling that occurs nearer the surface.

Aside from the physical characteristics and derivation of soils, their biological nature may be divided into three main divisions, based upon the size of the organisms present. These include the *microbiota,* the *mesobiota,* and the *macrobiota.*

The microbiota are microscopic and consist of algae, protozoa, bacteria, and fungi. Functioning in the decomposition of litter to its mineral constituents, or in the synthesis of new organic material in the case of the algae, the microbiota are preyed upon by higher level organisms within the food pyramid of the soil.

The mesobiota include larger animals, some of which are permanent residents of the soil and some of which leave the soil, such as the larvae of certain beetles. It is not unusual to find more than 10 million mesobiotic organisms living in 1 m² of soil. Although a large number of individuals, the total biomass of these individuals would be about 20 g. Primarily they feed upon detritus and bacteria, although some feed on other mesobiota and others feed on plant roots and on soil algae.

The macrobiota, as the name implies, includes the largest residents of the soil, including the roots of plants. Perhaps the best definition for this division of soil life is that the macrobiota can be readily segregated by hand and include such things as roots, earthworms, and burrowing animals. Depending upon habitat, it is not unusual for roots to represent the major macrobiota present in an average square meter of ground surface. Digging down as far as the root system extends in a mature prairie, for example, about 1 kg of roots (dry weight) may be collected. In a forest, more than 3 kg may be found (Figure 3.12).

With respect to the animal life in the soil, the macrobiota, such as earthworms, are important in mixing and aerating the soil. They are also important in initially breaking up the litter, making it more readily decomposable by the microbiota. In fact, the biomass of organisms in 1 hectare of forest surface may exceed 1 ton, of which about 0.2 tons would be earthworms (Figure 3.13). Furthermore, although life in the soil represents only about 1% of the dry weight of a sample of meadow soil, the presence and ecologic balance of this system is crucially important to the long-range ecologic stability of the terrestrial ecosystem as a whole.

The essence of this examination rests with the fact that soil, whether rich or poor, is a complex environmental system that provides a nutrient sink of mineral constituents that enrich and support much of the terrestrial ecosystem in which we live. In particular, the soil, in combination with climate, is a primary determinant for the type and character of vegetation that often dominates our perception of a terrestrial habitat.

From an ecologic point of view, the four factors that are most important in defining a terrestrial habitat include:

1. Temperature
2. Availability of water

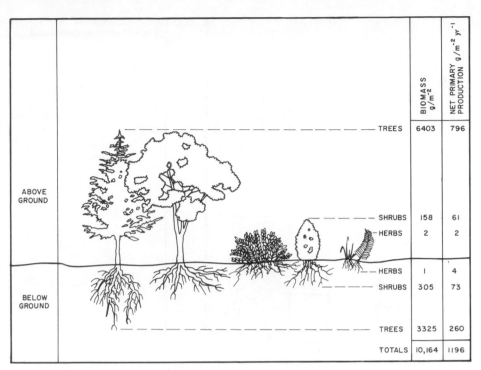

Figure 3.12. Shown here is the distribution of productivity in the environment of a young oak-pine forest. (Based upon data from Whittaker and Woodwell, 1969, from an interpretation by Odum, 1977.)

3. Mechanical–physical factors
4. Chemical constituents in the soil

All these are important factors in determining the ecologic distribution of vegetation within the terrestrial environment, and none of them is particularly interdependent in its presence or absence in a particular location. Radical or slight differences may be present within a relatively small geographic area with respect to any or all of these factors. Those differences, however slight, will often result in marked effects upon the type and character of the vegetational distribution, or *phytogeography* of that particular location. For example, the vegetation on the northern slopes of a hill may differ from that on the southern slopes because evaporation of water on the northern, and therefore more shaded, slopes will be less than on the southern slopes. If ground water is available to both slopes, however, the distribution of plants is not determined by water availability but may be influenced by other abiotic factors, such as temperature, mechanical factors present, or chemical constituents. Taken together, these factors comprise an environmental setting that usually grades imperceptibly from one point to another and is somewhat ambiguously reflected in the changing population of plants that are able to inhabit the land in question.

Gradient modeling of population distribution is an important ecologic

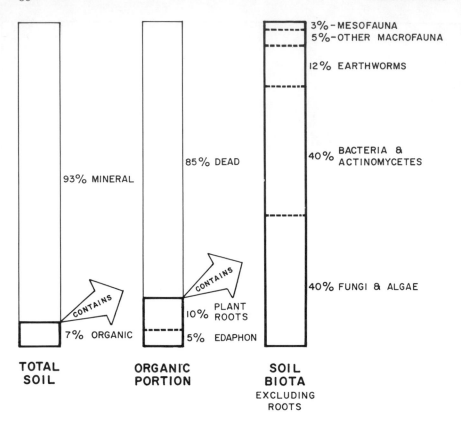

Figure 3.13. The living and nonliving components of soil make it a dynamic medium of crucial importance in determining the nature and ecological stability of those vegetative and animal communities which grow on and in it. (Diagram based on data from Tischler, 1955, *Synökologie der Landtiere*, Gustav Fischer Verlag, Stuttgart.)

concept useful also in providing a philosophic approach to our understanding of populations and community structures. Gradient modeling deals with the gradual changes and combination of ecologic factors that overlap, causing a particular point in the environment to fall within the niche of one or more species. As we consider different points that fall more within one species niche than in another, the favored species will naturally dominate in these more favorable environmental conditions.

From this point of view, the terrestrial landscape and the vegetation which it supports does not naturally consist of a distinct "checkerboard" of species distribution. One population usually grades into another and there are usually no sharp and discontinuous units of one life form as opposed to another in a natural environment. The gradient of abiotic factors in that environment reflects an equally dynamic biotic gradient, which leads to the concept of the *ecologic optimum*. That concept may be contrasted to the *physiologic optimum* for a particular species. The ecologic optimum of a species is represented by a natural environment in which the species is most abundant. The ideal conditions provided in the controlled and artificial environment of a laboratory, in contrast, represent a physiologic optimum that is virtually always different than the ecologic optimum for a particular species. Furthermore, the presence or absence of a particular species in a natural environment not only depends upon the capacity of the niche of that species and the inclusion of the subject environment within that niche capacity, it also relates to competition within the plant community.

Competition within a community of different species is measured by the relative success of one species to displace and ecologically dominate another species. This is called *interspecific competition,* whereas competition between members of the same species, such as one oak tree crowding another oak, is called *intraspecific competition.* Although the actual factors that cause one plant species to compete with another one are complex and involve all aspects of the mutual niches involved, the ultimate ecologic results make the final plant community better suited to survive the environment than may otherwise be the case. Intraspecific competition tends to crowd out the weaker members of a species population. Interspecific competition tends to result in a final population that is most successful (i.e., best adapted) to the environment that exists in that particular geographic location.

With respect to the broad vegetational zones of the earth, temperature and water and their seasonal patterns of fluctuation are especially crucial. These factors, as well as mechanical and chemical factors, combine with natural geographic barriers, such as mountains and oceans, to create the patterns of plant communities that we see today in the biosphere.

A unique characteristic of terrestrial ecosystems, as compared to aquatic ecosystems, is *transpiration*—the physiologic process by which water, usually absorbed through a root system in the soil, is lost to the atmosphere through leaf and other plant surfaces. A wilting plant is usually one in which more water is being lost to the atmosphere than is being absorbed from the soil.

From an ecologic point of view, transpiration, which requires a constant once-through circulation of water, leads to two primary observations. First, the niche capacities of plants have evolved to allow various species to successfully cope with a very wide variety of climates relative to water availability, from swamps to deserts. Second, transpiration causes plants to serve as living wicks that actively move water and its contained substances from one point in a plant to another point.

From a philosophic point of view, the basis for the natural success and dominance of a plant species within the terrestrial environment primarily lies in its consistently most efficient management of a water budget at a particular location within a particular period of time. That is, the physiologic processes of life that take place within a plant occur in an aqueous cellular medium. In terrestrial ecosystems, the balance of water intake and water loss is critical and those plants that can best manage this balance in a particular climate will be the dominant species of that climatic area. In fact, climatic and vegetational characteristics in the terrestrial environment are combined in defining the broadest ecologic divison of the biosphere into units called *biomes*.

These biomes are relatively vast areas of characteristic vegetational patterns (Figure 3.14). Supposedly, they are each in an ecologic climax as defined in Chapter 2. The biomes of the biosphere may be fairly well segregated into eleven types.

Tropical Evergreen Rainforests

The tropical evergreen forest biome usually has an annual rainfall that exceeds 200 cm, including at least one relatively dry season in which less than 12 cm of rain will fall each month. Combined with a mean annual temperature of about 25°C, the ecologic conditions are conducive to life because, at least from these abiotic considerations, the environment is not stressed, and vegetation proliferates. The upper dense canopy of leaves, exposed to the direct rays of the sun are desertlike. They are adapted to reduce transpiration loss of water and therefore they have a thick outer

Figure 3.14. These vegetational zones, or biomes, are much simplified without variations of soil or man influences shown. These generalizations include: The tropical evergreen rain forests (1), the semi-evergreen and deciduous tropical forests (2), the dry woodlands, and savannas or grasslands (2a), the semideserts and deserts (3, 7a), the chaparral or sclerophyllous biomes (4), the moist, warm forests (5), the temperate climate, deciduous forests (6), the steppes of the temperate zone (7), the boreal coniferous forests (8), the tundra (9), and the mountain biome (10). (From Walter, 1973.)

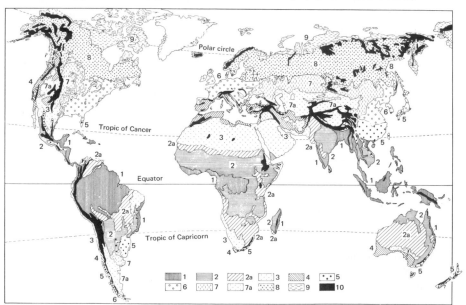

layer, or *cuticle,* and leathery leaves. These leaves form an upper canopy, which shades the forest floor and thus creates an entirely different, far more hospitable and homogeneous climate at lower levels in the forest. Between 0.1 and 1% of the incident full daylight which strikes the dense canopy of this biome reaches the forest floor. This light intensity at the forest floor is not much different than that found in a deciduous forest of temperate climates.

One of the most remarkable ecologic conditions of this biome is that the soils are extremely poor, in spite of the fact that the luxuriant vegetation contributes about 10 g of dry weight litter per day on each square meter of forest floor. Apparently, the nutrients required to support the forest are present in the living vegetation, with virtually no reserve present in the soil. The litter is mineralized so rapidly that nutrients are not leached out by percolation into the subsoil. In part, this theory is borne out by the fact that in this forest, runoff stream water is relatively pure, carrying virtually no salts.

Typically about 55 m high, the upper canopy of a rain forest is not continuous but is represented by scattered species reaching far above the other trees. A lower second canopy, and sometimes a distinguishable third canopy, are usually present with as many as 100 different tree species present in 1 hectare. Crowded in the forest, each tree usually has a small trunk diameter, and because annual growth rings are not present in the absence of seasonal changes, it is not easy to estimate the age of a forest. Based upon the growth rate of these trees, it is estimated that a mature and stable rain forest is about 200 or 250 years old.

Trees dominate the rainforest biome; about 70% of plant species there are trees. The prime ecologic danger of environmental manipulation of the rainforest is tree removal by natural or other means. The impact results in (1) removal of the nutrient reserves from the system, (2) exposure of the poor soil to erosion, and (3) alteration of the microclimatic conditions at the ground level. The virgin biome apparently cannot be reestablished after deforestation.

Semi-evergreen and Deciduous Tropical Forests

The semi-evergreen and deciduous tropical forests biome seems to be a result of a longer dry season and decreasing rainfall unequally distributed over the seasons. Experiencing an average of about 150 cm of rain, this forest type covers a relatively large land area. Humans have converted much of it to agricultural uses, for example, the savannas of Central Africa (Figure 3.14), where the original forests are burned during the dry season; this practice is relatively ineffective with the rain forest biome, which cannot be set afire. In the undisturbed forest, many of the trees lose their leaves in response to the seasonal decrease in water availability. Some species may drop their leaves for the duration of the dry season of 3 to 6 months, whereas the same species may retain its leaves if it is located in an area where ground water is present, such as a valley.

Apparently, the trees of these tropical deciduous forests lose their leaves as a physiologic adjustment to the fluctuating presence of water. Such a response is called *facultative*. Those species which do not lose their leaves often possess small, hard, leaves that appear desertlike and are able to survive the dry season. It appears that the metabolic activity of at least the soil organisms in the wet season is three times greater than it is in the dry season. As a result, the forest slows down during the dry season, producing a winter–summer landscape in response to water, not temperature.

Dry Woodlands, Savannas, and Grasslands

The dry woodlands, and savannas, or grassland biomes result when the naturally competitive and usually mutually exclusive niches of grasses and woody species overlap in an ecologically stable state, permitting the two groups to coexist. This results in savannas, a biome dominated by grasses but with woody species more or less evenly and sparsely distributed over the landscape.

In order for the savanna to survive ecologically, seasonal rain and deep loamy sand must be present, as they are in several areas of the tropics (Figure 3.14). The unusual balance between grasses and woody species in this biome results from complementary morphologic and physiologic conditions relative to their different root structures and water balances. Grasses have dense, shallow roots that draw water from the shallow depths; they respond to drought by most of the aboveground tissues dying. Transpiration is great in grasses when they grow, and within a short time after the end of the dry season, shoots sprout and primary production moves very rapidly. In the dry seasons, only roots survive.

Woody species have coarse roots that penetrate vertically and horizontally to a much greater degree than the grasses. In response to drought, the woody species reduce water loss through transpiration by closing the *stomata,* microscopic surface openings through which transpiration water vapor passes. As the drought continues, the leaves are dropped and only a relatively small quantity of ground water is necessary to carry the woody species through that drought.

The savanna survives because the climate, primarily water, stresses both grasses and woody species sufficiently to shift the dominance depending upon average seasonal rainfall. For example, in an arid region where 10 cm of rain falls, only grasses survive from one season to the next. As the rain falls, the grasses absorb it, preventing much from sinking below their shallow, dense roots (Figure 3.15). In areas where 20 cm of rain falls, more robust grasses grow with deeper roots than in more arid areas, but there is still insufficient water penetrating below the grass root level to support woody vegetation.

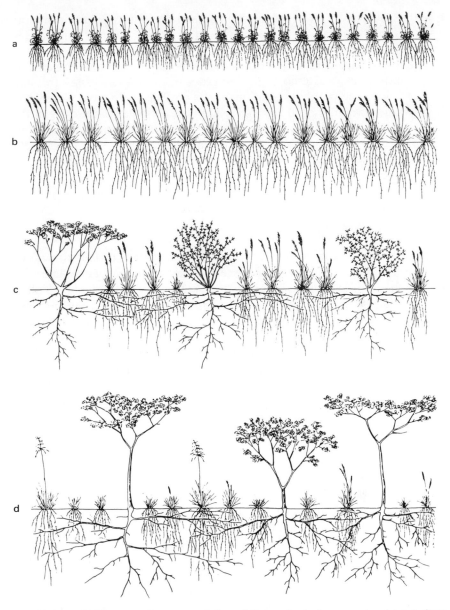

Figure 3.15. A diagrammatic representation of the ecological progression from a short or tall grass grassland (a and b) to a savanna (c) and to dry woodland (d), as explained in the text. (From Walter, 1973.)

When 30 cm of rainfall is available, some water does pass the grass root zone during the wet season. This allows some hardy, small woody species to survive the dry season by drawing upon the little water present in the deeper soil layers. If 40 cm of rainfall is available, more water passes to the deeper soil layers, allowing larger woody species to survive and forming a true tree savanna. A forest forms when greater amounts of rain can support a population of trees whose crowns touch, forming a canopy that shades out the grasses below.

An interesting disruption of this balance occurs when a savanna is overgrazed by cattle. Removing some grass reduces water losses through transpiration. The woody vegetation is thus provided with more water, and it grows more vigorously. Its seeds are rapidly distributed in the dung of the cattle and a scrubland is rapidly formed, replacing the savanna and thus eliminating the pasturage. The Texas mesquite thicket is a brush encroachment that has evolved from such unintentional disruption of the natural ecologic balance.

Semideserts and Deserts

Semideserts and deserts, whether hot the year round (in the tropics) or having relatively cold winters (in the temperate zone), are all characterized by the fact that evaporation is regularly much greater than total rainfall. The rainfall is characteristically less than 20 cm a year. They are also characterized by a relative sparseness of vegetation, partially determined by the availability of water.

Assume that an individual hypothetical species requires about 5.5 liters of water a day to survive, and that the prevailing hypothetical climate provides an average of 200 cm of rain each year. Therefore, 2000 liters of water each year, or about 5.5 liters a day, fall on each square meter of this hypothetical biome. Therefore, that species could theoretically survive at a concentration of one individual per square meter. If the rainfall is reduced in half and only 100 cm falls on each square meter, than the density of the hypothetical vegetation must also fall by half, which becomes possible if the species has the physiologic means to gather up the water it requires by drawing the water from 2 m² instead of from 1 m² of ground.

The water requirements of plants in the desert biome support physiologic processes that are fundamentally identical to the cellular processes of plant species growing in a rain forest. Although the external environments of the species living in these different biomes are markedly different, their internal life processes are the same. This circumstance results in the ecologic condition that the dryer the climate, the farther apart the plants grow. In arid environments, the root systems tend to spread out, resulting in more plant tissue below ground than above. The reverse is true in the hospitable climate of the rainforest, where water is not a factor limiting growth.

The genetic adaptations that have evolved to expand plant niches into arid environments seem primarily based upon methods of providing the most efficient use of water, rather than upon any alteration in the basic physiologic processes that require water. These plants are called *xerophytes* and they provide several interesting conditions of natural, as opposed to technologic, adaptations to a hostile environment. Some xerophytes have soft leaves, which wilt in dry conditions and drop off if the drought is extended. Others, such as the olive tree, have small, stiff leaves and are able to adjust their rate of transpiration to reduce water loss. Others close their stomates, reducing transpiration loss by that means. In all these xerophytes, some water is required throughout the year. The *succulents* are xerophytes that can store water and therefore need no external source for seasonally dry periods.

These typical desert species may have (1) succulent leaves; (2) succulent stems, such as cacti; or (3) succulent roots, such as asparagus or other species with large, underground tubers. Within this harsh environment, productivity is logically relatively low but the ecologic adaptations of the arid biome species results in their remarkable survival. For example, the oldest known living thing is the 4500 year old bristlecone pines *(Pinus longoeva)* growing in the very arid mountains of the southwestern United States. If even older species are discovered, they are most likely be found in the extreme environments, such as the desert, rather than the hospitable environments, such as the jungle.

Chaparral

The chaparral, or sclerophyllous biomes of the world represent an ecologic transition between desert and the semidesert conditions. They are dominated by trees or shrubs with hard, thick evergreen leaves, which actively grow during the wet winter season of approximately 6 months. With occasional winter frosts, an average temperature of about 17°C, and average annual precipitation of between 50 and 70 cm, the chaparral of central and southern California, the maqui of the Mediterranean, the renoster of South Africa, the mallee of Australia, and a similar ecologically stable transition zone in northern Chile are all sclerophyllous biomes.

Many of the plant species characteristic of this biome possess deep root systems, which may extend between 4 and 10 m into rock fissures, where water availability permits survival in the face of the summer dry season. Accumulation of litter in their dry conditions leads to the occurrence of natural fire by lightning. Fire appears to be an important factor in maintaining the character of the chaparral. The dominant species of the biome survive fire and the chaparral remains stable, provided fire sweeps it at least once in 12 years. If fires are less frequent, fire-sensitive species begin to encroach.

Moist, Warm Forests

Moist, warm forest biomes cover relatively few areas. They experience mild winters, for example, the redwood forests of California, the valdivion forests of southern Chile, and the Varri Forest of southeastern Australia. This biome is evergreen and is usually found on the eastern seaboards of the continents, where rising mountains cause trade winds or monsoon winds to carry up moisture-laden ocean air. The moisture condenses and falls along this elevated land, supporting the moist forest with a rainfall of between 70 and about 350 cm and a relatively high humidity all year round.

With approximately 1% of the available sunlight filtering through the trees to the forest floor, ferns, mosses, and other *hydrophylic* (water loving) plants produce a relatively dense, parklike environment. As in tropical rainforests, a large percent of the mineral content of this biome is present in the biomass of the standing crop. If harvested, the mineral content of the biome is diminished by an amount equal to the equivalent mineral content of the material harvested (i.e., removed).

Temperate Climate, Deciduous Forest

Temperate climate, deciduous forest biomes have relatively cold but not prolonged winters, with an annual precipitation of between 50 and 150 cm. This biome occurs only in the northern hemisphere and most forests contain both deciduous trees and evergreens. Unlike the leaf drop of deciduous trees in warm climates, which is facultative in response to drought, leaf drop in the temperate forests is obligatory because of some physiologic mechanism apparently triggered by the shortening days of fall. Such trees require a longer period of warm weather to sprout new leaves and survive than do the evergreens of this biome. The evergreens can survive with a shorter period because they do not lose their relatively small and dense needles, which can begin photosynthesis as soon as the winter breaks. These needle-bearing trees, the *conifers,* can survive in climates with only 30 days a year registering temperatures above 10°C, whereas deciduous trees require at least 120 days above 10°C. Therefore, as the average daily temperatures fall, the proportion of deciduous forest diminishes.

In its natural state the deciduous forest is a multilayered biome, just as are the other forest types discussed. It typically contains at least three layers, including one or two uppermost strata formed by the crowns of the taller tree species, an intermediate stratum of shrubs, and the lowest level, the ground cover or herbaceous layer. The lesser strata are adapted to grow in relatively low light levels because only about 0.6% of the incident sunlight reaches the forest floor in an oak forest about 10 years old, whereas about 2% penetrates a mature oak forest. The mature forest is more open than the young, crowded forest.

Relative to the productivity of this biome, an eastern European or an eastern United States deciduous forest may have a net primary productivity of 12 tons per hectare, of which 5 tons is wood. The amount of chemical energy contained in this biomass represents about 1% of the incident light energy striking 1 hectare in 1 year. About 33% of that light energy is dissipated in transpiration, and about 65% is dissipated as heat energy, eventually radiated back into space.

Temperate Zone Steppes

Steppes of the temperate zone are biomes that have a rainfall averaging between about 25 and 50 cm each year, with an average annual temperature of about 5°C and a growing season of about 190 days. They are characteristically plains with grasses on a soil derived from wind-blown deposits *(loess)*. The roots of grasses may penetrate to a depth of 2 m. Although there are transition zones between the forest and the steppe, it appears that the forest can compete successfully only when its location is well drained, through porous soils, whereas the steppe is successful in competing for space in which to grow on poorly drained and heavy soils. On the mature steppe, unless transplanted by humans, the seeds and seedlings of trees are not able to survive the crowding of the steppe grasses.

Stressed by a scarcity of water, grasses can survive more effectively than can trees. A savanna does not appear because the deciduous species types associated with a savanna cannot compete for space with the dense steppe grasses to get a foothold.

The prairie grasslands of the United States consist of tall grass prairie, mixed tall and short grass prairie, and short grass prairie. They are found in this order moving from east to west, which appears to be an ecologic response to decreasing precipitation. Here, the exclusion of trees seems clearly linked to (1) the difficulty with which tree seedlings compete with grass coupled with (2) the sensitivity of trees to fire. Fire favors grasses but, in its absence, the forest can encroach upon the prairie at the rate of about 1 m each 3 to 5 years.

A particularly interesting example of environmental management is seen in the natural interrelationships between grazing and grasses in this biome. In the absence of adequate grazing, the aerial parts of the grasses seasonally die and accumulate at such a rate that steppe felt, or tatch, develops to the point of reducing the regeneration of the dominant grasses. Weeds encroach and the preexisting steppe vegetation is altered. The natural grazing of antelope, horses, rodents, and insects was apparently sufficient to prevent the excess accumulation of litter. In the absence of such grazing, mowing the steppes every third year, or fire management, seems adequate to prevent the natural imbalance described above from occurring.

Boreal Coniferous Forest

Boreal coniferous forests form an extensive biome without a distinct transitional zone between the warmer deciduous forests of lower latitudes.

Between 25 and 50 cm of rain and snow fall each year in this biome, and there are fewer than 120 days each year with an average temperature above 10°C. The dominant type is the needle-leaved evergreen tree, including spruces, pines, and firs. These grow to form a dense and year-round canopy that prevents the colonization of the forest floor by all but the most shade-tolerant species. Furthermore, conifers are very efficient in absorbing nutrients and can apparently stress other vegetation by reducing the availa-bility of soil nutrients, such as nitrogen, to those other species. A rich although shallow humus soil is generated by the litter of needles and wood dropped to the forest floor each year at the rate of about 1000 tons per hectare in mature forests. About 950 tons a year is recycled through the natural decomposition of that litter.

Tundra

The tundra is a biome in which less than about 25 cm of precipitation falls and where only about 30 days a year have an average temperature above 10°C. The cold season lasts at least 8 months and during the growing season, the top several centimeters of ground thaws. As a treeless, wet arctic grassland the grasses, lichens, and sedges dominate the landscape, with occasional woody plants dwarfed by the harsh environment. The climate is humid even though precipitation is low because low temperatures reduce evaporative water loss from the air near the ground. Therefore, an excess of water, coupled with very slow microbial decomposition, leads to the formation of a thin peaty layer at the base of the growing plants.

Physical and mechanical factors influencing the habit of growing vegeta-tion relate to the abrasion caused by blowing snow and ice. An interesting adaptation by some plants to this environment has been development of a life cycle extended over several seasons. This allows the plant to grow in relatively short bursts, taking advantage of the brief summers so that seeds may be developed over a period of a few years to be released very early in the summer of an ensuing year. This capability increases the likelihood of the seeds' sprouting, growing, and accumulating sufficient food reserves to survive the winter. The primary net productivity of the tundra is about 14 times smaller than that of a deciduous oak forest.

Mountain

The mountain biome is a complicated region with a wide variety of climatic differences. Perhaps the only consistent feature of this biome is that the average annual temperature decreases with increasing altitude. Based upon this characteristic, coupled with the distribution of precipitation and soil conditions, the complex mountain biome usually is represented as a series of roughly parallel irregular bands ringing the mountain.

In this sense, the typical mountain may be a composite of several biomes, as previously described, with their location determined mainly by the climatic factors of temperature and precipitation. For example, creating a hypothetical mountain about 10,000 ft high, rising from a desert floor, we might encounter the desert biome extending up to an elevation of about 1000 ft, where cooler temperatures lead to the appearance of a grassland biome reaching up to about 3000 ft. With cooler temperatures and slightly more precipitation, a sclerophyllous scrub forest could lead up to 5000 ft, where a still cooler and wetter climate could support an evergreen, coniferous forest capable of climbing to 8000 or 9000 ft. There the climate may cause this forest to give way to a tundra biome, crowning the mountain in its coldest region.

Whatever its vegetational form, the terrestrial environment serves as a vertical transitional zone between the atmosphere and the earth as well as a horizontal transitional zone between the communities within the ecosystems of a biome. That characteristic is amply demonstrated by the preceding catalog of orderly divisions and it leads into the next section of this chapter, which treats a major nonterrestrial aspect of the biosphere, the aquatic environment, beginning first with fresh-water ecology.

Fresh-Water Ecology

From a philosophic point of view, ecology is a system of studies that attempts to make order out of the slow and natural autodigestion of the biosphere by our environment. The sequential processes that move elements through the biosphere may be physical, such as erosion of soil by water and wind, or biologic, such as digestion and assimilation by living organisms. Transformations occur and water is always directly or indirectly involved in both those physical and chemical transformations through which elements pass from one point to another in the biosphere.

As suggested earlier in this book, all life is a water solution. Our own bodies contain about 70% water. We require about 1.5 liters of water each day for our survival, and terrestrial ecosystems are directly dependent upon water as a determining factor in the structure and functioning of all the biomes on earth. In these senses, water has been called "the web of life," implying the interdependent and complex nature of life and water.

From an ecologic point of view, and aside from the physiologic requirements of life in which water participates, water serves the biosphere quite literally as a conveyor belt, linking elements and activities in one biome with those in another. The movement of water is therefore a crucial factor in ecological systems and its basic pattern of movement is composed of four major divisions.

1. Water evaporates into the atmosphere from water bodies.
2. From the atmosphere, the water vapor precipitates, some falling on the land.

3. On land, water percolates through the soil or runs over the ground, moving toward the lowest point of the watershed in question.
4. The water eventually returns to the ocean, from whence most of it had evaporated in the first instance.

Summarized on a global basis, the cycling of water through the biosphere is essentially balanced and there is no net gain or loss over a period of several thousand years. The actual movement of water through the atmosphere is quite rapid and complex and it creates problems that must be faced by an ecologist who seeks to trace the movements of water-borne materials in any ecosystem. For example, each year about 88 cm (3 ft) of water, or 100 million billion gallons, evaporates from the oceans of the world. In addition to the water vapor, a considerable quantity of salts also enters the air as a result of such surface phenomena as bursting bubbles and spray. By such means, about a third of the salts found in rivers seems to be derived directly from the sea salts returned to earth by the precipitation that supplies the rivers with their water.

Precipitation directly into the oceans accounts for about 78 cm (32 in.), or 91.5 million billion gallons, each year. On the land, about 44 cm (18 in.), or 16 million billion gallons of water, evaporates, and about 66 cm (27 in.), or 24.5 million billion gallons, falls as dew, fog, rain, or snow on the land. From the land, some water flows downgrade on top of the ground as *surface water* and some moves through the ground as *ground water*. The length of this journey may be very short or it may last hundreds or thousands of years. However, each year about 22 cm (9 in.) of water, or 8 million billion gallons, return to the oceans as surface water runoff, and about 1.2 cm (½ in.), or ½ million billion gallons, of water return to the ocean each year as ground water.

Some of these estimates may be larger or smaller by as much as 20%. However, the essential magnitude of water movement is correct and the very dynamic nature of that cycle is the important illustration drawn from this investigation. Temporary imbalance may occur if, for example, a new ice age were to develop, causing more water to be held in the glaciers than would otherwise be the case. If the reverse were to happen, and our present glaciers were to melt, we could expect the oceans to rise about 61 m (200 ft). That would result in an increase of 1½% in the water volume of the world oceans.

From an ecologic point of view, the movement of water through the atmosphere is important not only in studying climate and the hydrologic cycle of any ecosystem; water is also important as the universal solvent and as a scrubbing agent, which tends to move materials from one place to another. Gases, pollen, and other particulates are washed out of the atmosphere by precipitation. Remarkably, it takes about 12 days for all the water in the atmosphere at any one instant to be completely replaced by new water molecules previously in the oceans or on the land. However, because the process by which precipitation washes "impurities" from the atmosphere is not especially efficient, it is estimated that it takes at least several years to scrub out the impurities contributed to the atmosphere in any 1 year.

Let us, for the moment, consider only the surface waters within the terrestrial environment. Precipitation collects in streams and rivers, which may or may not enter ponds or lakes, in its movement to the lowest elevation of a terrestrial environment or to the seacoast. We can estimate that streams and rivers hold about ten times more water at any one instant than does the atmosphere. As a point of special interest, the Amazon River, fed by its surrounding watershed of tropical rainforests, carries about 20% of all the stream water on earth.

Ponds and lakes hold about ten times more water than the streams hold at any one instant. This pyramid of volumes is ecologically significant, especially as it relates to the renewal or flushing times involved in these different systems. Therefore, the atmosphere can recycle its water more quickly than can the streams of the earth. Similarly, streams are more rapidly renewed with water than are ponds or lakes. The oceans, in contrast, are never renewed because they serve as the final sink for all manner of materials leached from the atmosphere and the terrestrial environment by the waters moving down through those fresh-water environments to the oceans.

As soon as precipitation reaches the earth, it begins its migration in response to gravity (Figure 3.16). Although it may accumulate as snow and

Figure 3.16. The Terrestrial Water Cycle shows the pathway of precipitation divided between groundwater and surface water. The water migrates toward the ocean in response to gravity and is a primary factor linking one ecosystem with another.

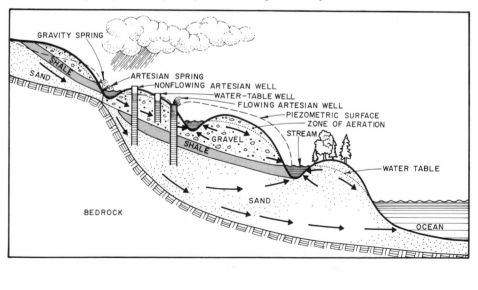

ice to form a glacier, more usually it moves as liquid water along the surface or through the ground, changing in chemical constituents as it progresses. It may take 135 years for ground water to move through 1 mile of sand, depending upon many factors, such as grain size and the pressure driving the water. During its passage through the various mineral constituents of the earth, water may dissolve or deposit some of those minerals through various geochemical processes. The specific natures of these chemical changes are of considerable importance to the responses seen in those ecologic systems that utilize this water. Just as with the terrestrial biomes previously described, an orderly sequence of aquatic environments can be constructed with respect to physically limiting factors imposed upon the ecologic niches evolved within those aquatic environments.

Yet the logic here obviously cannot revolve upon water availability, for, by definition, water is abundantly present in these aquatic environments. Instead, in this division of ecology, temperature, currents, and nutrient content are especially important considerations in the determination of those communities which form and become stabilized. In particular, the physical chemistry of water is a central factor in determining the ecologic nature of aquatic ecosystems, their sensitivity, or their stability relative to imposed ecologic alterations upon the system.

For example, water has a *high specific heat,* which means that a relatively large amount of heat is required to change the temperature of water. As a standard of measure, 1 gram-calorie (cal) of heat will raise 1 cc of water 1°C (between 15° and 16°C). Only ammonia and a few other substances have a specific heat factor any larger than that of water. From an ecologic point of view, this means that an aquatic environment tends to have a much more stable temperature than might a contiguous terrestrial environment, because fluctuations in the air temperature are much more rapid than are those of water.

Water also has a *high latent heat of fusion.* If 80 cal are removed from 1 g of pure water at 0°C, the temperature will not change but the water will crystalize into 1 g of ice. Conversely, if 80 cal are added to 1 g of ice at 0°C, the ice will melt to form 1 g of water at 0°C. In addition, water has the highest known *latent heat of evaporation.* One gram of water requires the addition of 536 cal for that water to evaporate into vapor with no change in temperature.

All three of these characteristics allow water to serve as a temperature moderator for the terrestrial environment immediately adjacent to the water body as well as for the aquatic environment itself. Its temperature fluctuation is greatly diminished by the heat sink capacity of the water.

Perhaps the most curious and ecologically important single characteristic of water is the fact that the greatest density of fresh water occurs at a temperature of 4°C. Above and below that temperature, water expands to become less dense. For this reason, lakes do not freeze solid and, furthermore, they freeze from the top down, not the reverse. As the density of water increases, consistent with a declining temperature, the water sinks, displacing lighter and warmer waters. If the water temperature drops below 4°C, that body of water rises and may ultimately freeze at 0°C, forming ice on top of the water body. Further drops in temperature causes the ice to expand, becoming less dense. If the water body is sufficiently shallow, it will eventually freeze through. Most often, however, once the ice covering is formed it insulates the remaining liquid and retards further heat loss.

Given such characteristics, it is usual for aquatic organisms to be more *stenothermal* (i.e., having narrow temperature range tolerances) than terrestrial species. For example, the antarctic fish *Trematomus bernocclai* can live within a range of only 4°C, between −2° and +2°C.

Another physical feature that is important in limiting the exploitation of the aquatic environment by plants and animals is transparency. The degree of cloudiness, or *turbidity*, within a water body not only affects the visual perception of certain animals in that environment, it also decreases the depth to which sufficient light can penetrate in order to support photosynthesis in green plants. The depth at which the rate of photosynthesis is equal to the rate of respiration is referred to as the *compensation level*. Light intensity at that level is usually about 1% of the full sunlight striking the surface of the water. For comparison, the intensity of light at the compensation layer is of approximately the same relative strength as the light that filters through the thick deciduous canopy of a forest to reach the forest floor.

The movement of water (i.e., current) is another physical factor that defines an important aspect of the aquatic environment. Currents not only physically limit the distribution of organisms, especially swift currents, but they also determine the distribution of gases and other elements within aquatic systems. Among these gases, respiratory gases and dissolved oxygen (DO) in particular are important to the ecologic status and balance of any fresh-water ecosystem.

Oxygen is primarily transferred to water bodies from the atmosphere, which contains about 20% oxygen. At 5°C and 760 mm mercury air pressure pure water will hold 8.96 ml oxygen in solution in each liter. In warmer water at the same atmospheric pressure, 6.42 ml of oxygen can be absorbed in a liter of pure water. If there is turbulence or if active photosynthesis within the water body is generating oxygen, a supersaturated condition in which more oxygen is contained in the water than can theoretically be present, is likely to hold for awhile.

The importance of the oxygen concentration in aquatic environments is particularly relevant to the distribution of certain animal species, which require oxygen to respire. Many have very specific ranges in which they can survive for long periods (Table 3.1). Fluctuations within the environment that cause these required ranges of oxygen content to change also generate corresponding dramatic and rapid changes in at least some of the animal populations within the ecosystem in question.

Although each fresh-water ecosystem must receive a site-specific evaluation, including some determination of the resident population, a general

Table 3.1. Dissolved Oxygen in Fresh-Water Systems in Parts per Million (mg/liter)

Tolerance range (ppm of O_2)	Fish species
7–11	*Salmo trutta* (trout)
	Phoxinus phoxinus (minnow)
	Cottus cobio (bullhead)
5–7	*Thymollus thymollus* (grayling)
	Leuciscus cephalus (chub)
	Lota lota (burbot)
4	*Rutilus rutilus* (roach)
	Acerina cernua (ruff)
0.5	*Cyprinus carpio* (carp)
	Tinca tinca (tench)

(Adapted from Nikolsky, 1963.)

guideline may be used; it calls for an oxygen concentration of 3 to 4 ppm for waters with an average temperature greater than 12°C (warm), and 5 ppm or greater for fresh waters with an average temperature of less than 12°C (cold). Under those conditions, one can expect the distribution of fish to not be restricted, at least because of oxygen.

Carbon dioxide is also an important gas in the fresh-water ecosystem in part because it reacts with the water to form carbonic acid (H_2CO_3), which in turn reacts with bases to form carbonates and bicarbonates. These factors provide a source of nutrients for plant growth and also buffer the aquatic environment, keeping the hydrogen ion concentration near a neutral pH level of 7. In addition to the dissolution of carbon dioxide from the air into the water, the gas is also added to the water by the processes of respiration and decay and from soil and subterranean sources.

In addition to these important gases in the aquatic environment, a number of crucial *biogenic salts* may be found which, by definition, are vital to sustained ecologic balance of the system under study. These are the nutrients that permit primary productivity to continue in the presence of adequate light; they are often quantified when an ecologic study is undertaken to evaluate the real or potential ecologic condition of a water body. In their general order of decreasing importance to aquatic ecosystems, six prime chemical factors include phosphorous (PO_4^-), nitrate (NO_3^-), potassium (K^+), calcium (Ca^{2+}), sulfur (S^{2+}), and magnesium (Mg^{2+}).

In fresh-water systems these and other chemicals contribute to a weak salt solution, seldom exceeding 0.5 part per thousand. They constitute the environment surrounding the organisms in fresh water and, compared with the much higher salt concentrations encountered in approaching sea water, with 32 parts per thousand (32%) of salinity, fresh-water organisms have problems of *osmoregulation*. They require highly evolved physiologic mechanisms of control. This compensation for osmosis is necessary because in fresh water, the body fluids of an organism are usually more salty than the surrounding water. Therefore, water will tend to diffuse into the organism, causing it to swell. Such body fluids are *hypertonic* to the surrounding aquatic environment.

Those organisms which have evolved to a point at which they can control this influx of water are able to survive. However, many organisms cannot compensate for osmosis and they remain restricted to salty environments, including the oceans, where their body fluids are *hypotonic* (less salty than the environment). There, it seems physiologically easier to excrete excess salt and/or to retain water in order for the organism to survive with a stable internal environment.

From this fundamental information, we can define fresh-water ecology relative to a hypothetic middle plateau between the terrestrial environment and the marine environment. In general terms, in the fresh-water environment, water is not a limiting factor; neither does temperature widely or rapidly fluctuate as it does in terrestrial systems. Other factors may be chosen to use as guides in categorizing the aquatic environment similar to some of the classification schemes used to describe components of the terrestrial environment. Some main differences seen between terrestrial and aquatic ecology rest with the buoyancy and currents present in a homogeneous aquatic environment compared with the unbuoyed, heterogeneous nature of the terrestrial environment.

The concept of biomes in the terrestrial environment cannot be satisfactorily transferred to aquatic environments because (1) individual aquatic environments are much smaller than a biome, (2) aquatic systems are present within the biomes, and (3) the genesis of the ecologic analysis of terrestrial versus aquatic environments seems to have evolved from different human directions. That is, it is uncommon indeed for a terrestrial ecologist to know much about aquatic ecology, and vice versa. Yet, certainly the fundamental principles of each are identical. Therefore, because the scientific logic of terrestrial ecology is identical to the scientific logic of aquatic ecology, the practical differences must reside in the scientist, not in the science.

In any book by any single ecologist, his or her specialty virtually always dominates in quantity or in quality by comparison with his or her nonspecialties. For a student to avoid this unilateral imbalance, he or she must make a very conscious effort from an early start. The best corrective training is to seek out opportunities to investigate applied projects involving ecological linkage between terrestrial and aquatic systems. Building a confidence that crosses the environmental boundries involved forces the recognition that a continuum of ecologic principles can be transferred from one system to the other and that the analytical sequence followed in one can be applied to the other. From the point of view of environmental structure, for example, the aquatic environment is ostensibly identical in fundamental ecologic structure to all terrestrial systems.

For example the distribution and basic functioning of organisms in a fresh-water environment may be grossly divided into trophic levels which includes first the autotrophs, or green plants, which produce food from minerals, water, and light. Second, there are heterotrophs in the aquatic system, including the phagotrophs (macroconsumers, which directly or indirectly feed upon the producers) and the saprotrophs (microconsumers, which serve to decompose the dead organic materials within the ecosystem).

There are five broad categories of life forms whose habits define the trophic levels within fresh-water environments, beginning at the bottom of any aquatic environment and at the interface between the bottom sediments and the overlying water. Organisms that live in the bottom and may or may not extend up above the bottom are called *benthos* and are subdivided into *phytobenthos* (plants) and *zoobenthos* (animals).

The zoobenthos may be further subdivided by feeding habits. Those which pump water and sift out food particles are *filter feeders,* such as clams and oysters, and those which extract food from the sediments are *deposit feeders,* such as certain snails. Because the benthos are in direct contact with the sediments and because they usually feed upon those sediments or particles suspended in the water, they tend to reflect the chemical nature and overall ecologic stability of the particular aquatic environment in which they occur. As will be pointed out in more detail in subsequent chapters, examining the benthos is often a useful means of evaluating the general status of an aquatic environment.

The next category of organisms is called *periphyton* (or *Aufwuchs*). It includes both plants and animals that are not directly imbedded in the bottom of an aquatic environment but that are attached to any surface, such as benthos, racks, or pilings, that extend off of the bottom. Like the benthos, these plants and animals reflect the ecologic state of the surrounding aquatic environment and, depending upon the nature of the ecologic information sought, the periphyton may be used as a diagnostic summary of the productivity or nature of the community.

The third category of plants and animals in the aquatic environment is the *plankton* (from the Greek, to wander). The plankton consists of those organisms which are generally carried about by water currents, although some have weak swimming abilities. However, none can effectively overcome being transported by the currents, although many can move vertically, apparently in response to light. The plankters are usually small, often microscopic, and are subdivided into two size-related categories. *Net plankton* includes all those species which are retained in a net with a net size of 200 mesh per inch. *Nanno* (dwarf) *plankton* consist of those organisms which are not retained. They are usually collected by taking a sample of water, which is then filtered through fine-pored membranes in the laboratory.

Plankton is an important indicator of the ecologic state of an aquatic environment because it reflects the availability of nutrients and the ecologic productivity of that environment. It may also be more conveniently collected than other organisms and it lends itself to statistical as well as chemical analysis.

The fourth category of organisms in this environment is the *nekton,* which are relatively larger than the plankton and which have strong swimming abilities. They are usually quite able to avoid being trapped by a plankton net. Fish represent the main body of organisms in the nekton.

Within a large regional area, the nekton certainly can be felt to represent an aspect of the ecologic state of that area. The fundamental problem with an ecologic evaluation of any mobile population is that, by their nature, these populations may represent a state foreign to the one in which they are recorded. Whether conscious or subconscious, it appears that the most reliable and site-specific ecologic analysis of a community is primarily based upon an evaluation of the *sessile* (i.e., stationary) populations. Most often that sessile population is dominated by plants, not animals.

The fifth and last category of organisms defining the aquatic environment is the *neuston.* These include all those organisms which float on the surface of the water or which are dependent upon the surface film of water. Mosquito larvae are associated with the bottom of the surface film, whereas the water strider is associated with the top of the same surface film.

The distribution of these five categories of life forms is determined by the physical conditions present within the habitats of the aquatic environment. Most often, these categories overlap in a habitat and we can recognize that any particular habitat is more or less an expression of some basic ecologic considerations that define two fresh-water habitat types: (1) ponds and lakes and (2) streams and rivers. All other fresh-water habitats are variations of these two types and, therefore, the basic *morphologic* (i.e., form, structure, and development) *nature* of the fresh-water environment can be defined by determining which characteristics of the two basic habitat classes are present in the subject habitat.

The first two characteristics pertain primarily to bodies of fresh-water, ponds, lakes, and bogs and some aspects of swamps and marshes. These habitats are collectively called *lentic communities* because their dominant feature is standing (i.e., still) water (*lenis* = calm). Within these communities, there are two groups of zones related to water depth and light penetration. The *littoral zone* is relatively shallow, with light penetrating to the bottom at sufficient intensity to support rooted green plants in the benthos. Ponds may be defined as fresh-water bodies containing only the littoral zone.

In lakes, a *limnetic zone* is paired with the littoral zone. It is the open water zone that descends from the surface to as deep as the lower limit of the adjoining lower limit of the littoral zone (Figure 3.17). That lower limit of both the littoral zone and the limnetic zone is called the *compensation level* where, with about 1% of the incident sunlight impinging on the surface of the water, the net productivity of green plants is zero. That is, the energy produced through photosynthesis is exactly balanced by the expenditure of energy required for the respiration of the green plants in question.

The littoral zone, therefore, is usually dominated by rooted vegetation, whereas the limnetic zone is dominated by plankton, nekton, and neuston

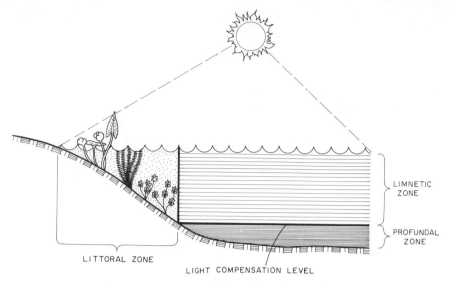

Figure 3.17. The lentic environment is subdivided into three zones as shown above based upon the depth of water penetrated by sufficient light to support plant growth. That depth is the *compensation level* and may or may not coincide with the depth of the thermocline of a summer stratified lake in temperate climates.

communities. This open water complex of suspended plants and animals intermixes with the shallow littoral zone. A physical relationship based on the compensation level is used to designate two zones describing fresh-water ecosystems with respect to the potential availability of light. The *euphotic zone* includes the littoral and the limnetic zones. The other zone includes the waters below the compensation level, referred to as the *profundal zone,* where respiration exceeds photosynthesis.

The second class used to define fresh-water habitats refers to moving water bodies, including streams and rivers. These are *lotic* (*lotus* = washed) communities, which are further defined by the possible presence of *rapids zones* and *pool zones*. The rapids zone consists of shallow and swift currents that prevent the accumulation of silts or other light materials. A hard bottom usually supports a benthos that does not penetrate into the bottom (i.e., substrate). The pool zone represents deeper, slower moving water, where silts and detritus can settle out of the water column to form a relatively soft, penetrable bottom. These characteristics usually allow such a habitat to support a substantial burrowing benthos which, in turn, may support a nekton population dependent upon that benthos and the accumulating sedimenting materials for food.

Regardless of the specific fresh-water environment that is being investigated, an ecologic overview of the biotic elements of that environment can be constructed as a guide to perceiving the relationships between various aspects of a specific community. If these characteristic biotic factors are ranked relative to their dominance and importance to the ecologic balance of a system, the following generalizations can be made.

1. The flora of a fresh-water system is ecologically more important than the fauna and is usually represented by the *algae* (i.e., aquatic plants containing chlorophyl and having no roots, leaves, or stems) and by aquatic seed-bearing plants (i.e., containing chlorophyl and having roots, leaves, and stems). In some cases of unnatural enrichment of the water, *phytoplankton* may dominate the system.
2. The fauna of a normal fresh-water system is usually dominated by the biomass of four groups of animals. These include mollusks, such as snails and clams; aquatic insects; crustaceans, such as shrimp; and fish. In addition, in some systems that have been ecologically disturbed or enriched by some nontoxic pollution, other groups of animals are likely to dominate the biomass. Usually, these upset systems become very restricted in their diversity and often contain a great many individuals of such groups as the segmented worms (annelids), wheel animals (the microscopic rotifers), protozoan, and/or roundworms (nematodes).

The saprotrophs of the aquatic system consist of bacteria and fungi just as they do in the terrestrial system. They are of primary importance in recycling the organic materials within the *detritus* of the ecosystem back into a pool of simple mineral nutrients available to green plants from their primary production of food. The saprotrophs are present in all aquatic systems, but their biomass is usually quite small, although their relative activity (i.e., respiration) may be very great.

When the plant and animal structure of the lentic ecosystem is examined further, divisions of biotic structure can usually be identified as concentric bands of vegetation around the pond or lake in question (Figure 3.18). The shallowest water, usually restricted to the shoreline, contains rooted plants that project above the water surface, with most of their leaf surface in the air. These *emergent* plants, consisting of such species as cattails, rushes, and pickerelweeds, provide a buffer zone between the water and the land. They provide protection and food for many animals, such as shore birds. Emergents also provide a ladder for larval aquatic insects by which they can leave the water environment to pupate on the stems and enter the aerial environment.

In somewhat deeper water, a zone of floating-leaf vegetation, water lilies, is often found in both lakes and ponds. The presence of such a zone provides surface area for the attachment of eggs by insects, snails, and other organisms. It also provides a habitat for certain nekton that seek shelter and food within and adjacent to the lily pads.

In the deeper water of the littoral zone, beyond the water lilies, a zone of *submergent* plants may be found. Usually with small and/or feathery leaves, which provide a large surface area for the physiologic exchange of gases and other materials with the water, these plants extend down as far as the compensation level and the end of the littoral zone.

Within the littoral zone, and extending beyond the littoral zone in lakes, the phytoplankton is usually represented by a variety of species, including the single-celled diatoms; green algae, including single-celled and fillamen-

Figure 3.18. The littoral environment is populated by three types of rooted water plants including the *emergents* which extend above the water, those *floating* with leaves on the surface of the water, and those *submerged* which do not reach to the surface of the water. Phytoplankton of the littoral zone either drift or are attached to the submerged stems and leaves of the rooted vegetation.

tons or colonial forms; and blue-green algae. The latter are of special interest because, like certain soil bacteria, these algae can take molecular nitrogen directly out of the atmosphere and incorporate it into the organics which it synthesizes.

With respect to the composition and distribution of animals within the lentic community, there appears to be a direct relationship between the diversity of the vegetative species and a parallel diversity of the animal species. In particular, it appears that the littoral zone has the greater diversity of animal life as compared with the other zones. However, unlike the horizontal zonation of concentric rings of vegetation, the zonation of animals appears to be much more vertical. The more or less sessile organisms, at least, tend to be associated with surfaces and sediment types rather than with depths or light intensities.

The periphyton, consisting of such organisms as snails and damselfly nymphs, forage for food on submerged surfaces, whereas benthic organisms, such as crayfish, shrimp, and certain larvae, reside and feed in close association with the bottom. A *subsurface benthos,* or *infauna,* is represented by clams, worms, and certain larvae that are found in relation to the characteristics of the sediments, burrowing into those sediments rather than moving on top of them. An interesting and relatively little studied benthic zone is the psammolittoral habitat, which consists of the microscopic community of organisms existing among the beach sand grains. Apparently especially sensitive to such disturbances as chemical pollution, the ecology of the psammon may become an important aspect of pollution monitoring.

The nekton of the littoral zone is usually restricted to those beetles, amphibians, and fish which nest and are protected within that zone. In the same general manner, neuston of the littoral zone consists of water striders, mosquitos, protozoa, and other small organisms that are associated with the surface film of water and that find some physical protection in the littoral zone as compared with the much more exposed physical and biologic environment of the limnetic zone.

The *zooplankton,* or animal plankton, of the litoral zone consists of small crustaceans, rotifers, some insect larvae, and other invertebrates and is often found in an inverse relationship to the phytoplankton. It appears that a large population of phytoplankton will support a growing population of zooplankton. As the animals feed on the plants, the zooplankton become dominant in place of the phytoplankton. If all else is equal, the final large population of animals and low population of plants is unstable because the animals soon decline in the absence of the food provided by the plants.

It is not at all unusual for the net primary productivity of the phytoplankton contained within the limnetic zone to exceed the productivity of the littoral zone of that lake. Continued production is possible provided the phytoplankton remains within the euphotic zone, which is achieved by the physiologic adjustment of buoying mechanisms. These include gas or oil bubbles and the coincidence of convection currents in the lake, which keep the synthesizing phytoplankton near the surface.

Under favorable environmental conditions, a rapid increase in the plankton may occur, which is called a *bloom.* Such an event often occurs in the spring and fall, when nutrient cycling and temperature coincide such that the phytoplankton of a lake produces a massive increase in biomass. Eventually at least one of the necessary nutrients is depleted or some toxic waste from the growing population grows in concentration in the water, causing the bloom to cease (See Figure 5.3).

Zooplankton in the limnetic zone of lakes is somewhat different from the zooplankton found dominant in the littoral zone. There are generally different and fewer species in the open water. Many of these species gather their food by straining the water through *combs of setae,* which look much like a microscopic hair comb and function much like a strainer. These animals feed on bacterial cells, phytoplankton, and detritus. Others of the limnetic zone are predatory and feed on larger plankters.

An interesting characteristic of this zone and its plankton is that a pattern of daily vertical movement is usually seen, similar to a phenomenon in the marine environment, called a *vertical diel migration*. The zooplankton moves as a diffuse layer into the profundal zone at the approach of dawn, returning to the surface waters the following night. The ecologic significance of this migration may be that it acts to reduce the chances of these small zooplankters being seen in the light of the euphotic zone by day feeding, predatory nekton. It may relate to some physiologic circumstance involving the dilution of metabolic products, aided by the migration, which would otherwise concentrate at one level, or it may relate to a need to absorb some material found in low concentration at one level and therefore requiring passage of the zooplankters through a number of levels in order to gain exposure to greater net quantities of the material. About the only agreement that can be readily had at this time is that the migration is diel (once a day), and that it appears to be triggered in response to light intensity.

The profundal zone of a lake is an interesting ecologic zone because it is a sink in which the inhabitants must utilize outside sources in order to sustain life over the long run. Because light does not penetrate this zone with sufficient intensity to allow for net photosynthetic productivity, the residents occupy niches that nearly all utilize the rain of detritus sinking from the euphotic zone above.

The majority of the resident life in the profundal zone of a lake is represented by bacteria and fungi that are concentrated at the sediment and water interface. Certain worms, larvae, and filterfeeders, such as clams, also may be found as benthos. However, the diversity and density of life is most often very much less than it is in the littoral zone under normal circumstances.

The most significant aspect of the lake environment relates to depth rather than to size. As indicated earlier, lakes, unlike ponds, have a profundal zone and usually the limnetic zone of a lake, instead of its littoral zone, is the chief producing region of that ecosystem. Furthermore, lakes tend to become physically divided by horizontal layers of water at different temperatures as a result of seasonal climatic changes.

This phenomenon is especially prominent in temperature climates, where the upper waters of a lake are warmed in the summer to form a distinct upper mixing cell, called the *epilimnion* (i.e., surface lake). A deeper, cooler, more dense, and noncirculating cell of water in the lake is called the *hypolimnion* (i.e., underlake). There is a marked discontinuity layer between the epilimnion and the hypolimnion where the temperature changes precipitously. This steep temperature gradient is the *thermocline;* it produces an effective sheer layer of density differences between the upper and lower cells of the lake that minimizes their mixing. In fact, if the compensation level for light penetration is above or at the depth of the thermocline, oxygen that might otherwise be generated by phytosynthesis in the hypolimnion is excluded from it. The oxygen that has been present in the hypolimnion before the formation of the thermocline and the ensuing period of *summer stagnation* is slowly depleted as respiration of those organisms in the hypolimnion continues. That process leads to an increase in the acidity of the water as an anaerobic (i.e., absence of oxygen) condition is approached.

As the fall season brings cooler air temperatures, the warm epilimnion cools and approaches the temperature of the hypolimnion (Figure 3.19). The thermocline becomes less distinct and disappears when the epilimnion reaches the temperature of the hypolimnion, allowing the whole lake to mix by wind-driven or density-driven convection. This is the *fall overturn* of the lake. The surface waters continue to cool, becoming more dense as they approach 4°C. If they become denser than the underlying waters, they sink and are replaced at the surface by those deeper waters. If the surface waters continue to cool, at some instant in time the whole lake will be at 4°C; continued cooling causes the 4°C water to remain at the lowest (deepest) part of the lake, whereas cooler waters float, to ultimately freeze on the surface.

As winter wears on, and especially with an ice cover preventing wind currents and reducing or eliminating light if there is a snow cover, a period of *winter stagnation* is likely to occur, reducing oxygen in the lake. Even though respiration is much lower in winter than in summer, some continues, leading to reduced oxygen and increased acidity.

In the spring, as the surface waters warm, they become more dense as they approach 4°C and they sink if the deeper waters are colder or warmer. If those deeper waters are themselves not at 4°C, a *spring overturn* occurs, bringing about a homogenization of the lake. Subsequent to this spring overturn, the summer stagnation develops. As a general rule, the deeper

Figure 3.19. The temperature induced stratification of a temperate climate lake is the result of density differences in the water which results in layering of three distinct zones in the summer. Winter mixing is usually much more complete. It brings nutrient rich waters from the deep zone to the surface. This process will enrich the waters of the euphotic volume of the lake which serves as a prerequisite to the spring bloom of phytoplankton. (Adapted from Odum, 1971 after Deevey, 1951.)

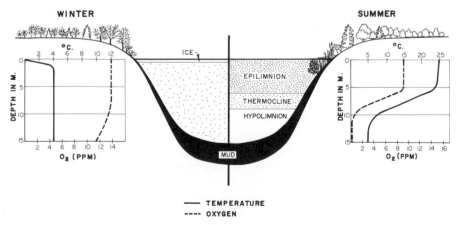

the lake is, the slower is the development of stratification and the thicker is the hypolimnion. With respect to oxygen depletion in stagnated lakes, we can demonstrate that it occurs in proportion to the amount of detritus on the bottom and the volume of the hypolimnion.

The stagnation and overturn of lakes is a key ecologic consideration to their overall condition or capacity for environmental change or adaptations to change. Therefore, the classification of lakes with respect to their mixing characteristics is another system of which we must make note. There are six categories, which include five that mix entirely. These five are called *holomictic* and they include:

1. *Dimictic lakes.* Those lakes of the temperate climates that have one overturn in the spring and one in the fall.

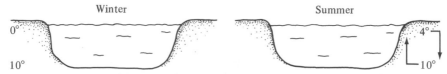

2. *Cold monomictic lakes.* Those cold climate lakes in which the water does not climb above 4°C. They have one weak overturn in the summer.

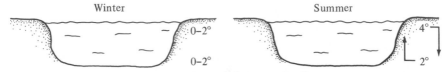

3. *Warm monomictic lakes.* Those lakes in the warm temperate or subtropic zones where the water is never below 4°C. They have one winter overturn.

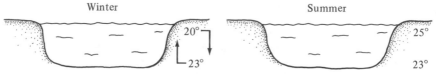

4. *Polymictic lakes.* Those of high altitude or that are equatorial. They are continually mixed by slight temperature gradients that are not seasonally distinct.

5. *Oligomictic lakes.* Those of the tropics, where temperature fluctuations are such that they are rarely or only weakly (slowly) mixed.

The sixth and final category of lake mixing is *meromictic,* which term refers to the absence of complete mixing, such as may be caused by a chemical alteration of density. The intrusion of a salinity gradient, or some other *chemocline,* as compared to the temporal thermocline, can cause a lake to be stratified and remain so regardless of seasonal temperature changes.

This classification of lakes based on their mixing patterns in response to climate is somewhat reminiscent of the scheme used to describe terrestrial biomes. In large part, temperature and rainfall were used as the key factors, determining what niches could succeed within the specified terrestrial biome. This scheme therefore gives the ecologist an hypothesis that can be tested in untangling the interrelationships of the natural environment and in trying to disassociate its pieces in order to better understand the structure of the whole.

With respect to fresh-water ecosystems, we have reviewed the basic structure and the associated distribution of organisms. In 1957, G. E. Hutchinson published a list of 75 lake types that was based upon the geomorphology and origin of differerent lakes. Perhaps this list is closest to a reasonably complete guide to identifying the basic groups of lakes and their fundamental differences and similarities.

However, for the basic purposes of this book, it is enough to indicate that there are three broad categories of lakes which adequately show the important ecologic concerns that should be borne in mind when evaluating the structure and interrelationships of the components of lakes and their significance. Those three categories are based upon characteristics which are generally described with respect to the availability of nutrients they contain and the profusion of plant life those nutrients support. The first and least productive of these types is called *oligotrophic* (i.e., few foods) because they have clear waters with relatively few nutrients entering the lake from the surrounding lands or from its nutrient-poor sediments. The next type of lake is called *mesotrophic,* because it does contain some nutrient factors which support plant proliferation, although not so much that lake productivity is excessive. The final major classification of lakes is *eutrophic* (i.e., good foods) in which nutrients are available and plant growth is prolific. From an ecologic point of view, this series can illustrate two important considerations.

In the first instance, the series of oligotrophic–mesotrophic–eutrophic may represent a natural sequence in which a lake ages, collecting detrital materials and thereby accumulating a reservoir of nutrients which leads from oligotrophy to the next stage of productivity, mesotrophy, through nutrient availability. The accumulation continues and eutrophication develops, further accelerating the availability of nutrients to the succeeding seasons of growth. Eventually, the excess accumulation of detritus can be such as to entirely fill the lake bed, producing a bog, and then a field, and ultimately a forest.

A second category of lakes has been called *special lakes*. It includes seven unique types which are fundamentally different than the typical lakes classified as oligotrophic, mesotrophic, or eutrophic. They include:

1. Dystrophic lakes, containing dark brown waters of humic and tannic acid derived from soil and vegetation, such as in a bog or swamp.
2. Deep and ancient lakes that are unique, with an endemic fauna, and are unusual because of their age but are otherwise no different, ecologically, than other lakes.

3. Desert salt lakes have a low diversity of organisms and a large population of those few species that can survive the high concentration of salts. They form where the rate of evaporation from the lake is higher than the flow of water into the lake.

4. Desert alkali lakes form as indicated above, but with drainage entering the lake from igneous areas. This results in a relatively high pH and a high concentration of carbonates.

5. Volcanic lakes found in active volcanic areas contain water often derived in part from the magma of the earth and presenting extreme chemical conditions and a consequently very restricted biota.

6. Chemically stratified lakes, or meromictic lakes, as previously noted, result from such circumstances as the intrusion of a salt tongue, which restricts both lake overturn and species diversity.

7. Polar lakes are categorized separately where they have year-round temperatures of less than 4°C.

The third and last category of lakes is called *impoundments*. They are manmade and are often characterized by their fluctuating water levels (which humans manipulate) and the relatively high turbidity (i.e., cloudiness resulting from a suspended sediment load). In addition to these typical differences, impoundments are interesting because water passing through such a system can be released either over the top of a dam or through a pipe from the bottom of the impoundment. We find that when water is released over the top, the impoundment mimics a natural lake and serves to trap sediments that might otherwise pass through the system. It also warms the surface water more than if the water passed the area in a stream. Therefore, this impoundment exports heat downstream.

If the same amount of water is released from the base of the impoundment, sediments are not retained well, but the main differences are that the cooler bottom waters are usually lower in oxygen, are more acidic, and contain more nutrients released from the sediments than surface waters. Bottom discharge from an impoundment, therefore, usually results in that impoundment serving as a heat trap and a nutrient exporter.

From a philosophical point of view, if one considers the similarities and differences between terrestrial and fresh-water ecology, and particularly as represented by the sink and source functions of impoundments as illustrated above, it appears that terrestrial biomes are apparently stable in their natural state. A climax community can be achieved and maintained indefinitely. Freshwater systems, however, such as lakes or ponds, are not stable. They are literally low points in the landscape which receive runoff water and other materials from the contiguous land. These water ecosystems serve as sinks for all manner of nutrient and nonnutrient factors generated within or without that water body. In that sense, their ecologic function and natural destiny is to be slowly or quickly filled and converted to a climax terrestrial community within the appropriate biome surrounding the original lentic water body. The lotic communities, in contrast, are substantially different in their basic structure and ecological functioning. They generally link the terrestrial environment with the lentic environment and with the marine environment.

The primary differences between the ecology of running water (i.e., lotic) environments and standing water (i.e., lentic) environments is summarized in four characteristics.

1. The relatively swift and constant currents of the lotic environment are much more controlling and limiting to biotic distribution than are the weak and often oscillating currents in the lentic environment.
2. The dynamic interactions of land and water and the actual material exchange between land and water are more rapid and of greater proportional significance in the lotic environment than in the lentic environment.
3. *Oxygen tension* (i.e., content) is usually higher and more uniform in the lotic environment. Generally, the lotic environment is more homogeneous than is the lentic environment. For example, there is no chemical or temperature stratification in streams or rivers.
4. The resident time that any volume of water spends in one section of a stream or river is relatively short, and, therefore, this water matrix of the lotic environment is continually replaced.

In consideration of currents, lotic characteristics are determined by (a) the slope (i.e., *surface gradient*) of the water course, (b) the roughness of the water course, and (c) the depth and width of the water course. These factors can be individually important in the site-specific evaluation of the *reach* (i.e., a stretch) of a water course. However, an initial evaluation of average current can be based solely on surface gradient if that information can be presented as units of elevation changes in the water surface along the water course. A good example validating the usefulness of such information was presented by Trautman (1942), who found that in a particular water course, smallmouth bass would select a reach with a surface gradient of between 7 and 20 ft per mile. Virtually none would select a reach with a surface gradient of less than 3 ft per mile or more than 25 ft per mile.

The dynamic interactions of land and water are critically important to the lotic community because the relationship of water surface to bottom surface in lotic environments is much smaller than in lentic environments. That is, streams and rivers are relatively shallow and long, lakes are relatively deep and short. A finite volume of water in a stream, therefore, comes into contact with more bottom sediments than the same finite volume of water in a lake. Similarly, the stream normally has a longer shore line than a lake that can hold the same volume of water as is in the stream at any one instant. Therefore, the stream is in much more intimate contact with the sediments and the contiguous terrestrial environment than is a lake of identical volume.

Despite this intimate contact with the nutrient stores of basic mineral requisites in the land and in the sediments, the primary productivity of streams is generally low compared to lentic communities. Streams and rivers have periphyton, such as attached filamentous green algae, encrusted diatoms, and aquatic masses, but the total biomass is relatively small and it cannot usually support the associated mass of primary consumers that can be typically found there. Clearly, the consumers must be in part supported by food subsidies delivered to the lotic environment from the lake or pond, or from the contiguous *terrigenous* (i.e., land), *sources* which literally feed water and its contained materials to the water course in question. These are *allochtonous* materials, derived from outside the system in question, such as leaves, sewage, or detritus. They are carried to the stream by water or wind, where they may realistically provide well over 50% of the energy requirements of that lotic ecosystem. *Autochthonous* materials are those indigenous to the system and they represent the materials generated by the resident biota in the presence of the allochtonous subsidy defined above (Figure 3.20).

A further examination of the characteristic structural components of the lotic community leads to a definition of the rapids and the pool environments. In each, a good starting place is the bottom types (i.e., substrate) found. These may be generally classified as being primarily composed of clay, sand, pebbles, rubble rock, or bed rock. The distribution of these bottom types is dependent upon the geomorphology of the region, including the surface gradient of the lotic system. However, we can generally evaluate that system without knowledge of specific conditions. Streams and rivers tend to erode the material of their bedding down to a base level; this causes the currents to diminish as the water course ages. Ultimately, a continuous channel is developed by this erosion. Such a channel habitat is intermediate between rapids and pools; its biota resembles the rapids community except that it tends to be fragmented, concentrating biota in the channel where favorable bottom conditions allow firm, stable attachment. Soft, shifting silts limit life to small, burrowing benthic forms.

The species found in the rapids community are usually entirely different from the pool species, and they are usually considered the representative lotic biota. Pool species, in contrast, often contain representatives one would find in the lentic communities, and so it is not an easy task to present one list of species that is especially diagnositc of all lotic communities. However, concentrating upon the rapids community, although recognizing the resemblance of the lotic pool to the lentic community, we can separate out some additional specific lotic characteristics.

For example, the diversity and stability of stream life may be generally categorized based upon the substrate characteristics of the stream bed. Firm and protective substrates provide a habitat for more species than soft, even substrates. Therefore, the following sequence can be constructed with least diverse and least productive first and most diverse and most productive last.

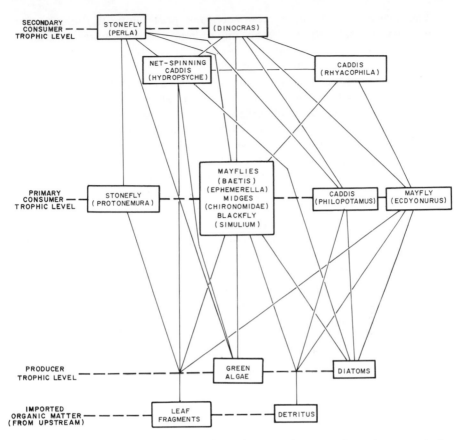

Figure 3.20. A typical food web of a lotic environment is energy dependent upon both allochtonous and autochtonous sources of food. If it were not for the supply of imported organic matter from upstream, it is very likely that the ecosystem would not survive containing the organisms illustrated here. (Redrawn from Jones, 1949.)

1. Sand or soft silt
2. Clay
3. Flat or rubble rock

With respect to animal distribution, benthic *invertebrates* (i.e., animals without backbones) are most prevalent in the rapids communities, whereas nekton and burrowing invertebrates congregate in the pool communities (Figure 3.20). In fact, the nekton will usually remain in a pool just below a rapids waiting for any material swept through or swept loose from the rapids upon which the nekton can feed.

Plankton is not endemic to the rapids, although some species may be stable and form a significant food source in larger pools. However, the plankton collected from the lotic environment is usually identified as allochtonous material in the lakes, ponds, or backwaters feeding into the lotic system.

The distribution of rapids, channels, and pools along the length of a water course usually results in the longitudinal zonation of stream life. In general, streams tend to reflect a progression of youth at the headwaters to relative old age at the mouth of the water course.

The young water course is small and possesses a steep surface gradient, dominated by rapids with occasional pools but few channels. The old water course is large and has a small surface gradient, dominated by meandering channels and few or no rapids or pools. Intermediate water courses have intermediate environments; changes in habitat are usually most pronounced at the upper and younger reaches, where the surface gradient, the volume of water carried, and the physical nature of the water is likely to change rapidly because of the intimate contact that the small, headwater stream has with the contiguous environment. In that sense, it is more usual to find community transformations and diversity along the first mile of a water course than it is to find a comparable quantity of change or diversity along the last 50 miles of the same water course.

Regardless of their specific distribution, the organisms whose niches include the lotic environment have evolved several mechanisms adapted to maintain their physical position in the currents. These can be summarized in seven different types of structures or behavior.

1. *Permanent attachment* to a firm substrate, such as by a holdfast, as represented by some algae, encrusting diatoms, masses of the genus *Fontinalis* sp., sponges, and some of the caddisfly larvae.
2. *Hooks or suckers,* used to anchor the organism to a firm substrate, as represented by the black fly larvae *Simulium* sp. and the caddisfly larvae, *Hydropsyche* sp.
3. *Sticky undersurfaces,* used to move over firm substrate as represented by snails and flatworms.
4. *Streamlined bodies,* resulting in the general egg-shaped organism which offers minimal resistance to water flow.
5. *Flattened* (i.e., compressed) *bodies* are prominent in those organisms which live in the rapids and find refuge in crevices and under stones where the current is minimal. Stonefly and mayfly nymphs are much flatter than are related species that are only found living in lotic pools, not rapids (Figure 3.21).
6. *Positive rheotaxis* (*rheo* = current; *taxis* = arrangement) is a behavior in which the organism orients itself facing upstream. The effect is similar to the objective of pilots taking off into the wind for better control. Species from the lentic community generally do not have this behavior when placed in the lotic environment. They are swept away.
7. *Positive thigmotaxes* (*thigmo* = touch, contact) is a behavior by which a stream animal clings close to the substrate, where the current is lower; by this means the organism reduces the surface area exposed to the force of the current.

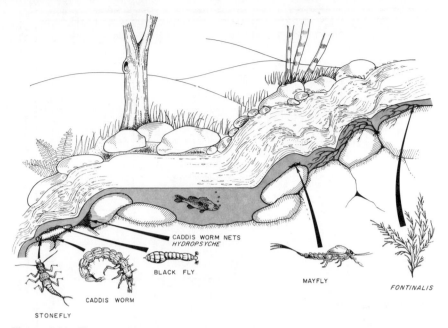

CADDIS WORM NETS
HYDROPSYCHE

BLACK FLY

MAYFLY

FONTINALIS

CADDIS WORM

STONEFLY

Figure 3.21. The organisms of a lotic environment are generally well adapted for life in the swift, oxygen rich waters. They are able to maintain their position. The predators and detrital feeders usually capture food carried past them in the current. Many species of fish seek out pools or eddies in the current when the currents drop sediments and food.

The lotic environment can be seen as serving to link terrigenous materials with the lentic environments. Water courses also ultimately enter the ocean serving as the drain for virtually all the terrestrial water-borne materials that leave the land to enter into the shallow coastal waters off the terrestrial biome of a particular region.

Of course, there are other means of transporting materials from the land-based ecosystems to the marine ecosystems, but the lotic environment seems to be the most efficient conveyor belt of those in operation. Wind, direct disposal of materials into the oceans by humans, and the coastal erosion of land by the ocean all contribute materials to the marine environment, but the contribution of the lotic environment is greatest. It delivers about 8 million billion gallons of water and all their contained dissolved and suspended materials each year to the oceans from the land.

Marine Ecology

The oceans cover about 71% of our planet's surface and they are relatively deep, with only about 8% of their floor any shallower than 200 m. Furthermore, the various basins of the ocean (Figure 3.22) are virtually continuous,

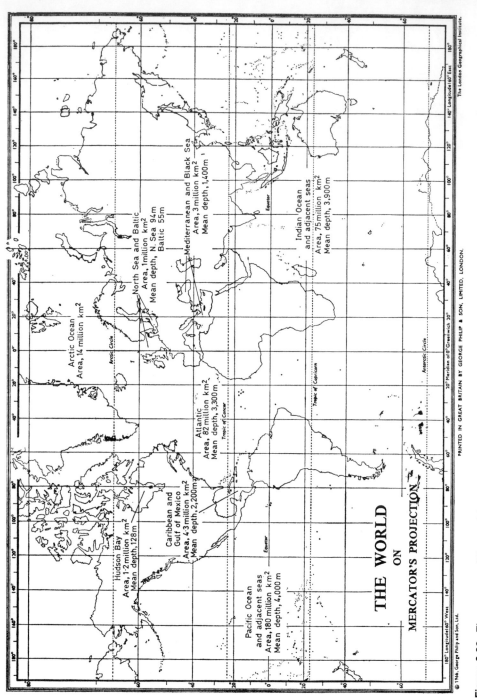

Figure 3.22. The oceans of the world, deep, interconnected, and with relatively constant physical and chemical characteristics defining their environments, account for about 71% of the surface of our planet. (From Tait and De Santo, 1972.)

with relatively gradual and small differences in temperature, salinity, and depth. However, these are the chief barriers to the free movement of organisms and of water from one basin to another. These barriers are relatively small when compared to the geographic barriers on land that tend to keep terrestrial systems insulated.

The air temperature differences between the poles and the equator create strong trade winds and, in combination with the rotation of the earth, they create surface currents, which drive deeper currents. However, it appears that most of the deeper currents result primarily from water temperature and salinity differences, which create *density gradients* impelled by the rotation of the earth *(Coriolis force)* and the configuration of the basin. Therefore, stagnation and associated anoxic conditions are rare in the ocean at any depth.

The *major currents* of the ocean (or *gyres*), such as the Gulf Stream, or the Humboldt (i.e., Peru) current, run clockwise in the northern hemisphere and counterclockwise in the southern hemisphere (Figure 3.23). In equatorial currents, surface waters move between 8 and 14 km per day, while some deep currents move between 2 and 10 km per day. Relatively fast, unusual currents, such as parts of the Gulf Stream, may move 180 km per day. A major pattern of water movement is based on the fact that low-latitude water warms in response to the intense rays of the sun, rises, and is moved to higher latitudes; there it is cooled, sinks, and tends to return to the equatorial zone (Figure 3.24).

As a result of these continuous mixing patterns, the oceans are kept quite well mixed. The general characteristics of the deeper areas allow particulate materials to settle out of the gently circulating sea water, with the basins serving as sinks for materials carried there. When currents move the deep, often nutrient-rich waters toward the surface, this *upwelling* enriches the surface water in somewhat the same way as dispersal of terrigenous materials and nutrients carried to the coastal waters enrich those waters by fertilizing them. This results in rich growth of the producers, which are fed upon by all levels of consumers.

From an ecologic point of view, because of the great volume of water in the ocean and because of the high specific heat of water, the ocean is a great heat sink. Therefore, it is a temperature moderator. Oxygen is well mixed and the capacity of sea water to buffer the environment is sufficient to keep pH remarkably stable. The buoying capacity of water, as in fresh-water systems, is also an important environmental characteristic in the sea. Taken together, the homogeneous and generally hospitable conditions of the marine environment are such that many marine plants and animals are very widely distributed. From another ecologic point of view, the characteristic stability and homogeneity of the marine environment has supported the evolution of many species whose niche requirements are very precise and whose ranges are controlled by very slight changes in such factors as salinity or temperature.

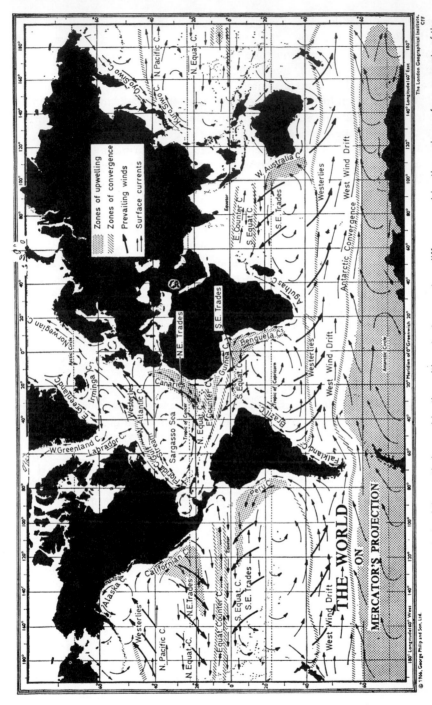

Figure 3.23. The prevailing winds of the earth, in combination with water temperature differences, drive the major surface currents of the oceans. Where these currents move deep, nutrient rich waters to the surface, such upwellings generally support unusually productive fisheries. (From Tait and De Santo, 1972.)

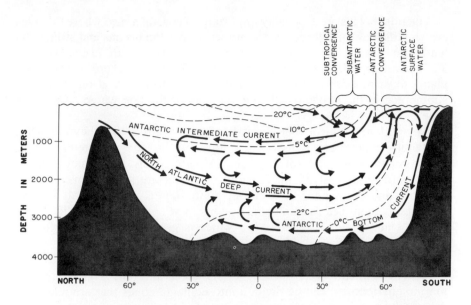

Figure 3.24. An idealized cross section of the ocean between the Arctic basin and the Antarctic continent shows the circulation of deep waters. Note the nature of the deep upwelling in the Antarctic surface waters which supports the rich fisheries found in that cold ocean.

The shallow parts of the oceans are uniquely dominated by waves. These waves have various periods of from less than a minute, for waves that lap or pound the shore, to months, for seasonal tide fluctuations, and days, for the tides. Ocean tides are caused by an interaction of the gravitational attraction between the sun, the moon, and the earth; they result in a bulge of water about 1 ft high at the equator, which has a period of about 12½ hours as it passes around the earth in the oceans. When the earth, moon, and sun are in a straight line (thus, ∘∘∘) the forces of gravity cause the fluid hydrosphere (the oceans) to form a larger bulge than when the earth, moon, and sun are 90° out of phase, forming a triangle (thus, ∘∘∘). The larger bulge creates the spring tides at the new and full moon, and the smaller bulge creates the neap tides at the quarter moon. In fact, a bulge is raised on each of the opposite sides of the earth, one in response to gravitational attraction and the other a compensating bulge as a result of the balanced centrifugal force, which causes the earth to remain in its orbit and which causes the water to pile up.

The tidal bulges are especially important in coastal areas, where the tides are amplified as the bulge of water moves across the oceans and strikes the land. The periodic submergence and exposure of the *intertidal zone* (Figure 3.25) is an especially rich and stressful environment where animals and plants must be adapted to live exposed to the drying air at one time and submerged in the strong oscillating currents of the tide waters at another time.

An additional critical characteristic of the oceans is that they are salty, containing about $3\frac{1}{2}\%$ of dissolved solids, mostly salts, reported as 35 $^o/_{oo}$ (parts per thousand). About 27 of the 35 parts consist of sodium chloride (table salt), and most of the remaining 8 parts are salts of magnesium, calcium, and potassium. In addition, approximately 60 elements can be predictably identified as small constituents in sea water. As a result of this complex mixture of chemicals, sea water is generally strongly buffered, with a pH of about 8.2.

Most of these characteristics of the oceans support the view that it is homogeneous. However, because of its great differences in depth, the marine environment can be divided into ecologic zones that are relatively distinct and are somewhat reminiscent of the concept of the terrestrial biomes. The fundamental difference is that water is not a limiting factor in the marine environment, whereas light, pressure, and turbulence are limiting factors. Based on the same general approach as used with fresh water environments, but on a grander scale, the marine environment includes several overlapping designations used to locate areas within the ocean with reference to depth, light, and habit.

The interrelationship of ecologic zones and geologic zones is of particular importance because of the significant influence that depth plays on the ecologic subdivisions of the marine environment. There are five central geologic features of special ecologic significance (Figure 3.26). Each of the continents is surrounded by a submerged margin, called a *continental shelf,* which slopes gently away from the coast at the rate of about 1 m in depth for each 1000 to 2000 m in width (i.e., 0.1° slope). Averaging about 70 km wide, the shelf ends at the *continental edge,* where the slope of the sea floor increases from 1 : 1000 or 1 : 200 to about 1 : 15 or 3 to 6°. This is an area called the *continental slope.* Although its average slope is about 6°, it may be as great as 45°. It extends from a depth of about 200 m to one of 3000 to about 4000 m. Here it grades into a gentler sloping surface of 1 : 100 to 1 : 700 (0.1–1°). This is the *continental rise,* which extends deeper, to about 6000 ft; here the deep ocean floor may be extremely flat as an abyssal plain and is usually less than 1 : 2000 in slope for hundreds of miles (less than 0.1° slope).

Certain areas of this deep ocean floor are interrupted with chains of *submarine mountains,* however, which may be more than 10,000 ft high, and *submarine valleys,* which may be more than 5000 feet deep. Even deeper features are called *deep sea trenches.* The Mariana Trench in the

a

b

Figure 3.25. The surface area of the intertidal zone may be either quite extensive or quite small since it is directly dependent upon the slope of the shore and the range of the tide. Here, a photograph of a small cove in Maine (U.S.A.) shows the intertidal zone created by a tidal range of about one and one-half meters. Such zones are generally rich areas for feeding shore birds and for the interactions of the sediments and the fluctuating water column. (a.) High Tide. (b.) Low Tide.

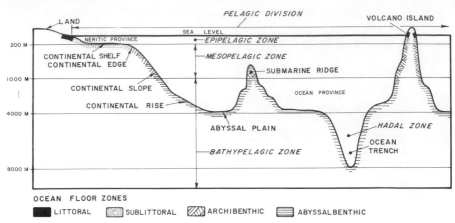

Figure 3.26. The divisions and main features of the marine environment are summarized here.

western Pacific, the deepest point in the oceans, is 39,198 ft (about 11,948 m) deep.

With general reference to these five geological features, or ocean realms, the associated ecologic zones may be constructed based on two broad divisions: (1) *aquatic habitats* and (2) *benthic habitats*. The aquatic habitat contains all those organisms which swim, float, or are suspended in the ocean and it is therefore called the *pelagic division*. The second type of habitat is the ocean floor and contains all those organisms which live in or on the bottom; it is called the *benthic division*.

These two broad habitats are further subdivided depending upon the essential physical characteristics of each which can be used to define more specific *subhabitats*. For example, the pelagic division is broken down into the *neritic province,* which includes all waters out to the continental edge, and the *oceanic province,* which includes all the waters of the ocean beyond the continental edge. The basic differences between these subhabitats of the pelagic division is that the neritic environment is generally more variable than the oceanic environment. Temperature, chemical composition, and turbidity all tend to be more variable in the waters closest to land than in the oceanic province, and these fluctuations make a difference in the ecology of the oceans.

In the deep waters of the oceanic province, the most significant environmental factor seems to be depth. Therefore, the province is subdivided into three pelagic zones based primarily upon the depth of water. The *epipelagic zone* extends from the surface down to a depth of about 200 m, the average depth of water above the continental edge. Compared with the ocean zones, the epipelagic zone is characterized by relatively sharp gradients of light intensity and temperature, which show both diurnal and seasonal changes. For example, thermoclines are often present in the marine environment. However, when present they are usually much less distinct than in fresh-water systems. In addition, currents may be far more rapid in this zone than in others.

The *mesopelagic zone* extends down from about 200 m to about 1000 meters. Within this zone, there is virtually no light penetration and temperature changes are slight, with only slight seasonal variations (5 to 15°C). The zone also usually contains a band of water containing very little dissolved oxygen. This *oxygen-minimum* zone is usually found between 400 and 1000 m. At low latitudes, water at 400 to 500 m has been found in which the oxygen content was nearly zero; the same water was nevertheless well populated. The natural range in sea water of dissolved oxygen is between 0 and 8.5 mg/liter, with the average overall mean value lying between 1 and 6 mg/liter. However, within the oxygen-minimum zone, where the mean value is virtually zero, it appears that in at least some of those instances, the respiration of the native population present depletes oxygen from the water just as rapidly as oxygen diffuses into the system and the system seems deceptively low in oxygen. Actually it is supplied with oxygen, which is then immediately utilized.

Another curious, and still unexplained, layer of water is often found at a greater depth, where maximum concentrations of nitrates and phosphates are found. This layer, found somewhere between 500 and 1500 m, contains nitrate nitrogen (NO_3-N) at a concentration of 200 to 500 μg/liter (surface concentrations are 1 to 120 μg/liter) and phosphate-phosphorous (PO_4-P) at a concentration of 40 to 80 μg/liter (surface concentrations are 0 to 20 μg/liter.

The deepest pelagic subdivision is the *bathypelagic zone,* which extends down from about 1000 meters to the bottom. Here, no surface light penetrates and the temperature is constant and low (less than 4°C). There is great pressure and virtually no current. Because the mean depth of the oceans is about 4000 m, most of the volume of sea water lies within the bathypelagic zone, where temperature of 4 to 0°C and blackness make individual productivity relatively low. In fact, the biomass associated with 1 m³ of water from this zone is about ten times smaller than the average biomass calculated for 1 m³ of water from the mesopelagic zone. Similarly, the diversity of life forms in the bathypelagic zone is low, including only about 150 different species of fish, whereas the mesopelagic zone supports about 800 species.

There are five zones in the benthic division, the habitats of all those plants and animals (collectively called the benthos) which live on or in the bottom. These included the littoral, sublittoral, archibenthic, abyssal-benthic, and hadal zones. There is some correspondence between these zones and those of the pelagic division, for they both relate to depth; however, the niches of each are distinctly different because one division is dominated by fluid alone (the sea water), and the other is dominated by substrate (the sea floor).

The *littoral zone,* very similar to the fresh-water zonation, refers to the shore line and shallow waters of the coastal marine environment. It includes the band of land known as the *splash zone,* which lies immediately above extreme high water and receives the sea spray of waves breaking upon the shore. From the shore, the littoral zone extends seaward to the extreme tidal low water mark. A central characteristic of this zone is that it is regularly exposed to the air as the tides ebb and flood. In general terms, the organisms that reside in this rather harsh environment must contend with rapid and regular extremes in moisture, salinity, abrasion, and light intensities as they are covered or exposed by the tides. In this intertidal zone, temperature may range between 5° and 25°C and salinity between 5 and 34 ppm. Tidepools are pockets of water trapped in depressions in the littoral zone as the tide recedes. In temperate and tropical tidepools, temperatures may reach 50°C before they drop rapidly as the tide returns. These stresses make tide pool organisms especially hardy.

The *sublittoral zone* extends from the extreme low water mark to the continental edge in about 200 m of water. This environment remains submerged at all times and is generally illuminated by sunlight filtering through the water. Relatively low pressures and swift currents are the basic considerations that define the ecologic interrelationships seen among the inhabitants of the sublittoral zone. Here the temperature may range between 5 and 25°C and salinity between 20 and 36 ppm.

The *archibenthic zone* is the next deepest benthic subdivision, beginning at about 200 m at the continental edge and extending down the continental rise to a depth of about 1000 m. In addition to increased hydrostatic pressures, this environment is characterized by gentler currents and less variable temperature (e.g., 5 to 15°C) and salinity (34 to 35.5 ppm) than is present in shallower circumstances.

The next and deepest benthic communities are found in the *abyssal-benthic zone,* which extends from a depth of about 4000 m down to the deepest parts of the ocean floor, at about 11,500 m. The deepest parts of the oceans, which drop below 6000 m in trenches, are refered to as the *hadal zone.* The area accounts for about 1% of the total area of the ocean bed. Here temperatures range from 1.2 to 3.6°C. It is impossible to separate all these zones clearly, and depth ranges must be recognized as being very approximate.

The relative calmness and constancy of all the ecologic zones within the marine environment, except for the turbulent littoral zone, is particularly well represented in these deeper communities, where about the only dynamic force present is current; and even there, the current is virtually always weak and constant (Figure 3.27). Yet all depths of the sea are populated and remain in ecologic balance so long as humans do not interpose rapid change to which organisms cannot adapt. Evolutionary adaptation is too slow to occur when those changes happen to fall outside the niche of a particular species found in the undisturbed habitat.

In general, the zones we have defined may be used to better identify and investigate ecologic or physical relationships that allow humans to manage and conserve the natural resources of the oceans. The need for such

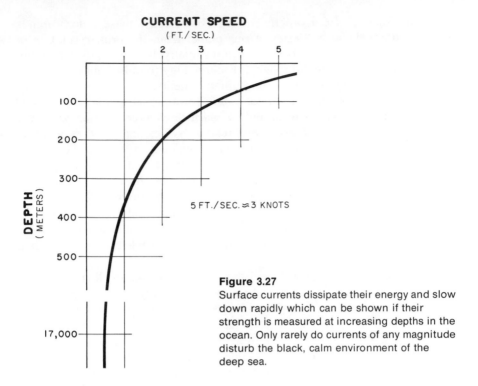

Figure 3.27
Surface currents dissipate their energy and slow down rapidly which can be shown if their strength is measured at increasing depths in the ocean. Only rarely do currents of any magnitude disturb the black, calm environment of the deep sea.

management should be intuitively recognized in the fact that about 75% of all of our harvest of aquatic resources, from both fresh-water and marine ecosystems, comes from the continental shelves of the world, an area that represents only about 8% of the surface of the earth. About 23% of our aquatic harvest come from fresh-water systems, and only about 2% of our harvest comes from the oceanic province, which represents 66% of the surface of the earth.

This division of productivity occurs because the shelf waters of our planet are generally far more productive than are the relatively desolate open oceans. Of course, the harvests cited above include not only the organic materials produced in aquatic environments, such as fish. They also include mineral and other chemical resources, such as the phosphorite and manganese nodules of deep waters and the detrital minerals, including sand and gravel, derived mainly from the land, which are shallow-water resources. The water itself holds great potential value because the complex chemical solution which is sea water contains many elements of mercantile value were we able to extract them. Every cubic kilometer of sea water contains about 40 million tons of dissolved solids which, in purified form, would be valued at many hundreds of millions of dollars.

From an ecologic point of view, one of the most important and challenging areas of marine ecology centers on the living resources of the oceans contained within the somewhat artificial zones that we have constructed for the convenience of our orderly minds. Historically, these orderly minds seek to unravel the mysteries of our planet from the primary motive of exploitation. Once understood, it becomes relatively easy for us to manipulate the environment and through our technology to utilize our new understanding in seeking new short-run human benefits.

The caution most of us try to use comes from the sense that uncontrolled exploitation of any resources, and of marine resources in this particular instance, can easily have disastrous results. An example will be given in Chapter 6 relative to overfishing, but here suffice it to say that the oceans, like all other ecosystems, have finite resources that must be recycled through the system. Harvesting, which removes those finite resources, ultimately will degrade the so-called renewable resources. The objective, therefore, of applied ecology is to understand a natural system sufficiently well to allow us to manipulate our intervention without reducing the finite resource base of the system while we harvest renewable resources for their utility.

The most direct purpose for all the naming and subdividing of ecosystems is the need of an ecologist to know where any facet of an ecologic system is coming from and where is it going. By knowing these routes and destinies, he can predict the ecologic interactions that the system will display. The problem we have in reaching this idealized goal is the confusion caused by our rather sketchy and at best incomplete knowledge about such systems as the oceans. However, one of the elementary factors in understanding any system revolves around the processes by which an ecosystem remains either homogeneous or heterogeneous. In simplest form, we should know what are the means of mixing in the ecosystem under study. In the case of the marine environment, this question concerns understanding ocean currents and mixing, some characteristics of which we have already outlined.

As a general overview, we can identify four aspects of currents that are of special importance to marine ecology and its associated environmental management.

1. Currents mix the water column and tend to homogenize its contents, including salinity, oxygen, heat, and any other additives.
2. Currents may return nutrients present in deep waters to the euphotic zone, where primary productivity can occur.
3. Currents tend to distribute larvae, eggs, and other plankton, including detrital food stuffs.
4. Currents cause erosion in those instances where the currents encounter erodible materials, and their absence causes sedimentation.

Of these four types of phenomena, perhaps the most dramatic one is the last, in part because shoreline processes are highly visible and are often very dynamic. Furthermore, the dispersion of near-shore pollution is directly related to currents generated by ocean waves, which cause both mixing and lateral movement. Waves cause water to move toward shore,

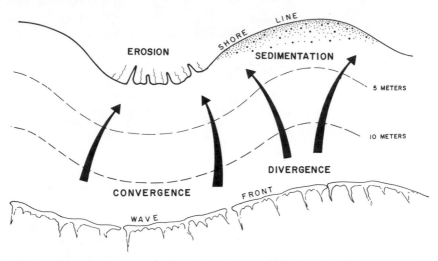

Figure 3.28. Waves passing over a shoaling (shallow) bottom tend to converge while those moving into bays or over deep water tend to diverge and weaken in their kinetic energy. The shore upon which the waves converge tends to erode while the shore on which the waves diverge tends to build seaward with sediments dropped on the beach by the waves.

away from shore, and/or along the shore, depending upon a number of readily definable factors or conditions.

In general, the shoreline of any coast tends to be eroded by the ocean such that it will become smooth and straight. Irregularities usually result from the presence of harder headlands that are not as erodible as adjacent land, which the waves cut and carry along more easily. As a straight wave approaches a coastline and begins to pass over a shoaling bottom, the wave front turns in, focusing upon the point of land from which the shoaling is derived. This *convergence* (Figure 3.28) concentrates the wave energy, tending to erode the shoal and the prominance on the coast that receives the brunt of the force of the pounding surf. The areas adjacent to the shoal, in contrast, are deeper and cause the approaching wave to diverge its energy. Whether over submarine canyons or in bays, this *divergence* lessens the erosion potential and encourages the deposition of suspended materials. This process tends to fill and extend the coastline seaward at these places.

When a wave breaks on the beach, some of the water slips back seaward under the approaching train of waves as a backwash, or *undertow*, carrying water and sediments out from shore for a relatively short distance (Figure 3.29). However, if the waves approach the beach at an angle, the breaking waves will carry water along the shore in the direction of the breakers. This movement may pile up the water along the shore, building a momentum that eventually overcomes the confinement of the landward movement of the waves. At that location, the water flows like a river out from the shore

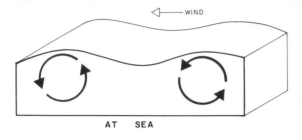

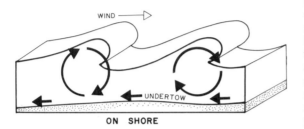

Figure 3.29
Caused by the winds, waves move over the surface of the water while the water itself moves up and down in a general circular movement. As the waves approach the shore, their bases are slowed by friction against the raising bottom. The top of the wave continues forward, over-reaching the bottom and breaks, rushing up the beach face. Returning under the approaching waves, the spent waves create an undertow which carries materials seaward, perpendicular to the shore.

through the breakers, forming a *rip current* (Figure 3.30). This current is rapidly dissipated and sediments or other materials swept out from shore drop from the current or are dispersed by the currents running across the seaward end of the rip current.

Rip currents are relatively isolated occurences and their importance as a mechanism for the movement of materials and organisms is relatively slight. The most significant factor of littoral marine ecology, in this respect, is the current formed by obliquely breaking waves, which carries the water parallel to the shore and provides the potential necessary to create a rip current. However, this *longshore current* is a regular phenomenon. Such movement is called *littoral drift* (Figure 3.31) and results from the zigzag pattern of sand grain movement along the beach in response to the oscillating movement of obliquely breaking waves.

Waves build or scour the beaches depending on the season. For example, winter waves in temperate and tropical climates, and storm waves, tend to be high with a relatively short distance (i.e., time) between one wave crest and the next. These *destructive waves* tend to pound the beach, loosening and eroding its materials and carrying them seaward off the beach in the strong undertow of each wave. *Constructive waves* are usually associated with summer conditions; they are relatively low waves and have long distances (i.e., times) between each wave crest. This form of wave plunges forward upon the beach, moving materials up the face of the beach and depositing them there with little backwash, or undertow, to carry materials seaward.

From both an engineering and an ecologic point of view, littoral drifts are important shoreline environmental factors; they are exceptionally difficult

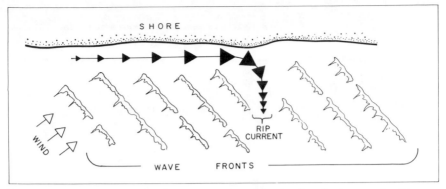

Figure 3.30. Rip currents may form as waves approach a beach at an angle piling up some of the water along the shore until, at some point along the shoreline, the momentum of that water (ie., its head) overcomes the landward movement of the waves. There, the current runs rapidly seaward perpendicular to the shore. Rip currents dissipate their energy rapidly and die not far off shore. If caught in such a current, a swimmer should swim parallel to the shore and not attempt to fight against the current which would tire one rapidly.

to predict, however, because of the oscillating and multifaceted nature the shallow waters of the coast. However, manipulations are often prac ticed in an attempt to modify the progression and consequence of longshore currents. In particular, jetties and breakwaters are often used to modify the movement of beach sands, especially along shoreline communities where the beach is an important recreational factor or where erosion threatens the safety of real estate previously built upon the unstable shore.

The principle of this type of manipulation is simple but its application is complicated by the need to know all possible repercussions to the natural environment before it is undertaken. The principle is that if an obstruction is placed in the path of a longshore current, materials being transported in the littoral drift will tend to accumulate at the *updrift* side of the obstruction. This is because of the slowing and eddying of current within the artificial bay formed by the obstruction. However, it is often possible to upset a littoral drift pattern, causing no net systemwide change in the transport of materials while causing decided shifts on a site-specific scale of concern. For example, the installation of a jetty perpendicular to the sandy beach of a North Atlantic community caused the deposition of sand against the updrift side of that jetty (Figure 3.32). It also caused the erosion of the *downdrift* side, where sand along 450 m of beach was removed by the waves over a period of about 20 years. In that instance, it appeared that the beach was relatively stable provided littoral drift replenished sand being carried downdrift. When the littoral drift was interrupted by the jetty, the sand that accumulated on the updrift side was unable to replace the sand being eroded from the downdrift side of the jetty.

In such an instance as this, there must have been a *net drift* greater than

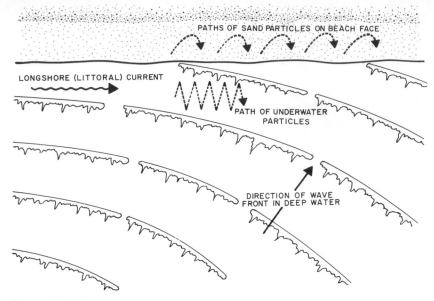

Figure 3.31. Longshore currents (littoral drift) carry materials in a zigzag pattern along the shore in the direction of the wave front movement. Particles are moved both on the beach and in the water.

zero that went undetected until the jetty demonstrated that sand was tending to move across the stretch of beach under consideration. The obvious implication is that at some point along the shore, or in the water upcoast from the jetty, there must have been an eroding source of the drift sands that were being trapped by the jetty, just as the downdrift sands near the jetty were being eroded and sent down coast.

In several instances, it has been shown that the net littoral drift of sand may be supplied by the erosion of sands several miles up coast from the study area. Interestingly, though, there are some rare coastlines, such as sections of Washington and Oregon, that seem to have no net littoral drift over a season. In these instances, there is a seasonal reversal of the direction from which wave fronts strike the beaches. The wave energy is thereby balanced, resulting in the littoral drift of material first in one direction and then, in the succeeding season, in the reverse direction. This results in oscillating, localized drifts which, in one case along a 30 km Oregon coastline involved about 620,000 m³ (810,000 cubic yards) of sand being moved each year back and forth parallel to the coast along the beaches. That shoreline is in *equilibrium* and will remain so provided no new obstructions are placed in the coastal waters to disrupt the balance.

From an ecologic point of view, littoral drift is of primary importance as it relates to the stability and grain size of the sediments that are affected by the currents. In general, when small grain sizes are present, sorted by the

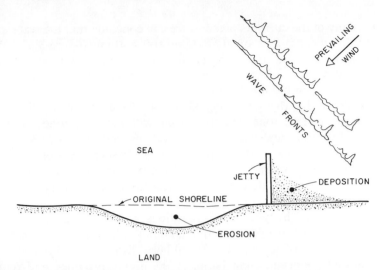

Figure 3.32. Jetties, breakwaters, and groins, all function to obstruct the ocean currents in the nearshore environment. If a net drift is interrupted, erosion and sedimentation (accretion) will appear and continue, usually until such time as a new equilibrium is reached. At that point, the net drift resumes its pattern.

Figure 3.33. The profile of a beach usually reflects the grain size of the material which comprises the surface of that beach. The smaller the grain size, the flatter the slope, which is an expression of the calmer surf of that beach. (Based on data and adapted from Moore, 1971.)

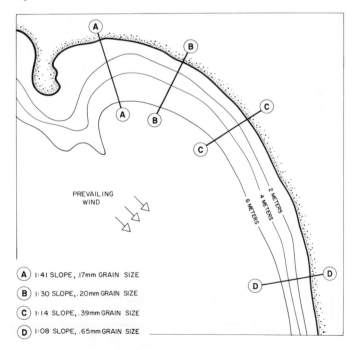

relative energy of the currents present, we can conclude that the waters are quiet and that there is a relatively flat and soft bottom (Figure 3.33). With large grain sizes present, the currents are more energetic and the bottom is consequently harder, often steeper, more abrasive, and generally less hospitable to its benthic inhabitants.

As suggested by the preceding sections, the ecologically significant aspects of any marine zone tend to overlap with adjacent zones, and it is therefore difficult to deal with isolated systems and still understand the interdependence of one system upon another. It seems easier to isolate terrestrial and fresh-water systems, perhaps because they are physically smaller and because they do not usually present the same scale of materials exchange as occurs in most marine systems.

As an example of the type of considerations an ecologist uses in studying this sort of exchange problem within the context of the marine ecosystem, two types of communities are introduced here. *Estuaries* and *marshes* are prominent and important subhabitats of the neritic province and require special attention with respect to management and protection of the coastal zone of every ocean. They are important as intermediary zones between the terrestrial and marine environments. They are highly productive in terms of both primary and secondary productivity. They serve as reservoirs and buffers for the dispersion of nutrients that support the high primary productivity of coastal waters.

Figure 3.34. Comparing the temperature and salinity of an estuary and a marine environ- ment produces a *hydroclimograph*. It illustrates the differences present with respect to the niche capacities required of the inhabitants of these two habitats. The graph records the ranges of temperature and salinity over the seasonal fluctuations characteristic of these habitats. (Adapted and redrawn from Hedgpeth, 1951.)

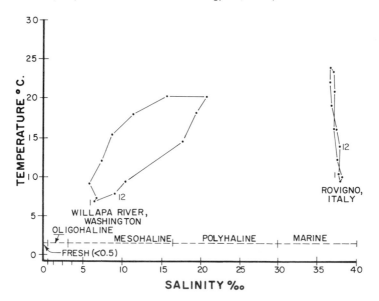

Estuaries are semienclosed, coastal water bodies that receive fresh-water drainage from the surrounding terrestrial environment and have a direct connection with the ocean. They have several characteristics, and prime among them is the fact that salinity and temperature may fluctuate widely in comparison with the waters of the open ocean. Fresh waters, with a salinity of less than 0.5 ‰, dilute the marine waters in an estuary, and the plants and animals within that estuary must be capable of adapting to these fluctuations (Figure 3.34). *Oligohaline* environments contain between 0.5 and about 3 ‰ salinity; *mesohaline* environments, between 3 and 16 ‰; *polyhaline* environments, between 16 and 30 ‰, and marine environments between 30 and 40 ‰.

Seasonal variations usually occur in the temperature and salinity charac-teristics of an estuary based on the amount of precipitation that reaches the sea each season. However, there are daily fluctuations as well, which reflect the ebb and flood of the tides. Depending upon the range of the tide at a particular estuary, the system may be considered a fluctuating water-level ecosystem that is *pulse* (i.e., cyclically) *stabilized;* it is highly produc-

Figure 3.35. The incomplete mixing of fresh water flowing into the marine environment often creates a stratified condition in estuaries. The lighter freshwater floats on top of the denser salt water creating a salt tongue lying on the bottom of the estuary. Plankton collected at a depth of 2 m in the Tyne estuary in England shows a characteristic distribution of organisms reflecting differences in salinity. (From Tait and De Santo, 1972.)

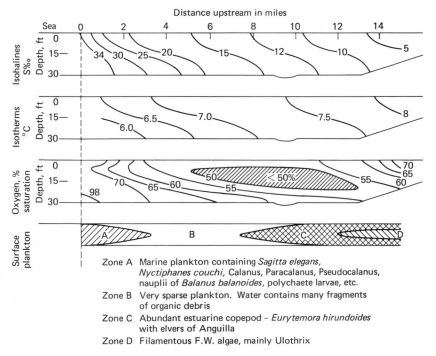

Zone A   Marine plankton containing *Sagitta elegans,*
          Nyctiphanes couchi, Calanus, Paracalanus, Pseudocalanus,
          nauplii of *Balanus balanoides,* polychaete larvae, etc.

Zone B   Very sparse plankton. Water contains many fragments
          of organic debris

Zone C   Abundant estuarine copepod – *Eurytemora hirundoides*
          with elvers of Anguilla

Zone D   Filamentous F.W. algae, mainly Ulothrix

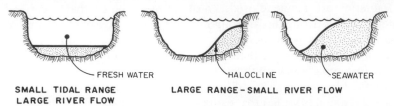

SMALL TIDAL RANGE   LARGE RANGE – SMALL RIVER FLOW
LARGE RIVER FLOW

Figure 3.36. As illustrated here, the relative amount of salt water to fresh water, the slope of the estuary floor, and Coriolis Force are all significant factors in determining the location of the halocline (ie. the salinity discontinuity) which characterizes every estuary to a lesser or greater degree.

tive as are most pulse stabilized systems, such as terrestrial flood planes. For the purposes of the marine ecologist, a hydrographic basis of estuary classification is useful because it centers on characteristics that effect the dispersion and deposition of materials, including pollutants, in the estuary. This approach is important to the successful use of sampling and monitoring technology, as will be discussed in Chapter 4.

This hydrographic classification includes (1) highly stratified, (2) poorly stratified, and (3) unstratified estuaries. Highly stratified estuaries usually have a relatively large fresh-water flow, and fresh water is less dense than salt water. Therefore, the fresh water floats on top of the salt water which remains lying on the bottom of the estuary as a *salt wedge,* or *salt tongue* (Figure 3.35). This condition creates a *halocline,* or salinity gradient, which may be depicted as a *salinity profile* of the estuary showing the heterogeneous nature of the cross-section of the water course (Figure 3.36). It is temporarily altered in response to the relative volumes of fresh water and sea water flows, which respond to seasonal and tidal cycles causing sharp demarcation boundaries to shift about. This affects the distribution of both plankton and benthos.

The *Coriolis force* acts upon estuaries as well, causing the salt tongue to be shifted to the eastern side of the water course in the northern hemisphere and to the western side in the southern hemisphere. It is caused by the rotation of the earth, which causes moving objects to turn right in the northern hemisphere and left in the southern hemisphere. If there were a theoretically perfect estuary of this nature, one could expect to find fresh-water species distributed along the shallow upper estuary above deeper marine species; however, such an idealized distribution has not yet been discovered and an intergrading of species is found. This intergrading does show a tendency toward the theoretical ideal defined above. In poorly stratified estuaries, the volumes of fresh water leaving and the tidal waters entering the system are nearly equal and the salinity gradients (i.e., salinity profiles) are less distinct (i.e., less precipitous) than those in highly stratified estuaries. This situation may create a physical *nutrient trap* (Figure 3.37) in a midwater mixing cell, which tends to retain nutrients and other entrained materials suspended in the estuary.

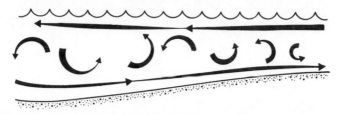

Figure 3.37. It is not unusual for poorly stratified estuaries, where volumes of fresh and sea water are essentially equal, to create a nutrient trap. Such a condition tends to retain nutrients and other entrained materials within the estuary being held there by the turbulent currents created by the upper layer of fresh water moving over the lower layer of marine water.

The third hydrographic classification of estuaries is the *unstratified,* and therefore homogeneous, type. It is characterized by a relatively large tidal range with associated vigorous mixing. This disrupts the formation of a salinity gradient. Although it is possible for a single estuary to pass through all those hydrologic classifications in one year, according to season, it is far more usual for one type to dominate and be reflected in the distribution of organisms and sediments within the estuary.

However, physically and ecologically, all estuaries tend to mature and age as the erosive forces of the oscillating sea water and the constant abrasion of the fresh-water flow tends to wear away the bottom and shoreline. As this process continues, estuaries tend to evolve as a result of (1) decreasing river flow and speed, if they were once augmented by melt water from ice age glaciers; (2) increased tidal velocity as a result of increased sea level; (3) increased width; and (4) decreased depth as a consequence of the erosion and sedimentation within the estuary (Figure 3.38). The ecologic implications of this is that the biogeographic distribution of organisms occurs in response to those physical factors which constrict the range of niche successes that exploit the marine environment in estuaries (Figure 3.39).

Marshes need not be associated with estuaries but they usually are. They may be defined as being tracts of soft wetlands, populated by low

Figure 3.38. Idealized cross sections above show the aging of an estuary which leads from the young, heterogeneous, and high flushing-rate river to the old, homogeneous, and low flushing-rate of a bay-like form.

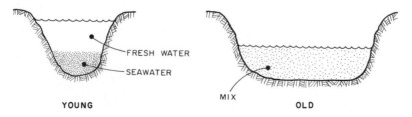

FRESH WATER

SEAWATER

MIX

YOUNG   OLD

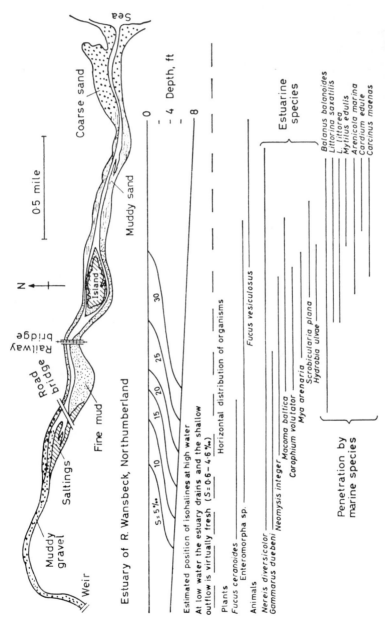

Figure 3.39. The biogeographical distribution of bentic organisms in an estuary in England reflects all the environmental factors which define that ecosystem. A combination of salinity and sediment types are of primary importance and must be presented in any thorough analysis. (From Tait and De Santo, 1972).

vegetation usually dominated by grasses, which are exposed to tidal waters regularly or occasionally. As an ecologic system, marshes serve as a continuum connecting terrestrial and fresh-water systems with the coastal marine ecosystem. They are clearly highly productive of both primary and secondary sources of food and usually generate a greater biomass than either terrestrial or aquatic systems.

The shoaling bottoms of a marsh, which are usually very broad and very flat, are either bare mud flats, when exposed at low tide, or are colonized by grasses that serve to baffle the kinetic energy of waves and currents (Figure 3.40). This allows suspended detrital material to settle out, with the result that such vegetated marshes act as both sediment and nutrient traps. Currents in the flooded marsh distribute food and also flush nutrients in and out, supporting the typically high productivity of the marsh and its sur-rounding waters (Figure 3.41). For example, it can be shown that approxi-mately 20% of phosphorus, a critical nutrient to plant growth, is absorbed by the salt marsh grass *(Spartina alterniflora)* in an experimental plot to which the nutrient is applied. Other studies have shown that the *Spartina* contains about 660 mg/m², whereas the sediments contain 5×10^5 mg P per square meter. This is a good indication that although the living biomass of the marsh may absorb nutrients from the water, the sediments serve as a sink and reservoir of these nutrients, such as phosphorus, which may be redistributed to the water from the sediments.

However, it cannot be concluded that these nutrients come initially from the fresh-water drainage leading to the marsh from the surrounding terres-trial environment. It is possible, and quite likely, that estuaries and their associated tidal marshes are fertilized by the ocean rather than by the inflow from rivers. Nonetheless, marshes clearly serve as buffers and they contribute at least four important functions to the stabilization of coastal marine ecology:

1. They serve as *physical and biologic filters,* which remove materials carried to them in fresh and sea waters.
2. They may serve as *biomagnifiers* (see Chapter 2) which concentrate elements in the ecosystem because of the relatively high productivity rates of marshes.
3. They may simultaneously serve as *sinks and sources of various chemi-cals,* both good and bad for life, which are metabolized (i.e., incorpo-rated into the biomass) and stored or released from this temporary storage in an altered or unaltered chemical state.
4. They also serve as a primary *nursery area* where the juveniles of many animals are provided with food, shelter, and a mild environment. They remain sufficiently long in the marsh to grow to survive in open coastal waters, where many of these species reach adulthood. It is estimated that approximately 90% of all sport fishing species are dependent upon tidal wetlands for the natural propagation of their species.

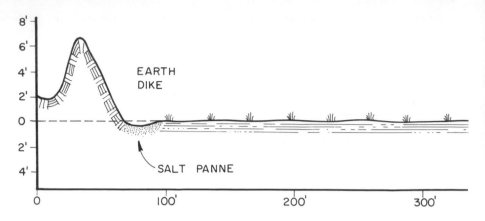

Figure 3.40. The cross sectional elevations taken across a typical salt marsh emphasizes the flat nature of this habitat. By that nature a large surface area of the shore is regularly inundated as the tide floods. Here, over a distance of 580 feet, an elevation change of only a few inches is evident. The weight of a man-made dike caused the marsh surface to slump and form a slight depression in which water is caught, evaporates, and creates a salt panne in which only highly salt tolerant vegetation can survive.

Considering the next level of the relationship between terrestrial and fresh water systems, and marine systems, we can look offshore from the marshes and estuaries to the coastal waters of a hypothetical ocean. Such an ocean is interlocked with all the systems previously described in this book. It is the ultimate sink for all the nutrients, pollutants, and sediments derived from the air, land, and fresh water that we have already considered or that we will consider. Mastery of understanding where these materials come from, how they function as ecologic facets in the biosphere, and what ultimately becomes of them is the fundamental premise upon which the philosophy of ecology is based. Perhaps we now have 10 or 20% of all the

Figure 3.41. Tidal marshes, which often form behind a barrier beach created by the littoral drift of sand, provide a protected and nutrient rich nursery ground for many species of wildlife. It is an interface between the sea and the land and represents one of the highest ecologically productive areas of the biosphere. The irregular pattern of tidal streams carry water, nutrients, sediments, and organisms throughout the system.

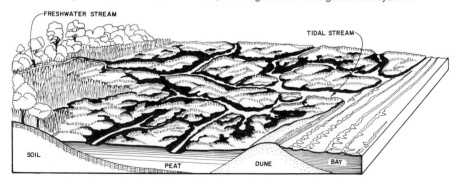

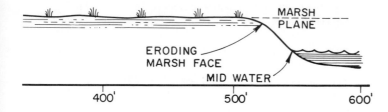

Figure 3.40   (Continued)

answers we are seeking. Nevertheless, our incomplete knowledge is enough to give us a good idea of the trends and major principles that we must focus upon and pursue. The challenge we face is to seek perfection while contending with the incomplete answers that we must use in any attempt to solve or control the environmental problems we have today. Doing as well as we can in our analysis and treatment of these problems requires the use of tools, whether concepts or hardware. We use such tools to define the ecologic state of an area of concern. Our interpretations of the findings we derive by using these tools are our chief means of dealing with ecologic problems.

Bibliography

General References

Allen, J. M., (ed.). 1967. *Molecular Organization and Biological Function*. Harper & Row, New York, 243 pp.

Allen, M. B., and Kramer (eds.). 1972. *Nutrients in Natural Waters*. Wiley, New York, 457 pp.

Asimov, I. 1963. *The Human Body, Its Structure and Operation*. The New American Library, New York, 320 pp.

Asimov, I. 1964. *The Intelligent Man's Guide to the Biological Sciences*. Washington Square Press, New York, 402 pp.

Baer, A. S., W. E. Hazen, D. L. Jameson, and W. C. Sloan. 1971. *Central Concepts of Biology*. Macmillan, New York, 385 pp.

Benarde, M. A. 1970. *Our Precarious Habitat.* W. W. Norton, New York, 362 pp.

Brachet, J., and A. E. Mirsky (eds.). 1961 *The Cell,* Vol. 2. Academic Press, New York, 916 pp.

Cohn, N. S. 1969. *Elements of Cytology.* Harcourt, Brace & World, New York, 495 pp.

Cox, G. W. 1969. *Readings in Conservation Ecology.* Appleton-Century-Crofts, New York, 595 pp.

Crowe, B. 1969. Tragedy of the commons revisited. *Science,* **166**, 1103–1107.

Davis, B. D., and L. Warren (eds.). 1967. *The Specificity of Cell Surfaces.* Prentice-Hall, Englewood Cliffs, N.J., 290 pp.

Gabriel, M. L., and S. Fagel. 1955. *Great Experiments in Biology.* Prentice-Hall, Englewood Cliffs, N.J., 317 pp.

Hammond, E. C. 1962. The effects of smoking. *Sci. Am.,* Vol. 207 No. (1); p. 39–51.

Hardin, G. 1968. Tragedy of the commons. *Science,* **162**; 1243–1248.

Kennedy, D. (ed.). 1965. *The Living Cell.* W. H. Freeman, San Francisco, 296 pp.

Simpson, G. G., and W. S. Beck. 1965. *Life, An Introduction to Biology.* (2nd ed.) Harcourt, Brace & World, New York, 869 pp.

Stein, W. D. 1967. *The Movement of Molecules Across Cell Membranes.* Academic Press, New York, 369 pp.

Veeton, W. T. 1967. *Biological Science.* W. W. Norton, New York, 955 pp.

Wallach, D. F. H. 1972. *The Plasma Membrane: Dynamic Perspectives, Genetics, and Pathology.* Springer-Verlag, New York, 186 pp.

Watson, J. D., and F. H. C. Crick. 1953. The structure of DNA. *Cold Spring Harbor Symp. Quant. Biol.,* **18**; 123.

Terrestrial Ecology

Alexander, M. 1961. *Introduction to Soil Microbiology.* Wiley, New York, 472 pp.

Black, C. A. 1968. *Soil–Plant Relationships.* (2nd ed.) Wiley, New York, 792 pp.

Brady, N. C. 1974. *The Nature and Properties of Soils.* (8th ed.) Macmillan, New York, 639 pp.

Braun, E. L. 1950. *Deciduous Forests of Eastern North America.* Blakiston, Philadelphia.

Brill, W. J. 1977. Biological nitrogen fixation. *Sci. Am.,* **236**(3), 68–81.

Buckman, H. O., and N. C. Brady. 1960. *The Nature and Properties of Soils.* Macmillan, New York, 567 pp.

de Beaufort, L. F. 1951. *Zoogeography of the Land and Inland Waters.* Sidgewick and Jackson, London, 208 pp.

Deevey, E. S. Jr. 1951. Life in the depths of a pond. *Sci. Am.* **185**, 68–72.

Elias, T. S., and H. S. Irwin. 1976. Urban trees. *Sci. Am.,* **235**(5), 110–118.

Fenton, G. R. 1947. The soil fauna: With special reference to the ecosystem of forest soil. *J. Animal Ecol.,* **16**, 76–93.

Hedgepeth, J. W. 1951. The classification of estuarine and brackish waters and the hydrographic climate. In: Report No. 11 of the National Research Council Committee on a Treatise on Marine Ecology and Paleoecology. pp. 49–56.

Good, R. D. 1964. *The Geography of Flowering Plants.* (3rd ed.) Longmans, London, 518 pp.

Jones, J. R. E. 1949. A further ecological study of a calcareous stream in the Black Mountain district of south Wales. *J. of Animal Ecol.* **18**, 142–159.

Kessell, S. R. 1976. Gradient modeling: A new approach to fire modeling and wilderness resource management. *Environ. Man.,* **1**, 39–48.

Moore, G. 1976. The natural role of fire. *Horticulture,* **54**, 44–50.

Vollmer, A. T., B. G. Maza, F. B. Turner, and S. A. Bamberg. 1976. The impact of off-road vehicles on a desert ecosystem. *Environ. Man.,* **1**, 115–129.

Vrajina, V. J. 1969. Ecology of forest trees in British Columbia. In: *Ecology of Western North America.* University of British Columbia Press, Vancouver 2, pp. 1–147.

Walter, H. 1973. *Vegetation of the Earth in Relation to Climate and the Eco-physiological Conditions.* Springer-Verlag, New York, 237 pp.

Whittaker, R. H., and G. M. Woodwell. 1969. Structure, production and diversity of the oak–pine forest at Brookhaven, New York. *Ecology,* **57**, 155–174.

Williams, R. E. O., and C. C. Spicer. 1957. *Microbial Ecology.* Cambridge University Press, London, 388 pp.

Wilson, J. T. (ed.). 1972. *Continents Adrift: Readings from Scientific American.* W. H. Freeman, San Francisco.

Fresh-Water Ecology

Fassett, N. C. 1960. *A Manual of Aquatic Plants.* (Rev. Ed.) University of Wisconsin Press, Madison, 405 pp.

Goldman, C. R. 1960. Primary productivity and limiting factors in three lakes of the Alaska peninsula. *Ecol. Monogr.,* **30**, 207–230.

Hunt, C. A., and R. M. Garrels. 1972. *Water the Web of Life.* W. W. Norton, New York, 208 pp.

Hutchinson, G. E. 1957. *A Treatise on Limnology.* Wiley, New York, 1015 pp.

Mackenthun, K. M. 1973. *Toward a Cleaner Aquatic Environment.* Publ. 5501-00573, U.S. Government Printing Office, Washington, D.C., 273 pp.

Morgan, A. H. 1930. *Field Book of Ponds and Streams.* G. P. Putnam, New York, 448 pp.

Nikolsky, G. V. 1963. *The Ecology of Fishes,* trans. by L. Birkett. Academic Press, New York, 352 pp.

Otto, N. E., and T. R. Bartley. 1972. Aquatic Pests on Irrigation Systems. Publ. 2403-0069, U.S. Government Printing Office, Washington, D.C., 72 pp.

Palmer, C. M. 1962. *Algae in Water Supplies.* Publ. PHS 657, U.S. Government Printing Office, Washington, D.C., 88 pp.

Prescott, G. W. 1954. *How to Know the Freshwater Algae.* Wm. C. Brown, Dubuque, Iowa, 211 pp.

Reid, G. K. 1967. *Pond Life.* Western, New York, 160 pp.

Stanley, N. F., and M. P. Alperts (eds.). 1975. *Manmade Lakes and Human Health.* Academic Press, New York, 495 pp.

Trautman, M. B. 1942. Fish distribution and abundance correlated with stream gradients as a consideration in stocking programs. *Trans. 7th North Am. Wildlife Conf.,* **7**, 221–223.

Marine Ecology

Barnes, R. D. 1968. *Invertebrate Zoology.* W. B. Saunders, Philadelphia, 743 pp.

Barton, R. 1970. *Oceanology Today.* Doubleday, Garden City, N.Y., 192 pp.

Bascom, W. 1964. *Waves and Beaches.* Doubleday, Garden City, N.Y., 268 pp.

Clark, J. R. 1977. *Coastal Ecosystem Management.* Wiley, New York, 928 pp.

Gordon, B. L. 1970. *Man and the Sea.* Natural History Press, Garden City, N.Y., 498 pp.

Green, J. 1968. *The Biology of Estuarine Animals.* University of Washington Press, Seattle, 401 pp.

Harvey, H. W. 1963. *The Chemistry and Fertility of Sea Waters.* Cambridge University Press, London, 240 pp.

Herring, P. J., and M. R. Clarke (eds.). 1971. *Deep Oceans.* Praeger, New York, 320 pp.

Komar, P. D., J. R. Lizarraga-Arciniega, and T. A. Terich. 1976. Oregon coast shoreline changes due to jetties. *Am. Soc. Civil Eng. J. Waterways, Harbors, Coast.* Eng. Div., **102**(WW1), 13–30.

Lauff, G. H. 1967. *Estuaries.* American Association for the Advancement of Science, Washington, D.C., 757 pp.

Moore, J. R. (ed.). 1971. *Oceanography.* W. H. Freeman, San Francisco, 417 pp.

Paskansky, P. F. 1977. Net drift in an atypical estuary. Long Island Sound. *Environ. Man.,* **1**(4), 331–342.

Redfield, A. C. 1972. Development of a New England salt marsh. *Ecol. Monogr.,* **42**(2), 201–237.

Riley, G. A. 1967. Mathematical model of nutrient conditions in coastal waters. *Bull. Bingham Oceanogr. Coll.,* **19**, 72–80.

Ross, D. A. 1970. *Introduction to Oceanography.* Appleton-Century-Crofts, New York, 384 pp.

Schroeder, P. M., R. Dolan, and B. P. Hayden. 1976. Vegetation changes associated with barrier-dune construction on the outer banks of North Carolina. *Environ. Man.,* **1**(2), 105–114.

Smith, G. L. 1977. The failure of success in fisheries management. *Environ. Man.,* **1**(3), 239–247.

Spoczynska, J. O. I. 1976. *An Age of Fishes.* Charles Scribner's Sons, New York, 152 pp.

Sverdrup, H. L., M. W. Johnson, and R. H. Flemming. 1946. *The Oceans, Their Physics, Density, and General Biology.* Prentice-Hall, Englewood Cliffs, N.J.

Tait, R. V., and R. S. DeSanto. 1972. *Elements of Marine Ecology.* Springer-Verlag, New York, 327 pp.

Zim, H. S., and L. Ingle. 1955. *Seashores.* Golden Press, New York, 160 pp.

4 Selected Tools of Ecologic Analysis

The practice of medicine and ecology is much the same. It seems that the need for the medical general practitioner increases rapidly while the supply diminishes. It is a symptom of our evolving and expanding culture as discussed in Chapter 2 of this book, where we considered the apparent ecologic trends seen developing within our technologic growth. The skills and information that physicians must comprehend logically pressures those professionals into pursuing specialties instead of the general practice of medicine. There is too much to learn for any one person to become expert and self-sufficient in the whole field, so instead of becoming a generalist, many physicians in our "advanced" societies seek the confidence provided through being expert in a small sector of the whole field.

Precisely the same condition prevails in ecology, which is a considerably broader field than medicine. The same frustrations exist with respect to learning too much in too short a time. The same skill in the art of diagnosis and prognostication is required. Furthermore, as is the case with any good professional, the same concern for being objective, accurate, and correct holds true for the applied ecologist as for the physician. Perhaps the ecologist should be something more of an attorney than the physician because much of what the applied ecologist does, relates directly to the law. Permit requirements, contractual obligations related to environmentally sensitive issues, and forensic ecology all require the central participation of an applied ecologist in order that the ecology be effectively combined and interpreted for use by an attorney.

Studying ecology, then law, is one approach to this problem, and assuming that it results in the development of a career specializing in the practice of law with an ecologic bias, it is a useful and increasingly marketable career. However, such an attorney recognizes the need to call upon the applied ecologist when the problem at hand requires the credentials and experience of a fulltime ecologist as opposed to the sole involvement of that fulltime attorney, "parttime" ecologist.

Of course, the basic tool of ecologic analysis is education; a good education, at least theoretically, prepares a student to strike off into a career with more momentum and sophistication than were that education lacking. Therefore, virtually every school offers a so-called interdisciplinary program leading to some formal or informal certification which proclaims that participating students are ecologists once they successfully complete their courses. This educational tool is certainly helpful but the ideal structure for such a program is often far from attained. The difficulty probably arises because traditional education and its institutions are very formally, and perhaps necessarily, divided into departments that, by their nature, cannot be effectively combined within something called an interdisciplinary study program. It seems that individual academic departments are forever and always individual and separate. Although members of the different departments may wish to come together in team teaching, their tendency is to teach as individuals nevertheless. More often, interdisciplinary programs offer a student various course options involving a *potpourri* of departments—each department selecting those courses it feels provide some appropriate facet to the education of the well-rounded ecologist.

These approaches work, provided the students are able to assimilate all this interdisciplinary learning as something more than an assortment of classes and class notes and textbooks. This is a major task at which virtually every present-day institution fails because the solution lies in the individual student, not the institution. The student must be trained to recognize that he or she is presented packages of information in relatively discrete categories. Education is taught in separate and clearly definable classes. It must be the student's responsibility to assimilate this diverse education so that she or he can solve the interdisciplinary problems encountered in the real world. But that ability is not effectively taught because, by its very nature, the traditional academic institution is insulated from the real world, especially in its molding of undergraduate students. That insulation may be necessary because it permits a more efficient cramming of information into the brains of students than might otherwise be possible in a nontraditional interdisciplinary learning process. A broad base of specific facts must first be amassed before a student can efficiently draw upon that library of facts to piece together the very specific solution of some real ecologic puzzle.

If we examine a few hypothetical problems, the limits of our present educational systems can be seen and the strain placed on the budding ecologist can also be better understood. For example, virtually all of the problems with which an applied ecologist deals relate to (1) environmental law, (2) engineering projects, or (3) some natural or social science involving ecology. In fact, regardless of the main aspect of the problem, the other two subordinate aspects are present to a lesser or greater degree (Figure 4.1).

If the construction of a sewer line is proposed, the major emphasis of the preliminary work necessary involves the engineering design. Although the legal agreements and controls on the project are important, and social

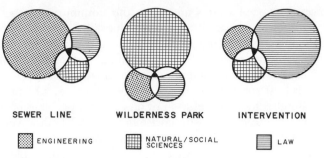

Figure 4.1. The predominating areas of concern which are present in any ecological problem with which an applied ecologist must deal include the above disciplines. One discipline may be clearly more significant than the others in a particular study. The ecologist should know enough of those other fields in order to identify the interrelationships between the fields and the role which he can best provide when called upon. The overlapping area of concern is at the center of most environmental problems and is an arena in which the applied ecologist must be experienced.

needs and restraints must be considered, the majority of the work concerns the engineering disciplines. If creation of a wilderness park is proposed, most of the work effort is within the natural sciences, which are used to define the concept and to collect the basic information necessary to support the proposal. Engineering and law are also necessary but are usually less dominant in such a project. Finally, if a project is opposed by some individual or group, this *intervention* requires an emphasis on the law as it pertains to the project. Obviously, the intervenors often require a considerable amount of engineering and natural or social science input, but the lead in that sort of project is clearly legal.

The fact remains that in all these generalized examples, the three fields overlap. That area where natural and social sciences, law, and engineering are superimposed is of prime concern to the applied ecologist because it is his or her job to interpret ecologic perameters as they effect or are affected by engineering practices, law, or the natural and social sciences. Therefore, the ecologist's job always directly or indirectly addresses those analyses which involve human impacts. This is often quite different from the work and interests of the theoretical ecologist, whose concern for ecology does not necessarily mean an interest in humans. Nevertheless, most of the tools used by the theoretical ecologist are identical to those used by the applied ecologist. The difference is that applied ecologists are usually concerned with presenting pertinent ecologic information in a format that can be used by nonecologists to weigh information in making a political (i.e., socioeconomic) decision. Several countries are formalizing requirements which aim to insure that political judgments made have access to ecologic considerations, for example, legal requirements directing that all pertinent human activities be thoughtfully planned, including prominent attention to ecologic sensitivity.

It seems quite obvious that the success of this formalized policy mainly depends upon our social values and upon our commitments to the costs versus the benefits of our environmental values. Its success is not especially dependent upon absolute, well-defined, and explicit goals because applied ecology often requires human judgments of relative values and goals. This dilemma of human interest in managing well our stewardship of the biosphere is best summed up in generalities because, when reduced to specifics, the approach to the solution of an ecologic problem becomes far more debatable. In 1969, the National Environmental Policy Act (NEPA) of the United States of America proclaimed its purpose " . . . to declare a national policy which will encourage productive and enjoyable harmony between man and his environment; to promote efforts which will prevent or eliminate damage to the environment and biosphere and stimulate the health and welfare of man; to enrich the understanding of the ecological systems and natural resources important to the nation. . . .''

There should be no question that these objectives are good and useful, but the means by which they can be achieved are not so clear. Therefore, the judgments that must be made should follow a pattern that insures that all the available and pertinent information is considered and that a conscious effort is always made to protect and conserve all environmental factors. In those countries in which guidelines for making these decisions involving applied ecology have been promulgated, those guidelines usually call for a written analysis of all environmental considerations for any significant project under the direct or indirect regulation of the government. Although such analyses have slightly different names, their purpose is to present an ideally unbiased report of the present and predicted future state of the total environment as it might be effected, or impacted, by the proposed action.

Often called an Environmental Impact Statement (EIS), each analysis represents at best a philosophy of dealing thoughtfully and evenhandedly with our short- and long-term uses of the natural and human resources for which use are steward. At worst, the EIS is thought of as a cookbook—a matrix—which, when filled in tells whether a project is good or bad. The specific nature of this analysis depends largely upon both the apparent social and the ecologic significance of the proposed project. The objective is to define the present environment in which the proposed action is to occur, to evaluate all possible and prudent alternatives, to identify all possible and probable eventualities, to insure that all negative impacts are minimized, and to demonstrate that the proposed project has been appropriately announced to all interested parties so that their concerns can be considered.

This approach can be applied to either large or small, simple or complex projects and the specific definition of what areas of concern should be emphasized must be provided by those supervising the preparation of the EIS. Although this book deals solely with applied ecology, it should be helpful to see this field in the context of the others that are necessary in the preparation of an EIS. Therefore, the following sample table of content is given in its entirety and in a sequence which seems to this author particularly logical, regardless of the natural or the human environmental emphasis for which a particular project may call.

As an approach to the challenge of interpreting the possible environmental impact of a proposed project, the preceeding outline can be expanded with emphasis on the various aspects of the natural or the human environment. Such an emphasis may be perfectly appropriate if the project is predominently natural or it may not be as appropriate if the project is predominantly human in its components. In either case, the analysis of the project may be divided between a natural environment team and a socioeconomic team.

Table 4.1. EIS—Basic Content Guideline

TABLE OF CONTENTS

   I. SUMMARY
  II. DESCRIPTION OF THE PROPOSED ACTION
     A. Project Area
     B. Project Design
 III. ALTERNATIVES TO THE PROPOSED ACTION
 IV. ENVIRONMENTAL IMPACTS OF THE PROPOSED ACTION
     A. The Natural Environment
       1. General
         a. Setting and Description
         b. Meteorology and Climatology
         c. Water Regime
         d. Land Regime
       2. The Specific Environment
         a. Physical Parameters
         b. Site Biology
           (1) General
           (2) Habitat Inventory
           (3) Organism Inventory
     B. Human Environment
       1. Land Use
       2. Economics
       3. Displacement (relocation)
       4. Traffic Movement
       5. Air and Noise Pollution
       6. Esthetics and Other Values
       7. Public Facilities and Services
   V. ADVERSE ENVIRONMENTAL EFFECTS AND THEIR MITIGATION MEASURES
  VI. THE RELATIONSHIP BETWEEN LOCAL SHORT-TERM USES OF THE HUMAN
     ENVIRONMENT AND THE MAINTENANCE AND ENHANCEMENT OF LONG-TERM
     PRODUCTIVITY
 VII. ANY IRREVERSIBLE AND IRRETRIEVABLE COMMITMENTS OF RESOURCES
     WHICH WOULD BE INVOLVED IN THE PROPOSED ACTION, SHOULD IT BE
     IMPLEMENTED
VIII. COORDINATION WITH AGENCIES, ETC.
APPENDIX
     1. Technical Reports
     2. Raw Data
BIBLIOGRAPHY

Table 4.2.   Guidelines for the Ecologic Documentation and Environmental Analysis of a Proposed Action. (Biologic Emphasis)

SUMMARY
- A. Significant impacts (positive and negative) that might result from the project
- B. Unquantified or unquantifiable impacts related to the project
- C. A graphic summary (as appropriate) of project impacts

I. DESCRIPTION OF THE PROPOSED ACTION
- A. Project Objectives
  1. Basic purpose of the proposed project
  2. Project nature relative to public policy/plans (long and short term)
  3. Current project justification based on need, absence of alternatives, including no development alternative (also see section II)
  4. Projected demand or need for the project based on documented methods and/or studies
  5. Cost effectiveness of the project from both an economic and a resource management point of view
- B. Project Description
  1. Project Location
     a. Mesoscale (i.e., regional) and microscale (i.e., local) map location (2 maps)
  2. Project Context
     a. Future expansion or modification
     b. Other interdependent existing or proposed projects or plans
     c. Existing projects in the area influenced by or influencing the project, including their map locations
  3. Project Activities
     a. Short Term
        (1) Construction activities that are temporary
     b. Long Term
        (1) Permanent structures
        (2) Ongoing (i.e., operational) activities

II. ALTERNATIVES TO AND OF THE PROPOSED ACTION
- A. Alternatives to the action that are prudent and possible, including the no-development decision; alternatives that may meet the same basic needs; and the subsequent main effects
- B. Alternatives to the action that are prudent and possible, including different locations and different component variations, such as design changes.

III. ENVIRONMENTAL SETTING AND IMPACTS OF THE PROPOSED ACTION
- A. The Natural Environment
  1. General
     a. Project setting and description (i.e., Biome and community)
     b. Meteorology and climatology as appropriate to an overview
        (1) Precipitation
        (2) Temperature
        (3) Wind (i.e., wind rise and overview of its significance)
        (4) Air quality in and about the project area
     c. Storms/Floods
  2. Terrestrial Environment
     a. Physical Components
        (1) Geomorphology
        (2) Bedrock geology
        (3) Surficial geology (excluding soils)
        (4) Soils of the project area (including a soils map)
        (5) Associated hydrology including existing erosion/sedimentation relative to the proposed project

(a) Mass movements (landslides, earthflows, mudflows, creep, solifluction, including bearing capacity of the soils as appropriate)
(b) Sheet erosion
(c) Channel erosion
(d) Channel, levee, and floodplain deposition
(e) Lacustrine (lake) deposition
(f) Riverine (river) deposition
(g) Palustrine (marsh) deposition
(h) Wave, current, or tidal erosion or deposition
(i) Aeolian (wind) erosion or deposition
(6) Soil characteristics relative to overall productivity (including fertility)
(7) Mineral resources (including sand and gravel deposits)
(8) Seismic or other geologic hazards
b. Biologic Components
(1) Biogeography of the project area (plant and animal)
(a) Plant and animal composition (including a biogeographic map)
(b) Definition of existing habitats (vegetation, migratory routes, den areas, water supplies, etc., mapped as necessary)
(c) Species diversity and abundance including resident and transient species
(d) Plant and animal dynamics (seasonal and stage of succession and migration)
(e) Plant functions in productivity, nutrient cycling, provision of habitat, and substrate stabilization
(f) Observations of wildlife interrelationships; territoriality, prey–predator relationships, etc., as related to the stability of the study area
(g) Hunting pressures
(h) Observations of other functions or capacities (plant–animal relationships, productivity in relation to theoretical carrying capacity, value of the vegetation in terms of wildlife potential)
(2) Highly productive and/or rare site communities
(a) Net productivity based on seasonal biomass and/or comparative evaluations
(b) Relative rarity compared with contiguous or nearby areas
(3) Rare, threatened, endangered species of plants and animals on or contiguous to the site
(a) Report those whose range theoretically extends into the study areas
(b) Report those which could probably exist in the study areas
(c) Report any that have been actually located in or near the study area
(4) Significance of undisturbed vegetation and/or wildlife on or contiguous to the project site
(a) Potential sensitivity of this vegetation and wildlife in terms of disturbances caused by construction or operations-related activities (i.e., air pollution, water pollution, trimming, thinning, grubbing, or clearing)
3. Aquatic Environment
a. Physical Components
(1) Project relationship to its watershed including a map (drainage area, runoff factors)
(2) Project relationship to surface and ground waters (including recharge potential)

Table 4.2 (continued)

(3) Existing downstream hydrologic conditions, including constrictions in flow.

(4) Project susceptibility to water-related hazards including floods, waves, etc.

(5) Existing water quality in and about the project area. Considering the following as appropriate.

    (a) pH (acidity)

    (b) BOD (biologic oxygen demand)

    (c) Plant nutrients (nitrate nitrogen and soluable phosphate)

    (d) DO (dissolved oxygen)

    (e) Chlorophyll a

    (f) Coliform bacteria

    (g) Trace metals (mercury, lead, zinc, cadmium, copper, etc.)

(6) Tides (tidal cycle curve—high water/low water, including range)

(7) Current measurements (direction and speed at 0.5 m from surface and 1 m off bottom)

(8) Sediments

    (a) VS (volatile solids, i.e., organic material)

    (b) COD (chemical oxygen demand)

    (c) TKN (total Kjeldahl nitrogen, free ammonia)

    (d) Hexane solubles (oil and grease)

    (e) Heavy metal

        1. Mercury (Hg)

        2. Lead (Pb)

        3. Zinc (Zn)

        4. Cadmium (Cd)

        5. Copper (Cu)

        6. Arsenic (As)

        7. Chromium (Cr)

        8. Nickel (Ni)

        9. Vanadium (V)

    (f) Insecticides, polychlorinated biphenyls (PCB), etc.

(9) Turbidity (measure at given depths 1 m and 3 m, etc., below surface at different tide stages, river flows, or effluent values)

(10) Salinity (in parts per thousand; multiple measurements as above)

(11) Temperature (multiple measurements as above)

(12) Color (multiple measurements as above)

(13) Depth (adjusted to mean low water)

b. Biologic Components

(1) Biogeography of the project area (plant and animal)

    (a) Plant and animal composition, including a biogeographic map

    (b) Plant and animal dynamics (seasonal, stage of succession, and migration, including anadromous fish movements)

    (c) Vegetation (productivity, nutrient cycling, provision of habitat, and substrate stabilization)

    (d) Definitions of existing habitats (vegetation, migratory routes, den areas, water supplies, etc., mapped as necessary)

    (e) Estimation of diversity, numbers of species present, and their populations, including resident and transient species.

    (f) Observations of wildlife interrelationships: territoriality, predator–prey relationships, etc., as related to the stability of the study area

    (g) Hunting or fishing pressures (commercial or recreational)

    (h) Observations of other functions or capacities (plant–animal

          relationships, productivity in relation to theoretical carrying
          capacity, etc.)
         (i) Value of the vegetation in terms of wildlife potential
      (2) Highly productive and/or rare site habitats
        (a) Net productivity based on seasonal biomass and/or comparative
           evaluations
        (b) Relative rarity compared with contiguous or nearby areas
      (3) Rare, threatened, or endangered species of plants and animals on or
        contiguous to the site
        (a) Report those whose range theoretically extends into the study area
        (b) Report those which could probably exist in the study area
        (c) Report any that have been actually located in or near the study area
      (4) Significance of undisturbed vegetation and/or wildlife on or
        contiguous to the project site
        (a) Potential sensitivity of this vegetation and wildlife in terms of
          disturbances caused by construction or operations-related
          activities (i.e., air pollution, water pollution, trimming, thinning,
          grubbing, or clearing)
  B. The Human Environment (addressed in detail by the appropriate specialists from
    disciplines other than ecology)
    1. Land Use
    2. Economics
    3. Displacement
    4. Traffic Movement
    5. Air and Noise Pollution
    6. Esthetics and Other Values
    7. Public Facilities and Services
IV. ADVERSE ENVIRONMENTAL EFFECTS AND THEIR MITIGATION MEASURES
  A. Short- and long-term impacts resulting from:
    1. Vegetation removal and its specific mitigation
    2. Erosion/sedimentation and its specific mitigation
    3. Solid waste generation and its specific mitigation
    4. Chemical pollution and its specific mitigation
    5. Special areas of concern (e.g., herbicides, fire, microwaves, etc.)
V. THE RELATIONSHIP BETWEEN LOCAL SHORT-TERM USES OF THE HUMAN
  ENVIRONMENT AND THE MAINTENANCE AND ENHANCEMENT OF LONG-TERM
  PRODUCTIVITY
VI. ANY IRREVERSIBLE AND IRRETRIEVABLE COMMITMENTS OF RESOURCES THAT
  WOULD BE INVOLVED IN THE PROPOSED ACTION, SHOULD IT BE IMPLEMENTED
  A. Significant loss of habitat
  B. Significant alteration of contiguous areas
  C. Significant "injection" of wastes into the environment
VII. COORDINATION
  A. List all agencies and other contacts made with respect to documentation
    collected and interpreted for this environmental review or for informational
    contacts
  B. Comments received concerning the report from any reviewer
APPENDIX
  Technical reports and raw data. Include any maps, photos, figures, technical
  reports, etc., that are crucial to the collection and interpretation of the information
  contained in this environmental review
BIBLIOGRAPHY

This is often the approach used in the commercial field of preparing environmental documentation, where specialties are divided and the final product results from the contribution of several disciplines and several individuals. As an example of one such option, the following guidelines were developed for use by a team of ecologists specializing in evaluating potential impacts upon the natural environments. Therefore, their approach deemphasized socioeconomic factors, such as noise, human community structures, employment, esthetics, and public services. Although these and related areas are obviously of considerable importance to any human evaluation of a proposed project, here we are considering the predominating interests of applied ecologists, who are primarily biologists, not sociologists, and this bias is reflected in the guidelines.

The execution of those sections of an EIS which should be controlled by an applied ecologist requires two phases of work. These phases can be subdivided into a total of ten tasks. The complexity and length of each one of these tasks is naturally directly related to the specific proposal being evaluated. However, regardless of the specific project, it is certainly possible to characterize each task in the following manner.

Phase I. Initial Stages of EIS Preparation

1. From the point of view of the natural environment, the ecologist responsible for project management must first consider the proposed project or action in light of its characteristics as an ecologic manipulation of the site-specific ecosystems. Usually based upon very gross information, the ecologist must define what general sorts of impacts and what relative magnitude of activities are likely to be associated with the project. His or her skill at making this first deduction is almost entirely based upon experience and intuition, supported by a few hints or guidelines that may often be extracted from preliminary engineering documents and interviews.

2. With some self-confidence that he or she knows the general magnitude and location of the project, the ecologist can make a preliminary definition of the data base necessary and the data base available, from which information about the proposed action and its environmental setting can be extracted. Such information about the setting may be present in the professional literature or in such fugitive literature as government publications, regional libraries, or private reports concerning some aspect of the proposed project area. At this early stage in an EIS preparation, the ecologist must become superficially conversant with the available data in order to proceed to the next task which includes determining the specific laws or regulations that relate to the proposed action and the associated EIS.

3. By further discussion with community representatives and individuals involved in the formulation of the proposed action, the ecologist must

determine the general nature of the community concern that is or will be focused on the proposed project. Part of this includes collection and review of the specific laws, directives, administrative procedures, and all associated application forms that may be pertinent when permitting procedures are involved. Virtually every time the ecologist is called upon to participate in the analysis of a proposed action, the client's primary motivation is compliance with the law rather than some particular concern that he or she may have for environmental protection.

However, whatever the reasons for his involvement in the project, the ecologist must determine exactly under what constraints he or she is working. This includes knowing the laws to which his or her work must answer, the "muscle" which that work lends to the success or failure of the project, and the time frame available for completing the EIS. Therefore, an important aspect of this task includes prediction of the level of effort necessary for preparing the EIS. This is determined by the level of technical information necessary to satisfy the level and nature of the EIS review process. That judgment requires technical meetings, which take place in the next task.

4. Meeting with representatives of the regulatory agencies, and with anyone else who may represent antagonistic or protagonistic views of the proposed action, allows the ecologist to construct a good profile of those regulations and social and environmental objectives that the proposed action must obviously meet. However, great care must be exercised to insure that all areas of real or potential concern have been addressed and that no area has inadvertently been missed. It is impossible to guarantee such unerring coverage and this task must remain an ongoing effort. However, most if not all of the associated findings must result from its early initiation. Its ongoing involvement in the project requires relatively little manpower, even though it provides an important safety factor. It helps insure that the final EIS is fully responsive to the central and critical issues of the proposed action, even if new or revived issues arise subsequent to the completion of the main efforts in this task.

5. Based upon the technical requirements of the EIS, the nature of the available data bases, and the time and budget provided for the study, the chief ecologist must designate a project team and she or he must assign the responsibilities of each team member. Highly controversial projects, requiring much technical information, or projects that have a tight time schedule can often be accommodated by increasing the division of labor within a larger environmental team. Provided that close coordination can be maintained, use of a larger team can theoretically produce more information in a shorter time. This concept may not produce the expected results, however, if the chief ecologist (1) does not delegate responsibility, (2) misassigns responsibility, or (3) if members of the team cannot effectively work as a team. The most common problem that arises to thwart the successful completion of complex

environmental analyses concerns interpersonnel relationships. These problems may form within the study team or between members of the team and other groups or individuals from which information must be obtained. Therefore, an effective ecologist has the responsibility of recognizing these human problems and seeks to correct them as far as possible. Diplomacy is an extremely important aspect in the successful practice of ecology as a profession.

6. The last task in this initial phase of work requires that the chief ecologist structure his or her involvement in the project so that he or she will always know what type of information and data are being sought and for what specific purposes. Especially when the team size exceeds three members, when the complexity of the analysis is great, or when the chief ecologist must direct more than one project or more than one team, there is a considerable danger that communication between team members will decay. Inexperienced team members may not recognize these circumstances and their work will suffer if they go off on a wrong tangent or needlessly duplicate the work of others. If the chief ecologist keeps abreast of the technical progress of each team, these potential problems can be avoided or minimized.

Phase II. Final Stages of EIS Preparation

7. With all the preceding information collected and organized, the chief ecologist, in consultation with the appropriate team members, must finally determine what applications, including permit applications, must be completed. These may include permits required for proposed actions that may take place in environmentally sensitive areas, such as wetlands, or they may relate to prospecting rights or any other regulated activities that involve environmental issues.

This task also requires that assignments be made to the staff of those individuals who will be responsible for preparing each required application. Ideally, each senior staff member is kept aware of all final drafts of these applications throughout the project, regardless of whether or not input from that particular team member ceases before the project is completed. The objective of this procedure is to provide as broad a view as possible of project work before that project is released as a final and supposedly complete document.

8. The use of outside consultants may be necessary for projects that require expertise beyond the capabilities of the resident team, or in those cases where additional staff is necessary for short-term work for which hiring of new staff is not appropriate. In either of those instances, the responsibility for project management rests on the chief ecologist, and either the chief ecologist or an assigned senior staff member must very closely observe the procedures and products issu-

ing from the outside consultant. The need for understanding all the steps undertaken by the outside consultant, and the need to insure a long-term continuity of work methods and interpretations, requires that at least one staff member of the project team be completely aware of all consultant activities, reviewing them at regular intervals with the chief ecologist.

9. This task calls for the integration of the objectives of the EIS (which is an overall documentation, interpretation, and mitigation of the proposed action), with the associated permit applications that may be called for. Both aspects of environmental management should occur simultaneously, provided that there is a free exchange of information between the EIS process and the permit process. Although this is not always the case, preparation of an EIS will ideally contain information that provides the technical framework on which required permit applications depend. In any event, at a minimum, tasks 3 and 4 result in a definition of the formal concerns that both EIS and permit applications must address. Once the technical work of answering those concerns is underway, a conscious effort should be initiated to coordinate the technical and the philosophic efforts of creating an EIS that is useful in assisting any authority to better judge the proposed action and all associated permits or other controls.

10. This final task involves assembling the EIS and any associated applications that may be completed simultaneously. If legal proceedings result from any of these submissions, it may be necessary, at that time, to reevaluate or reenforce the conclusions drawn from these documents. Such legal consequences may or may not be anticipated. Therefore, as a general safeguard, the EIS for any proposed action should contain as complete a bibliography as possible. It should be appropriate to the main issues and the peripheral issues involved in consideration of the proposed action.

If subsequent, legal review of the EIS is called for, it is likely that a comprehensive bibliography will be valuable. Furthermore, it is not unlikely that in the absence of such a bibliography contained within the published final document, copies of references will be lost because of the fast pace and potential confusion that always exists in the fields of applied ecology and EIS preparation.

Because of that fast pace, the packaging effort in this task is especially crucial. Because an EIS often results from the contributions of many authors, special care must be taken in assuring that the final document is consistent in format and style and that each section is in the proper sequence and is logically presented relative to the other sections. Because the contribution to the EIS made in the natural environment field is only a part of the statement, the overall responsibility of reviewing, final editing, and packaging of the EIS usually rests with an engineer or planner, as opposed to an ecologist. In such cases, the responsible ecologist must coordinate his or her report with the constraints of time and format determined through consultation with the overall EIS project manager.

Regardless of the specific proposed action being evaluated, there are four critical rules of caution that must be rigorously upheld by each team member involved in environmental assessments. In their absence, the project is likely to lose direction, conclusions will be difficult to identify, information will not be well organized or presented, and the inherent confusion present will not serve the best interests of environmental management and conservation.

Rules of Caution for EIS Preparation

1. As a responsible ecologist for the analysis of a particular proposed action, confirm the currentness of any forms, procedures, and timetables that define the project and determine what is expected of the natural environment section of the EIS. Subsequently, insist on written documentation if and when any of these project constraints are modified in any way.
2. Do not give free reign to data collection. Anyone responsible for the collection and/or the interpretation of data, regardless of its nature, must understand why the data is needed and what level of comprehensiveness and detail is required. In the absence of careful control over this function, great quantities of time, let alone budget, can be needlessly spent with no useful conclusion resulting from the work. The essence of applied ecology is solution or mitigation of real problems through the artful use of prudent and possible, environmentally sound methods of analysis and protection.

    Collecting data without knowing the possible, useful result of that effort in solving the immediate problem at hand is an egregious error of judgment in the field of applied ecology. Collecting such data may be an asset in the context of theoretical ecology, however, which does not confront a deadline; which does not, or should not, involve legal implications; and which is generally free of the constraints requiring the treatment of real problems through real interdisciplinary communication.
3. Maintain maximum flexibility in casting the final format and content of the EIS or application. Any project that is analyzed as outlined in this book should be viewed as a learning experience because no two projects are ecologically identical. Therefore, no two EISs can be identical, and the associated interpretations, emphases, mitigations, and results are different. One of the greatest dangers that confronts the applied ecologist is that of creating a rut from which innovation is hard pressed to escape. Because a set of guidelines has worked well in orchestrating the

analysis of a particular proposed action, it cannot be simply assumed that the same guidelines are going to work for another action.

Bearing this attitude in mind, it is often possible to fairly painlessly alter the content or format of an EIS at a late date in response to new insight into an analysis of the problem at hand. Dictating format at an early stage necessarily means committing team resources to a structured response to information not known at the time the format is finalized. Changing that format is virtually always a difficult, if not impossible, task when deadlines are tight and when large, complex projects are involved.

4. Do not assume that any agency, or individual antagonist or protagonist, knows his or her responsibilities, authority, or the limits of that authority. Perhaps most of us indeed do know and speak our limits in these areas of responsibility and authority. However, much of the human world, as much of the animal world, involves a serious game of self-confidence and bluff. A political leader may be kind or cruel, honest or crooked, sane or insane; however, he or she is never unsure and is likely to bluff—a lot or a little, but he or she will bluff.

In collecting information, evaluating environmental issues, and sensing the legal implications of the associated proposed environmental management, the applied ecologist must be sensitive to the motives and behavior of the humans involved in this process. Even with the best intentioned individuals, bias is inevitably introduced. Bias must be accepted and properly evaluated by the applied ecologist, who seeks as much input to his or her base of data as may be possible given the realistic limits of time and budget. Bias is only harmful to the preparation of an EIS if it consciously or subconsciously misdirects the applied ecologist in his or her preparation of a supposedly unbiased and professional analysis.

The problem of unraveling the lines and responsibilities of governmental authority is a somewhat different matter, often requiring legal interpretations by an attorney or someone intimately familiar with the process by which particular applications or other environmental documentation is processed. Great care must be taken in interpreting and verifying the information collected in task 3 above. Once verified, and kept current, such a specific listing of regulations provides sound guidance as a checklist to confirm that the technical data present in the EIS does or does not satisfy the formal requirements of the regulatory powers controlling the outcome of the application process.

In general, the attention of an applied ecologist is divided in three directions. First, he or she must define (or at least verify) the environmental questions which his or her work must address. Second, she or he must collect the data or other information that is needed upon which to base state-of-the-art answers. Third, the data must be presented and interpreted in a manner that allows its use by the decision makers who will judge the proposed action. Therefore, the objectives of data collection is usually to define what is present at the study area. The interpretation usually concerns a prediction of what will happen should the proposed action be implemented and superimposed upon the existing environment.

A starting point in defining the overall ecologic functioning of an area is to catalog the organisms that are present. Once the necessity for this information is clear, some discussion must follow to indicate the level of detail to be sought. At this time, two approaches may be used: One seeks to identify the populations present, and the other also determines the biomass of each population. Both approaches usually require at least some new field work and considerable effort is normally associated with the determination of biomass.

With respect to population identification, historical data may be sought from libraries, literature searches, local institutions, or other probable repositories of information about the natural environment of the study area. Although such searches usually produce only superficial or doubtful information, it may be all that can be gathered in the time available.

Ideally, the most specific source of current and pertinent information is accessible through the relatively greater expense and effort of *site-specific field surveys*. These are designed to accommodate the needs of the particular environmental documentation at hand. In utilizing such a survey, a major consideration must be to make certain that the survey does not inadvertently miss significant populations because of temporal or spatial variations at the site.

Design of the survey is highly variable, but it usually results in some level of mapping to show the geographic distribution of at least the dominating populations in and about the site of the proposed action (Figure 4.2). Such *biogeographic mapping* is a relatively superficial identification of habitats. It may lead to a more detailed investigation of how individuals within the identified zones or habits are distributed as a population density.

Whereas the construction of a biogeographic representation of the study area is qualitative, determination of population density is quantitative and the associated statistical and procedural problems of assuring accurate results are much greater. The usefulness of population density data is at least twofold in an EIS study. First, it tells the reviewer the quantitative and immediate impact upon the existing populations present that can be expected from manipulating the project site. Second, it provides information on the range of movement (i.e., *territory*) required by the animals present (Figure 4.3). For example, if deer are reported at a density of between 1 and 10 kg per hectare, one to ten 100-kg deer can be expected to range over 100 hectares in this hypothetical habitat. The problem of collecting statistically meaningful data on plant population density is much simpler for the simple reason that virtually all plants are sessile, whereas many animals are not.

These two procedures of biogeographic and population density surveys

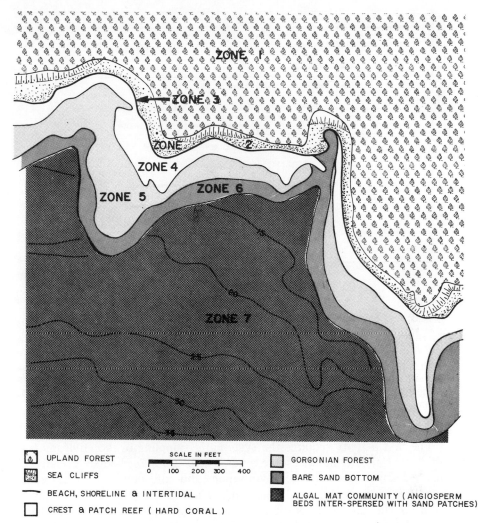

Figure 4.2. This biogeographical map was prepared based upon approximately twenty man-hours of effort collecting data in the field. Bathymetric contours are reported in feet at mean low water.

can form the base of all field interpretations of the natural environment. Most usually in the preparation of an initial ecologic evaluation, biogeographic procedures precede and dominate any quantitative analysis, and this is the approach presented in this book. Such an approach requires the two activities of sampling the environment and its populations and of identifying the materials sampled. They are two distinctly separate activities and can be undertaken separately, although for reasons of logistics and quality control it is of considerable advantage to have both occur simultaneously.

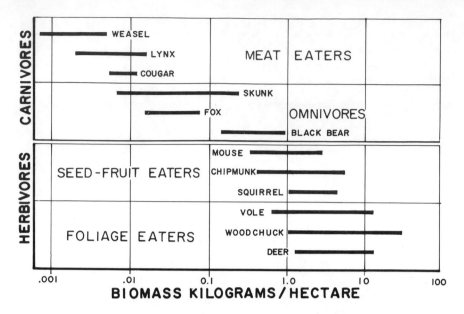

I HECTARE = IOO x IOO METERS = 2.47I ACRES , I KILOGRAM = IOOO GRAMS = 2.2 LBS

Figure 4.3. The distribution of population density is reported here as the biomass of a particular species which is likely to be found, on the average, in each hectare of the preferred, and stable habitat of that particular species. Organisms dependent upon relatively scarce food (ie. meet eaters) are not nearly as dense, relative to their biomass per hectare, as are less restrictive eaters (ie. foliage eaters). (Redrawn from Odum, 1971, based on data from Mohr, 1940.)

The first activity requires the systematic sampling of the environment, which is usually limited to some type of *grab sampling*—taking relatively small, discrete pieces of the environment and considering those as representative of the uncollected, surrounding area. The temporal and spatial nature of this procedure is such that the likelihood is great of its not being representative of the actual environment and community present. This is so because the ecologic philosophy supporting the use of grab sampling is that if the sampling sites are selected to be representative of the general area of interest, if the time and frequency of sampling spans productive and migratory cycles of the community, and if the appropriate sampling device is used, then the resulting samples will be representative of the larger area of concern. Assuring compliance with these three conditions often requires as much art as it does science, especially if the work is practically limited by time and money (Figure 4.4).

Grab sampling may involve not only the retrieval of organisms but also geologic samples or physical measurements of the environment. Therefore, the designation of sampling points becomes of prime importance in instituting any type of field survey program from which these several types of data are to be gathered. Part of these considerations must include the practical constraints of field work that often temper the theoretically ideal design defined above. For example, sampling points must be logistically accessible and their specific locations must be reproducible so that repeated sampling at precisely the same location can be undertaken if necessary. Furthermore, the sampling point should be vandalproof because, especially in urban locations, the validity of environmental sampling can be greatly influenced by vandalism to destructible markers, for example.

Whatever the study area, if grab samples are to be used as a means of defining the state of the environment, there are three basic patterns of sampling that may be used singly or in combination. The use of *random grids* as sampling points seems to contain the least potential bias of the three possible patterns, but it is much more time consuming. The traditional approach to random grid sampling has been to divide the study area into a projected matrix of equal sized grids and to sample the grids at random, covering a relatively small percentage of the total grids but resulting in a

Figure 4.4a. Grab sampling requires equipment which allows one to collect some type of quantitative piece of the environment. Here a trawl is shown which is used to collect benthic organisms and nekton swimming near the bottom, a Hensen plankton net to collect plankton in the water column, a Petersen grab to retrieve a bite of sediments from the bottom, and a snapper grab which takes a smaller bite of the sediments. Grab sampling may also be used for terrestrial surveys and often includes grab samples of air in order to estimate air quality.

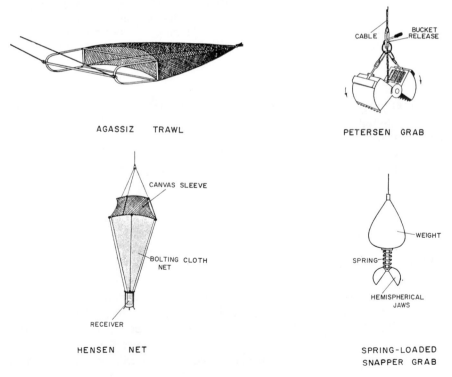

AGASSIZ   TRAWL

PETERSEN   GRAB

HENSEN   NET

SPRING-LOADED
SNAPPER GRAB

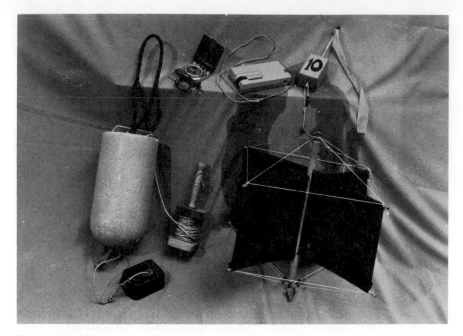

Figure 4.4b. Shallow currents in lentic and marine environments can be measured by drogues. A simple set of the required equipment is shown here including (clockwise from the upper left): compass, range finder, 10 foot drogue float, drogue, marker buoy and its line and anchor block. The drogue, slightly negatively buoyant, is released at a point marked by the simultaneous release of the buoy. After an appropriate length of time (five seconds to a few minutes) the direction and distance between the marker buoy and the drogue float is recorded and the marker buoy and the drogue are recovered and released elsewhere.

representative sample of the whole area. However, in the case of most environmental surveys, the type of species present must be first known in order to determine the most suitable frequency of random sampling. In addition, the topography of the survey area may be variable, which distorts the surface area of the projected matrix in unknown ways.

To overcome these problems, it is possible to superimpose a sinuous or zigzag course within the study area. Along such a track, grab samples can be collected randomly with the condition that a certain total number of grab samples is to be taken. That number is determined by the size of the area to be covered and the detail of information necessary. The general order of magnitude of samples taken in most analyses may be estimated at between one and five per hectare, which produces a standard error of about ±2%. That is, one can expect this procedure to result in data that distort the distribution of populations and ecologic conditions in the field by a factor of about 2%. This is limited to only certain surveys, such as those associated with large species and "normal" communities (Figure 4.5).

Continuous transects are usually favored over random grid sampling

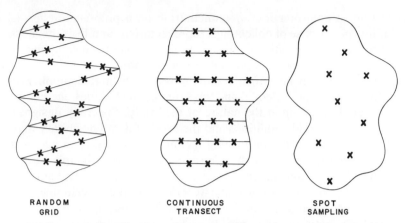

Figure 4.5. The location of stations for field sampling in ecological analyses may follow one of three patterns, in general. Such patterns may be used in either aquatic or terrestrial studies. Various modifications may be applied to these procedures in order to suit them to the specific requirements of the study at hand.

because the continuity of data generated by the procedure is often more accessible to interpretation. However, this pattern requires some basic knowledge of the distribution of habitats in the study area and some experience with similar analyses in similar environments. The technique requires that the study area be transected by a series of tracks, along which equally spaced grab sampling stations are designated. The number of tracks and the number of sampling stations along each track is determined by the size of the study area, the diversity of the environment, and the level of detail required. The primary danger in the use of this technique rests with the experience of the ecologist applying it in the field. It is possible to miss sampling significant populations or physical conditions if the transects do not pass through those areas or if the distance between sampling stations on a transect is such as to miss such an area.

Spot sampling is the last general pattern of environmental sampling used to define the site-specific environmental characteristics of a study area. It is the most difficult method to apply successfully because it ideally requires considerable background knowledge and experience relative to the study area in question. It is most often used only after preliminary analyses of the area have been completed using random grid or continuous transect sampling. Spot sampling usually involves relatively few stations and therefore the risk of missing significant existing conditions is very great. However, if good base information is available and sampling stations can be located that are indeed representative of the larger area, spot sampling is the most cost effective method available to the applied ecologist for the expeditious collection of field data.

As has been mentioned, the temporal nature of grab sampling is as important as the spatial distribution of stations. Although most sampling for the preparation of environmental documentation on a proposed action is short term, for the purpose of collecting a representative sample of the most productive or the most sensitive ecologic conditions long-term sampling or long-term monitoring are also important considerations.

When such long-term investigations are possible, a limited number of representative stations are selected and the sampling techniques and equipment are selected based upon the nature of the study. Specially designed equipment to suit the field conditions and the level of detail necessary may be required. One of the most interesting techniques of long-term ecologic monitoring at a particular site is the *bioassay*. The basic principle of this method is that living organisms may be used to reflect the general state of the environment. We can examine the reactions or the development of representative specimens living in the environment in question. Especially with respect to aquatic, sessile animals as bioassay materials, a cage or some other means is used to confine the assay organisms to a particular

Figure 4.6. Bio-assay cages are usually constructed of corrosion resistant materials. This one was designed for marine use and is made of plastic coated wire mesh placed over an oak cage weighted with several pounds of concrete. The coiled, plastic lines are stretched along the bottom of the sea perpendicular to the shore. The cage is recovered by grappling for these lines. A surface float is much easier to use but will often attract vandals to the cage. Sedimentation traps in this cage are at right angles to one another and record the amount of sediment which can settle out of the water in the stilled environment at the sight of the trap. They are seen here as two white boxes in opposite, bottom corners of the cage.

location in the environment (Figure 4.6). If an identical set of specimens is simultaneously kept in a controlled environment, comparisons of growth and death rates, physiologic functioning, etc., can be made between the experimental group and the control group; such information may be helpful in interpreting the long-term effects of the project being studied upon the environment and upon those organisms living in that environment.

Another often used technique for long-term monitoring of the biological environment is the *settlement plate* or artifical substrate (Figure 4.7). Usually constructed of some inert material, these devices provide a surface area on which organisms may settle and grow over a period of time. Settlement plates or artificial substrates are usually applied to aquatic surveys that have a duration in excess of 1 month. These procedures naturally favor the collection of benthic organisms but in combination with grab sampling within the other community levels, the analysis of settlement plates is the best practical technique for determining the characteristic fouling organisms (i.e., those that settle and grow on surfaces) in a particular, usually aquatic, environment.

Of all the possible communities an applied ecologist may be asked to analyze, perhaps the most difficult is the lotic environment. Swiftly moving waters have characteristics generally unlike those of lentic, marine, and terrestrial environments that complicate not only the ecology involved but also the logistics of carrying out the survey. In large part, these difficulties are caused by the fact that lotic waters mix in three dimensions, from top to bottom, from shore to shore, and along the length of the stream or river. If the environmental study is focused upon a point source of real or potential pollution, such as a pipe carrying sewage effluent to the water course, special consideration must be given to the three modes of mixing.

Figure 4.7
Settlement plates and artificial substrates may be used for relatively long term sampling of the types of fouling organisms and benthic community which are attracted to such surfaces. Seasonal variations and spacial differences in populations require sampling periods which ideally exceed several months. Inert material may be placed between settlement plates, as shown in this figure, or they may each be used separately suspended in the water or anchored to the bottom and in contact with the sediments.

These considerations are especially crucial in the lotic environment because *vertical mixing* is relatively rapid and is directly dependent upon turbulence. A vertical mixing cell, therefore, may be completely homogeneous within less than 100 m of our hypothetical sewage outfall. *Lateral mixing* is determined by turbulence across the water course, which may be created by sharp bends, rapids, or other agents that disrupt the forward movement of the water. Such lateral mixing may require several kilometers before it is completed.

The *longitudinal mixing* of a river is also a relatively unique characteristic of lotic environments. It must be considered in determining where, when, and how to sample for ecologic analysis. Water courses are literally conveyor belts with new water and its contained materials continually carried past any stationary sampling station. The ecologic characteristics of one neighboring station may be entirely different from those of another station because the longitudinal differences in water quality in a water course are usually very marked, especially when impacted by humans.

A sampling schedule to be utilized in the analysis of a water course based on grab samples should include the following:

1. Water samples taken at quarter points laterally across the water course, unless it is known that lateral mixing is complete at the sampling transect.
2. Tributaries that carry 10 to 20% of the water course under study should be sampled to determine the significance of any contribution they may make. Smaller tributaries must also be sampled if they represent a possible source of pollution or a possible sensitive receptor of existing or proposed activities along the water course under primary study.
3. The longitudinal distances separating sampling transects should in part be determined by calculating the *time of water travel,* the rate at which a volume of water flows down a river. Sampling station intervals of 10 to 14 hr are appropriate if the effects of a single point source of pollution is being studied. That is, the stations should be separated by a shoreline distance equal to the distance traveled by the water over a 10 to 14 hr period. If the ecology of the river reach is complicated by multiple sources of concern, the longitudinal distances between stations should be reduced.
4. Sampling of benthos, nekton, or any other biologic factors should be coincident with the water sampling transects defined above, but with biologic samples taken at each shore, at the two quarter points, and at midstream, assuming that the water course has a typical cross-section of regular form. Irregularities of geomorphology (form) require that site-specific judgments be made in order to locate sampling stations that are more likely to produce meaningful and representative results.
5. Sampling periodicity, the temporal nature of sampling, is a key to the success or failure of any field survey. It is particularly important in lotic

environment surveys, where the most stressful times usually correspond to relatively low river flow rates. Another constraint, shared partially by the lentic environment as well, is that the largest representation of benthic animals will be found in the fall, winter, and early spring, when many insects are in their aquatic stage of metamorphosis.

The fundamental issue to bear in mind when evaluating a lotic environment, more so than with most other environments, is that samples taken at a particular station usually represent the water conditions contributed from upstream. They do not represent conditions contributed from the environment at the exact location of the sampling station. Such a problem of perspective is not limited to the ecologic interpretation of lotic communities. In fact, the most useful attribute an applied ecologist can gain from experience is a sense of always trying to see environmental problems in the context of the whole ecosystem involved. The overview of a problem and the full ecologic context of its nature is a central objective that should be sought in any ecologic analysis. One tool for helping in that direction is the use of *remote sensing;* this is a developing science that offers great promise for large-scale ecologic data collection and interpretations. Looking too narrowly at an ecologic problem leads to missing the forest for the trees, whereas the danger in remote sensing lies in missing the trees for the forest. Avoiding that danger is simply related to the integrated use of remote sensing with other direct methods for the collection of ecologic data.

The principle of this growing technology of data collection is to seek information by noncontact through the use and interpretation of any portion of the electromagnetic wave spectrum. It often relates to interpretations of aerial photographs in the visible light spectrum. For example, biomass surveys of forests can be undertaken through the *photointerpretation* of low-level aerial photographs relative to the inventory of timber volume, giving an accuracy within 20% of the actual volume. This information is calculated based upon the ratio of the diameter of the crown of a tree to the diameter of its trunk. For example, western conifers have a ratio of 1 ft to 1 in., so that a tree with a crown observed in the aerial photograph of 30 ft is most likely to have a trunk diameter of 30 in. Eastern hardwoods have a ratio of ¾ ft of crown diameter to 1 in. of trunk diameter.

High-altitude remote sensing is primarily done with satellites (Figure 4.8). Their capabilities as ecologic tools are just beginning to be explored. Radiation from the earth's surface is scanned by the satellites, which transmit that information back to earth. Interpretations are made based upon the physical principles exhibited by electromagnetic wave generation associated with the processes being "observed" by the satellites. Because the spectrum these satellites can record is greater than the spectrum we see as visible light, certain types of information can be readily photographed by the sensors that would be invisible if recorded with ordinary film or observed by the unaided human eye (Figure 4.9). For example, remote

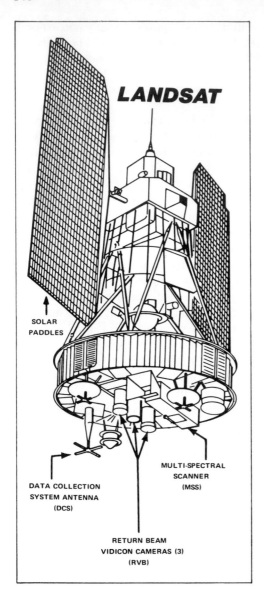

LANDSAT

SOLAR
PADDLES

MULTI-SPECTRAL
SCANNER
(MSS)

DATA COLLECTION
SYSTEM ANTENNA
(DCS)

RETURN BEAM
VIDICON CAMERAS (3)
(RVB)

Figure 4.8
A number of LANDSAT remote sensing satellites are circling the earth
transmitting multispectral images
and many other data to land stations
which reproduce and manipulate the
data. In 1977, the field of view for one
LANDSAT "photograph" was 185 ×
185 kilometers of the earth's surface.
(From Heaslip, 1976).

sensors in aircraft can easily discriminate temperature differences of 0.01°C
by sensing the transmittance of infrared radiation from quite small target
areas.

A third type of remote sensing is available to the ecologist via satellites
used to receive data from remote terrestrial- or water-based instrument
packages and to retransmit that information to a central receiving site
where the data can be interpreted. In this use, the satellite functions as a
relay station. The reliability of these stations, and the relatively low power
necessary to operate the instrument package, means that it is possible to

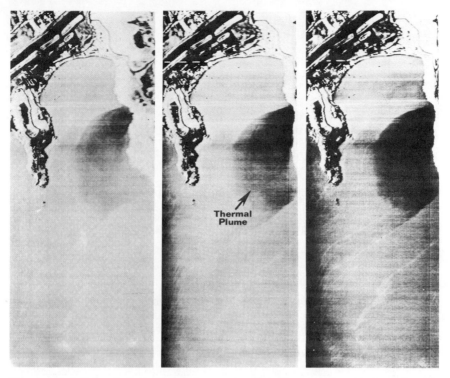

Figure 4.9. A thermal plume issuing from a desalinization plant into Lindberg Bay, St. Thomas, U.S. Virgin Islands is seen here in three thermal infrared images taken at the same time by a thermal IR scanner installed in an airplane. Each picture shows a different temperature differential with the smallest plume outline (dark color) depicting the greatest temperature differential. The solid black region in the upper left shows the runway, taxiway, and other paved areas of the Harry S. Truman Airport. (From Heaslip, 1976.)

place such packages in remote areas where, unattended by people, daily reports of data can be gathered for extended periods of time (Figure 4.10). Parameters that can be measured include: air temperature, dissolved oxygen, ground or water temperature, pH, precipitation, salinity, soil moisture, solar radiation, stream flow, tidal flow, wind speed and direction, snow depth, humidity, and seismic activity.

In virtually all these instances of remote sensing, the interpretations of the raw data collected must be validated through the initial use of *ground truths*. That is, observations are made of the events under study on the ground simultaneously with the recording of these events through remote sensing. By comparing the records of the known ground truth with the remotely sensed record, the reliability of interpretations of the remote sensing data can be developed. Once validated, interpretations based solely upon remote sensing become reliable.

The physical principle that energy is absorbed, emitted, scattered, and reflected by matter in a pattern and wavelength unique to the atomic and molecular structure of that matter is a key to the burgeoning field of remote

Figure 4.10. Remote sensing clusters of instruments can be placed in virtually any spot on earth to transmit environmental data up to a relay satellite which, in turn, re-transmits the data to a ground station for interpretation. (From Heaslip, 1976.)

sensing. Through our understanding of this principle, it will be possible within the relatively near future to collect an ecologic profile of the study area without any disturbance of the contained organisms or elements. An ecologic picture of the health or sickness of a community will be derived from a remotely sensed "photograph." Such an understanding will allow the applied ecologist to make a much more reliable prediction of probable environmental consequences than is now possible, even under the best circumstances of present-day technology. Such predictive modeling is a tool that is at a stage of development considerably more primitive than is remote sensing. In part this is because relatively few applied ecologists are conversant with that field of *ecologic modeling* and in part because the field is still largely an art frought with problems of validating predictive models with actual data.

Nevertheless, ecologic modeling is a tool that must be developed as a further aid in arriving at our goal of being able to predict the consequences of environmental management. We are still woefully short of that goal and ecologic modeling, theoretically, is the idealized, mathematical expression of that goal. Its present shortcomings seem to reside primarily with our incomplete knowledge of ecologic principles that would otherwise permit us to reduce these principles to the algorithms necessary for the derivation of a true mathematical description of an ecologic event and its ecologic consequence.

This field has developed with modeling as a tool to categorize changes in the environment, or to disect pathways within an ecologic system. Three basic strategies presently exist for the development of these modeling concepts.

1. Models that are real and specific but lack general applicability. Assuming that we have detailed, correct, and comprehensive information about a particular ecologic system, it is possible to construct a model that sacrifices general applicability to gain realism and specific precision about the system modeled.

2. Models that are general in their applicability and give specific predictions but are not real in their mimicry of the outcome of ecologic principles. Assuming that we wish only to empirically mimic the past and present environmental state and interrelationships of a particular system, a model can be constructed that ultimately sacrifices the realism of a full and true understanding of these interrelationships for an empirical and generally accurate mathematical description of the precise environmental consequences of the modeled system. However, such a model will not, except by coincidence, describe what will happen tomorrow, nor will it provide much insight into the mitigation of real or potential impacts.

3. Models that are real and are generally applicable but are not specific in their predictions. Assuming that we know enough about an ecosystem to describe its basic functioning, it should be possible to construct a model that is generally applicable to any ecosystem in which these same

Figure 4.11. Energy flows within an estuarine ecosystem are diagramed here for the Crystal River in Florida estimating the biomasses moving through the primary marine food chains involved. (From Odum, et al. 1977.)

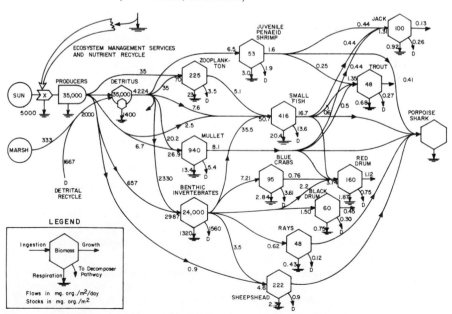

principles function. The problem with this sort of ideal model is that in virtually all instances, we do not understand ecologic interrelationships well enough to confidently predict the outcome of manipulations.

These problems of ecologic modeling can be appreciated in context if we consider the categories of information that must be fully understood if we are to model a sample area, such as the evaluation of aquatic environments and the nutrient chemical dynamics of those waters. In such an example, at least six areas are important. Some aspects are simple and some are hard to gather, whereas still others are still mysteries relative to their nature and interactions. These areas include: (1) the vertical mixing of the environment, (2) the nature of sediments as a sink or source of nutrients, (3) the nature of the recovery of the environment with respect of different types and loads of stress, (4) the relationships between physical parameters in the environment and the biotic responses that occur, (5) the interrelationships between land activities within the watershed and airshed and the aquatic environment, and (6) the dynamic biotic feedback within the variable environment present in any real ecosystem (Figure 4.11).

These requirements of this hypothetical model are formidable, and the ideal end of our formulation is still distant, to be found in *systems ecology*. The systems approach to mathematical modeling of ecologic events is a new and extremely challenging field. Probably more than any other branch of ecology, systems ecology always remains balanced on the edges of both theoretical and applied ecology.

Bibliography

Bragg, T. B., and A. K. Tatschl. 1977. Changes in floor-plain vegetation and land use along the Missouri River from 1826 to 1972. *Environ. Man.,* **1**(4); 343–348.

Cairns, J. Jr., and K. L. Dickson. 1971. A simple method for the biological assessment of the effects of waste discharges on aquatic bottom-dwelling organisms. *J. Water Poll. Contr. Fed.,* **43**(5), 755–772.

Canole, R. P. 1976. *Modeling Biochemical Processes in Aquatic Ecosystems.* Ann Arbor Science Publishers, Ann Arbor, 389 pp.

Caugley, G. 1977. *Analysis of Vertebrate Populations.* Wiley–Interscience, New York, 234 pp.

Edelson, B. I. 1977. Global satellite communications. *Sci. Am.,* **236**(2), 58–73.

Feldstein, M. 1977. A new modeling technique for air pollution control. *Environ. Man.,* **1**(2), 147–157.

Gaufin, A. R., and C. M. Tarzwell. 1952. Aquatic invertebrates as indicators of stream pollution. *Publ. Health Rep.,* **67**(1), 57–64.

Gevirtz, J. L., and P. G. Rowe 1977. Natural environmental impact assessment: A rational approach. *Environ. Man.,* **1**(3), 213–226.

Ghiselin, J. 1977. Analyzing ecotones to predict biotic productivity. *Environ. Man.,* **1**(3), 235–238.

Hall, C. A. S., and J. W. Day, Jr. 1977. *Ecosystem Modeling in Theory and Practice: An Introduction with Case Histories.* Wiley, New York, 684 pp.

Hanson, W. R. 1967. Estimating the density of an animal population. *J. Res. Lepidoptera*, **6**(3), 203–247.

Heaslip, G. B. 1976. Satellites viewing our world: The NASA Landsat and the NOAA SMS/GOES. *Environ. Man.*, **1**(1), 15–29.

Kessell, S. R. 1977. Gradient modeling: A new approach to fire modeling and wilderness resource management. *Environ. Man.*, **1**(1), 39–48.

Lewontin, R. C. 1969. The meaning of stability. Diversity and stability in ecological systems. *Brookhaven Symp. Biol.*, No. 22, 13–24.

Middlebrooks, E. J., D. H. Falkenborg, and T. E. Maloney (eds.). 1974. *Modeling the Eutrophication Process*. Ann Arbor Science Publishers, Ann Arbor, 228 pp.

Odum, H. T., W. Kemp, M. Sell, W. Boynton, and M. Leliman. 1977. Energy analysis and coupling of man and estuaries. *Environ. Man.*, **1**(4), 297–315.

Ouellette, R. P., R. S. Greeley, and J. W. Overbey II. 1975. *Computer Techniques in Environmental Science*. Petrocell/Charter, New York, 248 pp.

Oviatt, C. A., S. W. Nixon, and J. Garber. 1977. Variation and evaluation of coastal salt marshes. *Environ. Man.*, **1**(3), 201–211.

Paskausky, D. F. 1977. Net drift in an atypical estuary. Long Island Sound. *Environ. Man.*, **1**(4), 331–342.

Pielow, E. C. 1977. *Mathematical Ecology*. Wiley, New York, 385 pp.

Porter, J. W. 1972. Patterns of species diversity in Caribbean reef corals. *Ecology*, **53**(4) 745–748.

Rand, M. C., A. E. Greenberg, M. J. Tards, and M. A. Franson. (eds.). 1976. *Standard Methods for the Examination of Water and Wastewater*. (14th ed.) American Public Health Association, Washington, D.C., 1193 pp.

Stecher, P. G. (ed.). 1960. *The Merck Index of Chemicals and Drugs*. (7th ed.) Merck and Co., Rahway, N.J., 1641 pp.

Tait, R. V., and R. S. DeSanto. 1972. *Elements of Marine Ecology*. Springer-Verlag, New York, 327 pp.

Veziroglu, T. N. 1975. *Remote Sensing/Energy-Related Studies*. Wiley, New York, 491 pp.

5 Lethal, Sublethal, and Nonlethal Environmental Stress

A good half of applied ecology is usually devoted to determining what is contained within the existing natural environment. The other half generally deals with predictions of one sort or another, as specific as possible, about how the natural environment can be expected to respond to the proposed action or structure being evaluated.

A very large part of an applied ecologist's successful interpretation of those facts rests on the general experience that he or she has had with similar ecologic interpretations. Even though no two situations can be ecologically identical, similarities between them can obviously be seen and predicting one outcome is often helped through knowledge of similar previous experiences. Furthermore, because applied ecologists are often confronted with the doom and gloom of bad possibilities and *negative impacts,* their experience is usually dominated by tasks that attempt to quantify how ecosystems or organisms are or may be harmed by a proposed or existing action.

In this sense, one type of applied ecology involves the consideration of *lethal and sublethal stresses.* Such stresses may be biologic, chemical, or physical. They result in the death or weakening of individuals within the ecosystem, thus altering the ecology of communities through the imposition of that physiologic death or weakening. If such trouble or impairment concerns the internal physiologic balance of the vital functions of an organism within the ecosystem, some "disease" is likely the cause. The environmental health fields are most directly focused upon the environmentally induced diseases to which humans are heir. Much of our ecological experience, therefore, has been biased in the direction of human health, although many of the principles uncovered are just as applicable to the much broader diagnostic logic used to prognosticate environmental perturbations. These disturbances often afflict a whole ecosystem much as a disease debilitates a single organism. From the point of view of this book, whichever factors or combination of factors are involved, an environmental interpretation lends support to seeing how living and nonliving factors in the biosphere are all interdependent. The manipulation of one single factor will cause a remarkably wide circle of variable responses.

This principle may be called the *ripple effect,* suggesting that if a central factor is present that stresses a biologic system, the impacts of that one stress will have repercussions radiating outward and disturbing the system usually in diminishing magnitude, from the single, central impact. If an organism is injured, or becomes ill, all of its functions will be affected in some great or slight manner. If an ecosystem is stressed, it too is modified in accommodation to that stress. As a brief introduction to this sort of stress, as seen by the applied ecologist, it is useful to first see something of the personal human perception of our own stresses derived from some of the environmentally induced diseases which have caused us special concern.

The beginnings of our concern for occupational health can be traced back to the first century AD, when Pliny the Elder called for the use of protective masks by workers engaged in dust and fragment producing jobs of mining and grinding operations. More comprehensive was *De Morbis Artificium Diatriba* (The Diseases Of Workmen) which was written by Bernardino Ramazzini and published in 1700. It is generally considered the first treatise that clearly associates a person's occupation with his or her health and it therefore suggests that the health of an individual and the environment in which that individual lives are directly related.

That simple logic is one of the first steps taken by the applied ecologist in unraveling the specifics of a particular problem involving the *interdigitation* of biologic and physical or chemical disruptions of a complex system. Regardless of the particular system being investigated, it is not unusual for the pathway of interrelationships to be quite torturous, although in some instances they are easily defined and corrected. For example, *optokinetic nystagmus,* or conveyor belt sickness, is easily corrected. It was defined when potato chip factory workers became giddy, nauseous, mentally confused, and also suffered from fainting. Upon investigation, it was eventually discovered that involuntary contractions of eye muscles were caused when the speed of the conveyor belts they were watching increased above 32 ft per second. Workers trying to focus their eyes on moving objects at these speeds would suffer the symptoms.

Seldom has an applied ecologist so direct, controllable, or testable a circumstance as with optokinetic nystagmus. The agents encountered are often far more sinister, such as those associated with diseases grouped together as *pneumoconioses.* This is a disorder caused by dust retained in the lungs. It results in a reduction in the amount of the healthy, spongy tissue of the lungs, which is replaced by physiologically useless fibrous scar tissue. The scar tissue replaces the healthy tissue as dust particles between 0.5 and 150 μm in diameter lodge in the lungs and are incorporated into the tissue.

With respect to humans, experiments have shown that the mucous membranes and hairs of the nose normally remove particles greater than 10 μm in diameter from the air stream passing through. The surfaces and cillia of the bronchi and bronchioles leading to the alveoli filter out particles larger than 2 μm in diameter. Finally, particles that are smaller than 0.3 μm in diameter are not retained in the lungs but are exhaled. Therefore, we can predict that at least with humans, airborne particulates ranging in size between 0.3 and 2 μm in diameter are likely to reach the alveoli and once there, may cause problems. In fact, upon examination of lung tissue diseased by pneumoconioses, it was found that the average diameter of imbedded particles is usually about 1 μm.

Of the approximately 20 human disorders associated with pneumoconioses, the most common three are:

1. Silicosis, or miners asthma, caused by long-term inhalation of mineral dust containing silica (such as sandstone, flint, quartz, or many other mined minerals). Sand blasters, "mud hogs," and gold miners often suffer this affliction.
2. The black lung disease of coal miners, anthrocosilicosis, is basically similar and is caused by long exposure to coal dust.
3. Asbestosis is caused by the inhalation of asbestos fibers, which result in a diffuse fibrous infiltration of the lungs that apparently not only does physical injury but also has metabolic implications. Asbestos alters the internal cellular environment such that a cancerous growth is more likely to appear than were the imbedded asbestos fibers not present. Furthermore, the tumor may appear years after exposure to the asbestos.

If we seek to understand the health or illness of an organism or an ecosystem, we must at some time address the problem of developing an unambiguous, working definition of health versus illness. That is, where does health end and illness begin or is the boundary a continuum? Especially with respect to ecosystems, such a definition is very illusive. A complex system is seldom static and its changes are often virtually imperceptible. Changes need not be perceived as good or bad in the short run to prove bad or good in the long run of several seasons, years, or decades.

The ecologist may seek as a start to measure the health of a subject and the absolute impact upon that subject by some toxic material. The pharmacologic concept of toxicity has been developed to a fine art and will only be dipped into very gingerly here. We will select one point from which we can better diverge into examining ecosystems or communities as pharmacologists or medical doctors seek to diagnose the disorders of individual organisms. With respect to the relative toxicity of a substance, one simple measure commonly used is LD_{50}. This refers to the dose of a substance administered to a number of physiologically identical test organisms that results in the death of 50% of the organisms within the test period. Lethal Dose$_{50}$ is always reported as a number of milligrams per kilogram of organism body weight.

One further bit of information we will need relates to the general *synergistic effects* that must be accounted for whenever we seek to evaluate the effects of some disruptive and/or toxic material upon an organism or an ecosystem. Our increasing interest in occupational health has led us to numerous studies that attempt to codify and therefore standardize *thresholds* for the exposure of an individual to some toxic material which would result in no harmful repercussions below that limit (Table 5.1). Tests and case histories have been constructed and evaluated that provide support for various thresholds. However, the most significant findings and conclusions for applied ecology has been that the toxicity of a particular agent may be enhanced when one or more other toxic material(s) is present to stress the same organism at the same time. In simplest form, this suggests that if the LD_{50} of materials X, Y, and Z is 1 unit each when administered separately, the LD_{50} may be considerably different if they are administered simultaneously. That is, the LD_{50} in that circumstance may be 0.5 each for the materials in combination. This is a synergistic effect.

This seems certainly a logical condition with which we all have some experience. If we are not in good health because of one disease, we may fall victim more readily to some other disorder. If we sustain one onslaught it is less likely that we can equally well sustain a simultaneous onslaught. Of course, if we recover from the initial encounter before the second begins, it is not unlikely that the experience of the first will better fit us for our encounter with the second. In many respects, the disruption of ecosystems by toxic agents follows much the same logical pattern.

Ecosystems are even more complex than single organisms, simply because ecosystems consist of not one organism, but of many. Yet we do not know every aspect of the niche of even just one of the organisms of an ecosystem. In fact, as with all science, it seems that the more we know of ecology the more intricate becomes our confusion and the harder we try to keep our findings straight and tidy.

The fundamental issues at hand are:

1. Life in an organism or in an ecosystem is a complex of physiologic interrealtionships between individuals and/or cells and their mutual environment.
2. Our technology has allowed us to fabricate and manipulate both physical

Table 5.1. Toxicity Spectrum

	LD_{50}	Dose
Extremely toxic	<1 mg	taste
Highly toxic	<50 mg	teaspoon
Toxic	50–500 mg	an ounce
Moderately toxic	0.5–5.0 g	a pint
Slightly toxic	5–15 g	a quart
Nontoxic	>15 g	>quart

and biologic aspects of our environment well in advance of understanding the full ecologic meaning of our actions.

3. There is no divine providence to intercede and prevent us from suffering the consequences should we make a mistake that extracts an ecologic penance of lesser or greater magnitude.

Perhaps these clanking chains of doom are not as ponderous as they seem. However, in 1968 one link weighed in at approximately 120 billion pounds—the production in the United States of some 9000 synthetic chemical compounds that are in wide commercial use. Of all the more than two million chemical compounds that are known to exist in the biosphere, very few have been examined comprehensively enough to allow us much confidence in predicting their toxicity and/or their synergistic impact upon ecologic systems. The problem is further complicated by technologic experimentation, which produces approximately 1000 new commercial chemical compounds each year. These enter the environment at various points in various quantities, both intentionally and unintentionally.

Although the problems confronted by applied ecologists are confusing because of the number of different agents involved, their different effects, and the synergistic interrelationships that one agent may have with another, we can still seek a simplification of the overall field by examining a few typical examples. These reinforce the modern consciousness that we live in an artificially altered environment, requiring that we seek to understand complex ecologic stresses if we are to successfully manage that environment and our fate in it.

Of the most interesting are polychlorinated biphenyls (PCB's) (Figure 5.1) because, like few others, these substances represent many characteristics of the conflicts between modern technology and ecology, precipitated by our incomplete knowledge of both. Prior to 1971, PCB's were widely used in plastics, epoxy glue, lacquer resins, paints, varnishes, fire retardents, inks, carbonless copy paper, and a dielectric insulating fluid in electrical equipment, such as transformers. In the environment, PCB's have been discovered to be virtually nonbiodegradable; they can only be destroyed by incineration above 2700°F.

Even at fairly low concentrations in the environment, these *persistent* chemicals have been shown to cause skin lesions, reproductive failure, and liver cancer in laboratory animals. Yet their actual functioning in these

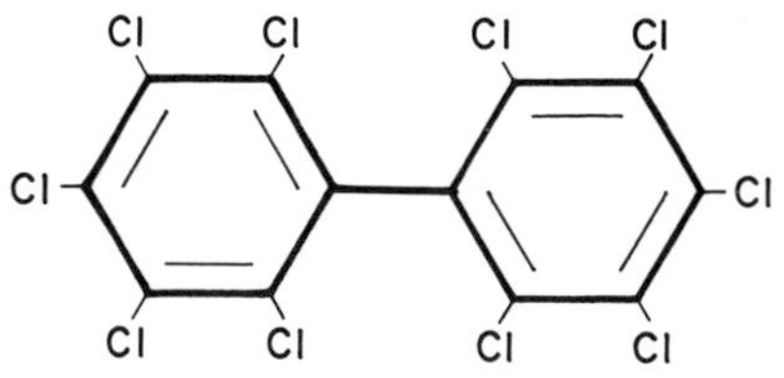

Figure 5.1

The Polychlorinated Biphenyl molecules (PCB) are made up of as many as ten chlorine atoms bonded to two phenyl rings. They are highly resistant to physical, chemical, and biological breakdown and they are, therefore, virtually indestructible in the environment.

living systems remains little known. Although excellent as plasticizers and dielectrics, PCB's have several unwanted attributes that can stress the environment. These escaped initial detection because the original development of the chemical was directed solely at technologic advancement through the empirical synthesis of materials, without a parallel concern for ecologic implications.

The sales of PCB's rose from zero in about 1930, when they were introduced, to about 34,000 tons a year in 1970. Estimates suggest that between 1000 and 2000 tons of PCB's are in the atmosphere as a result of the natural degradation of plastics and although it is known that PCB's alter liver tissue and a variety of enzyme functions, the threshold levels for such effects are not known. In fact, as with many agents that stress an individual or an ecosystem, the toxicity spectrum is a continuum and there is no clear ecologically safe exposure rate. An environment containing a low concentration of the material in question is not as bad as an environment containing a higher concentration.

Perhaps the primary example of the classical strain between technology and ecology is represented by the epic story of DDT (Figure 5.2). Dichlorodiphenyltrichloroethane (DDT) was first synthesized in 1874 by Othmar Zeidler in Germany. Not until 1939 did the Swiss chemist Paul Muller discover the value of this synthetic compound as an insecticide. He recieved the Nobel Prize for his work. Subsequently, its wide use led to a peak production of more than 400,000 tons in 1 year.

The basic uses of DDT are (1) the protection of food and fiber production against insect attack and (2) public health protection by control of the insect vectors, such as mosquitos, of certain diseases. Since 1955, several international organizations and various countries have undertaken campaigns to eradicate malaria through the widespread spraying of DDT in an attempt to eliminate their target, the *Anopheles* mosquito. In fact, the use of DDT can be credited with the control of at least 30 major diseases of man, including plague, cholera, and Rocky Mountain spotted fever. With respect to food and fiber production, DDT has been credited with doubling yields. However, the very characteristics that make DDT a good insecticide make it a stressful substance in the environment.

It is a *broad-spectrum insecticide,* which means that it not only kills what may be considered harmful insects, the target for which it is used, but

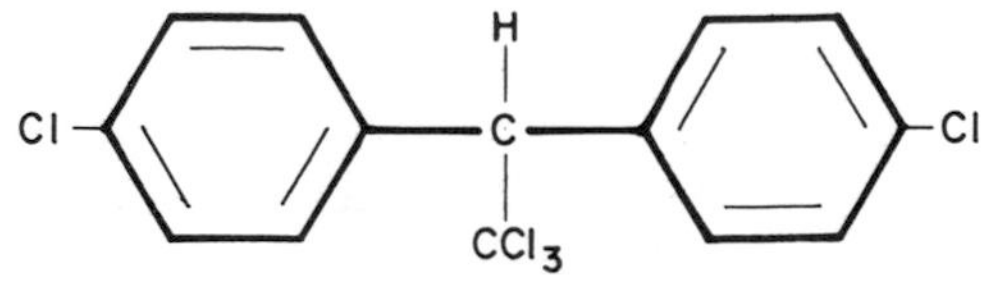

Figure 5.2
Dichlorodiphenyltrichloroethane (DDT) is an organochloride, or chlorinated hydrocarbon. It is highly resistent to biodegradation and concentrations may be found 15 or 20 years after a single application of the chemical.

it also kills beneficial insects. This often causes stress in the ecosystem when preexisting stable balances of organisms are upset. DDT also is not readily decomposed so it lasts a relatively long time (approximately 20 years) following one application. Its low toxicity to man and its high toxicity to insects, coupled with its low cost, have also encouraged its widespread use. Its actual cycling in the environment will be discussed further in the next chapter.

In general terms, we can use PCB's and DDT to portray a major part of the philosophic problem that usually confronts an applied ecologist in the course of his work. He or she is often called upon to correct a problem but with the stipulation that agents that may be causing the problem not be removed. Therefore, most of his or her efforts in such an instance are devoted to determining what are the best means available to maintain sound management of the environment through *mitigation* of any negative environmental impacts associated with the project. The applied ecologist usually does not have the option to eliminate the end product, though he or she may modify the means of producing that product. If a client wishes to grow crops and increase their yield through the use of an unsound insecticide, it is unlikely that an applied ecologist will succeed in redirecting this client unless the ecologist can understand and manage the problem while simultaneously educating the client to the long-range benefits of sound environmental management. A client whose sole motivation is short-term benefits is seldom interested in ecologic management practices, which are virtually always designed to avoid environmental stress. Sound environmental management often means that short-term productivity is not as great as is technologically feasible by use of unsound methods producing high returns.

Usually in the evaluation of an ecologic puzzle, no single piece is found that can be confidently identified as the keystone of some lethal control over the health of that system. Toxicity, thresholds, physiologic state, and many more attributes of the interrelationship between a stress and an ecosystem must be fully understood before a reliable interpretation or prediction can be made of the ecologic state of any stressed ecosystem. And although we have just considered a few of the lethal and sublethal stresses associated with PCB's and DDT as agents of modern technology, nonlethal and quite natural environmental stresses must also be introduced as an equally important facet of the applied ecologist's interests. These stresses concern chemical, physical, or biologic factors that are nontoxic but that cause stress by altering the fundamental stability of an ecosystem. Physical cycles, predation, and overfertilization are three examples of nonlethal stress that can change a stable community rapidly and dramatically.

One way in which to view this sort of ecologic nonlethal stress is that it serves to *throttle* the general state of the ecosystem in question. Stress withholds some freedom that would otherwise permit the relatively unimpeded exploitation of the full niche capacities of those organisms present.

Liebig's "law" of the minimum is a good starting point for the introduction of the concept of nonlethal stress as a major controlling factor in ecosystem dynamics. In the mid 1800s, Justus Liebig reviewed the effects of various environmental factors upon the growth of plants. He deduced that plant growth was not usually limited by the supply of those factors, such as water and carbon dioxide, which were needed in large supply (the *macroelements*) but that growth seemed most often limited by those factors which were required in minute quantities (the *microelements*) and which were generally scarce in the soil (such as boron or phosphorus).

The ecologic law that has evolved out of his work, beginning in about 1830, states that the total crop of any species will be determined by the availability of that single substance which, in relation to the physiologic needs of that species, is least available in the environment. Liebig investigated the proportions of various elements required for terrestrial plant growth. By comparing the relative amounts of these materials to phosphorus, in plants and in the earth's crust, he found that very few elements were in lower proportions than phosphorus. He empirically knew the need for phosphorus by growing plants. Today we know much more about that need, including the critical participation of phosphorus in the synthesis of adenosine triphosphate (ATP) and the structural integrity of desoxyribonucleic acids (DNA) and ribonucleic acids (RNA). It therefore seems quite logical for the absence or scarcity of phosphorus to serve as a major *limiting factor* in the growth of a very broad spectrum of species and to throttle that growth in direct proportion to its scarcity.

Unfortunately, it is never so simple that an environmental problem revolves around one factor, although one factor may predominate in a particular set of circumstances. More usually, because any ecological community is the living consequence of many organisms and ecologic facets, the importance in that community of *factor interaction* is very great. For example, in a few instances of ecologic study, the balance and change seen between certain limiting factors and their interaction on a community can be predictable and cyclical. One of the best examples is seen in the spring and fall *blooms* of algae in most temperate aquatic environments. There, both physical and chemical limiting factors combine to produce a seasonal pulse in the crop of plankton present at any one time. In the simplest cases, it appears that regular fluctuations in the limiting nutrients in surface waters combine with changes in illumination and temperature to generate the changes seen in productivity.

In the winter, when temperatures are low and sunlight is weak or absent, even though the surface waters are rich with such nutrients as phosphorus and nitrogen brought up by convectional mixing, the amount of plankton present is at a low ebb (See Figure 3.19).

As the waters warm in the spring as a result of stronger sunlight, and with the winter's high concentration of nutrients brought up from deep waters, there are that many fewer limiting factors present and the phytoplankton bloom begins. As it increases in magnitude and the waters continue to warm, thermal stratification tends to isolate the surface waters and nutrient replenishment from the deeper waters is stopped. Nutrients are rapidly assimilated by the bloom in the surface waters. This phytoplankton bloom of April and May is usually accompanied by a slightly lagging bloom of zooplankton, which grows as its supply of food, the phytoplankton, grows. Then, as the zooplankton increases in population, the phytoplankton declines rapidly (Figure 5.3).

In the summer season, nutrients in the surface waters have been depleted through assimilation and although temperatures are relatively high and sunlight is extended, the phytoplankton population is relatively low. The zooplankton may reach a peak at this time, which is followed by its rapid disappearance because of starvation toward the end of the season.

In the early autumn, as the surface waters begin to cool and as the thermocline begins to weaken and finally disappear, nutrients from the deeper waters are carried to the surface by convection currents. Nutrients, such as nitrogen in marine systems and phosphorus in fresh-water systems, are no longer limiting and, although temperature and light are rapidly diminishing, there is often enough time remaining for an autumn bloom. However, when it does occur, it is of a much smaller magnitude than the one in the spring. With shortening days and diminishing temperatures, that bloom of phytoplankton and then zooplankton decreases and the cycle enters its winter sequence.

Dealing with ecology from this point of view of limiting factors and the

Figure 5.3. The seasonal cycle of plankton productivity in the aquatic environment is dependent upon a number of coincident limiting factors. In temperate climates, there are usually two major blooms as illustrated here with one occurring in the spring and the other in the fall. These are triggered by the presence of an adequate concentration of nutrients in the surface layers of water and of adequate physical factors of light and temperature.

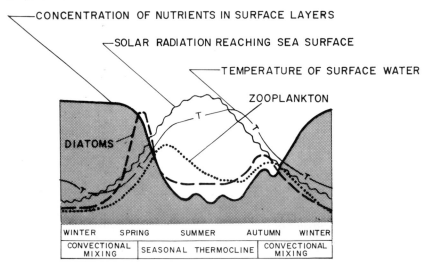

stresses that regulate populations and communities, we are led to the predominant theme of applied ecologists, *toleration ecology*. In 1913, V. E. Shelford developed the concept of the law of ecological tolerance, which includes five basic principles. These we can consider to be general characteristics of the niches of organisms that we may study and that we try to fit together in the environmental analyses described in the preceding chapter.

1. Organisms may tolerate (i.e., survive) a wide range of one factor and tolerate only a narrow range of some other factors.
2. Organisms that can tolerate a wide range for one factor or for a group of factors usually have a relatively wide distribution.
3. If an organism is stressed in its tolerance of one factor, its tolerance to stresses of other factors is likely to be lowered. For example, if grass is stressed by drought, it becomes less able to survive a disease than otherwise.
4. Organisms usually exist within a balance between stresses and tolerances. A community is subject to stresses consisting of competition for space, food, and other factors; of predation; of parasitism, and still others. For example, brine shrimp expend least metabolic energy in normal sea water. However, they are not found in the sea but in salt lakes, where they are stressed by high salt content and where they expend much more metabolic energy than would be required in sea water. The ecologic reason for their "preference" of brine seems to be that the predators of brine shrimp cannot survive in the brines. This allows the shrimp, tolerating that stress, to exclude their predators and they survive successfully at the expense of metabolic energy.
5. Organisms may have variable tolerances depending upon their physiologic state. Often, tolerance to stress is lower for an organism during its reproductive stages when it becomes more sensitive to stress.

In considering the ranges of tolerance within the niche of a particular species, relative measures include narrow *(steno)* ranges or wide *(eury)* ranges. These may relate to tolerance of any chemical, physical, or other limiting or stressing factors, such as stenothermal or eurythermal relative to temperature, stenohydric or euryhydric relative to water, stenohaline or euryhaline relative to salinity, stenophagic or euryphagic relative to food, and stenoecious or euryecious relative to habitats. With this approach in mind, the ecologic success and relative stability of a niche is controlled by two factors: (1) the types and quantities of all materials and physical factors that the niche (i.e., organism) requires and (2) the tolerances (i.e., living range) that the niche (i.e., organism) can successfully survive.

Therefore, ecosystem analyses may be directed at identifying the primary limiting factors with reference to the specific tolerances to these factors by the subject populations. In those cases, the predicted impacts, monitoring, and mitigation of impacts is concentrated upon the implications drawn from the often single-minded pursuit of the concepts of toleration ecology and limiting factors. Although the principles of toleration ecology seem sound, retrieving solutions to environmental management problems from their use is tedious and often requires much time. The usually imperceptible ecologic changes in a real system must be measured in order to solve, or even detect, a problem. So some aspects of toleration ecology are exceedingly nebulous.

For example, with respect to human ecology, it has been estimated that between 60 and 90% of all cancers (often called urban cancers) are the direct result of environmental factors created by our technology and the associated inadvertent release to the environment of *mutagens* (i.e., agents that increase the rate of mutations) and *carcinogens* (i.e., agents that increase the rate of cancer). Furthermore, it appears that there is no safe threshold that can be determined for these substances because, physiologically, their presence in any amount increases the associated stress, i.e., the potential for cancer. A small amount is not as bad as a large amount, but both are bad. Determination of the statistical significance of small versus larger amounts is especially difficult because in at least some actual instances of environmentally induced diseases, it may take between 15 and 40 years after exposure for a person to develop cancer resulting from the carcinogen (as with asbestos fibers).

In other instances, and usually in those most commonly dealt with by the applied ecologist, the limiting factors of primary concern are far less sinister than those associated with the defining factors in urban cancer. In fact, the concepts of toleration ecology and limiting factors are most understandable when they are used to define the ecologic structure of an environment stressed by some distinct single factor. In those cases, that one factor clearly dominates the system and there is little confusion possible in sighting the main aspects of the problem and in describing the apparent ecologic consequences of associated activities. For example, the dilution, dispersion, and decomposition of sewage discharges into lotic environments provides a model for both the stresses on an ecosystem caused by biodegradable (e.g., organic) materials and the transition or succession from one specific ecologic state to another within the same ecosystem. If we can assume that the river or stream in question is uniform in its ecologic condition and if we then construct a point source of pollution by having a sewage treatment plant empty its effluent into the stream, we find the following kinds of stress expressed in the natural environment.

After partial treatment by most sewage facilities, and assuming that no exotic or toxic material is significantly present in the sewage, the effluent is rich in chemicals that are essential to plant growth. Including nitrogen and phosphorus, which are usually limiting factors in the environment, the sewage effluent can often serve as a fertilizer, whether wanted or unwanted. If an ecosystem is overenriched by an unwanted fertilizer, problems arise because of the accumulation of these nutrients in one form or another. If the effluent enters a clean stream, there is an initial *zone of mixing and degradation* immediately downstream of the point source. In this zone, darker and more turbid than the clean upstream waters, bacteria begin to multiply in the presence of the effluent nutrients. This initial decomposition requires oxygen, which is removed from solution in the water, lowering the dissolved oxygen content of that water. If the effluent is heavily laden with nutrients, all the available oxygen may be burned up in support of the oxidation and utilization of the nutrients. This results in a virtually dead, anaerobic condition. In addition to the removal of oxygen from the *water column* present in this zone, fine solids present in the effluent may settle out over time to form *sludge banks*. As these banks oxidize, they will remove oxygen from the overlying waters if any is present. If the sludge banks are fiberous and cohesive, as are the matted banks formed in the vicinity of paper mills, which generate fibrous wastes, the evolution of gases within the banks may cause caked sediment to become dislodged and carried downstream.

This first zone is followed by the much longer *zone of active decomposition,* which, depending upon the size of the stream or river and the amount of nutrients emptied into the environment, may extend for miles. Its length is equivalent to a number of days of stream flow. Within this zone, those factors of the sewage effluent that have not settled out or reacted in the zone of degradation are removed or acted upon by biologic processes. It is an anaerobic zone of septic conditions, often accompanied by noxious odors and floating sludge, with no fish and only the most tolerant invertebrates surviving (Figure 5.4).

Biogeographic analyses of these sublethal environmental stresses show that starting with a naturally stable and diverse environment, pollution loading is often reflected in a marked reduction in the numbers (i.e., diversity) of species present. It simultaneously results in a marked increase in the population of the tolerant species. Therefore, depending upon the pollutant present and its degree of toxicity, there may be a tremendous increase or decrease in the individuals present of particular species. In particular, a tremendous increase is found when the species are able to uniquely tolerate the stressed environment and can also utilize the virtually limitless nutrient supply. For example, sludge banks may support a population of 50,000 slugworms for each square foot of surface. This is equivalent to 25 tons of slugworms per acre. These organisms are well adapted to burrowing into the sludge, extracting nutrients from the sediments and oxygen from even the very low concentrations in the overlying waters. If these physical conditions are found not to support a rich *monoculture* of one or a few tolerant species, one should suspect that there are toxic substances which have altered the normal pattern of decreased diversity and increased biomass.

Downstream of the zone of active decompostion, a *zone of recovery* can be found. In it, the water chemistry and the overall ecology of the water course is returning to its original state. It may extend for miles and it includes physical, chemical, and biologic interactions that are clearly reflected by increased dissolved oxygen, increased species diversity, and decreased biomass. However, nutrients associated with the organic pollution of the upstream waters are still present in forms available to fertilize an

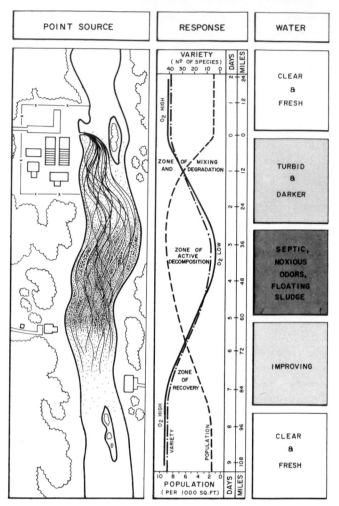

Figure 5.4. If over fertilization (ie., cultural eutrophication) of a stream or river results from pollution by sewage effluent, a number of changes can be predictably found downstream of that point source of the pollution. In general, this sort of organic pollution limits the diversity of the ecosystem present although the limited number of species adapted to live in the pollution may be present in exceedingly high numbers. (Adapted and expanded from Eliassen, 1952.)

abundance of growth. Caddisfly larvae and mayfly nymphs, characteristic of unpolluted waters, have been collected in the zone of recovery at concentrations of 1000 per square foot and 300 per square foot, respectively.

This zone is followed by the *zone of clear water,* theoretically defined as beginning at the point in the water course where there is no longer any effect attributable to the subject pollution source. The physical, chemical,

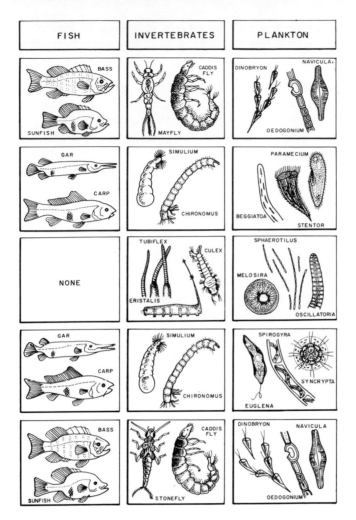

Figure 5.4. (continued)

and biologic features of the watercourse are similar to those upstream of the pollution source.

This consideration of stress is related to the *overenrichment* of the environment. Another equally important type of consideration of human imposed stress relates to *overharvesting,* which is something of an ecologic antonymn for enrichment. We have already mentioned some of the ecologic consequences of cutting down rain forests or overgrazing grass savannas. The ecologic accommodation to these types of harvesting manipulations can also be very clearly shown in the results of some fishery studies.

Given the growing efficiency and the ranging ambition of our commercial fishing industry, it is clearly possible for us to exploit a species to extinction; examples are whales, sardines, lobsters, and many others. When a stock of fish is underfished (underharvested), the majority of the catches contain older fish and evidence of disease or undernourishment with respect to an ideal stock of the same species. Because of crowding and lack of natural predation, the stock is less robust than it might otherwise be.

Increasing the fishing stress on the stock would thin the population as one would thin a garden or forest. This would result in the eventual harvesting of a younger, more robust stock. However, accelerating the fishing beyond this optimal level would result in *overfishing* the stock. Too many young fish would be caught, resulting in undersized fish, which would bring poor prices to the fisherman and stress the species through a serious depletion in the reproductive stock.

The enforced and virtually absolute prohibition of marine fishing in the North Sea during the First and Second World Wars provides a dramatic illustration of overfishing impacts. Plaice are flatfish that live as a *demersal* (bottom living) species in the northeastern Atlantic (principally the North Sea), where between 150,000 and 170,000 tons are currently being landed each year (Figure 5.5).

Figure 5.5. The graphic representation of the landings of plaice caught in the North Sea by first class English steam trawlers provides an excellent illustration of the impact of overfishing (predation) upon a large population when extended over a number of years. In particular, note the relatively rapid decline in population once fishing resumes subsequent to the war years. (Redrawn from Tait and De Santo, 1972.)

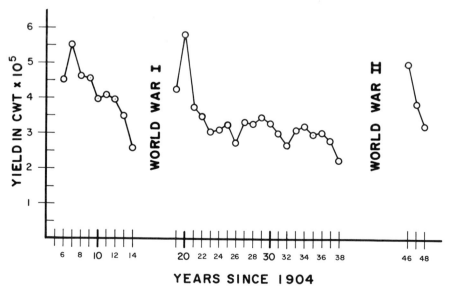

YEARS SINCE 1904

Plaice reach a maximum size of about 40 cm in 6 years; only those specimens 25 cm or longer may be kept while the rest are protected by law. However, such means of conservation were not generally practiced in the past and the resulting fluctuating population reflects stress by declining yields comprised mainly of young and small fish. When fishing ceased during war, the subsequent catch showed the potential for ecologic recovery. Resumed stress removed the recovery within a few years.

Whether environmental stress results from lethal, sublethal, or nonlethal factors, the results may be evaluated in terms of the tolerance by involved niches to the stress and their ecologic response to that stress, as suggested in Figure 3.1. Stress reflected in the progression of seral stages leading to a climax community (Figure 2.7) or stress reflected in the lack of species diversity seen in the zone of active decomposition in an effluent stream (Figure 5.4) are both perceptions of stress as an expression of toleration ecology.

However, in order to construct a complete understanding of any one ecologic analysis, the interrelations between abiotic and biotic factors must be defined more comprehensively than through a study of only limiting or stressing factors. We have seen how the major divisons of the biosphere are artificially segmented for our convenience into terrestrial biomes and the aquatic branches of ecology. We have seen how a complex ecosystem can be dissected into its principle components and how they can be evaluated, and we have touched upon the problems of environmental stress. We are led to conclude our examination of ecology by looking into some of the pathways that are involved in ecologic cycling of materials within an ecosystem. Considering the abiotic and biotic multilevel interactions of a few representative processes and principles, we can review the facts and techniques useful in understanding and predicting the ecologic outcome of the cycles. This concept of ecologic cycling is the essence of sound environmental management. It is, therefore, the central tenet of applied ecology.

Bibliography

Ballinger, D. G., and D. G. McKee. 1971. Chemical characteristics of bottom sediments. *Water Poll. Contr. Fed.*, **43**(2); 216–227.

Elias, T. S., and H. S. Irwin. 1976. Urban trees. *Sci. Am.*, **235**, 111–118.

Eliassen, R. 1952. Stream pollution. *Sci. Am.*, **186**, 17–21.

Gaufin, A. R., and C. M. Tarzwell. 1955, Environmental changes in a polluted stream during winter. *Am. Midland Naturalist*, **54**(1), 77–88.

Goldman, C. R. 1960. Primary productivity and limiting factors in three lakes of the Alaska peninsula. *Ecol. Monog.*, **30**, 207–230.

Hutchinson, G. E. 1973. Eutrophication. *Am. Scientist*, **61**, 369–379.

Lean, D. R. S. 1973. Phosphorus dynamics in lake water. *Science*, **179**, 678–680.

MacKenthun, K. M. 1973. *Toward a Cleaner Aquatic Environment*. Publ. 5501-00573. U.S. Government Printing Office, Washington, D.C., 273 pp.

Muus, B. J. 1974. *Collins Guide to the Sea Fishes of Britain and North-Western Europe*. Wm. Collins Sons, Landon, 244 pp.

Rohl, A. N., A. M. Langer, and I. J. Selikoff. 1977. Environmental asbestos pollution related to use of quarried serpentine rock. *Science,* **196,** 1319–1322.

Shelford, V. E. 1913. *Animal Communities in Temperate America*. University of Chicago Press, Chicago.

Smith, S. L. 1977. The failure of success in fisheries management. *Environ. Man.,* **1**(3), 239–247.

von Liebig, J. 1855. *The Relations of Chemistry to Agriculture and the Agricultural Experiments of Mr. J. B. Lawes*. (transl.) Albany, 23pp.

6

Pathways of Selected Ecologic Factors

A major requirement helping to shape the character of applied ecology is that the ecologist is challenged to draw upon all his or her knowledge and intuition in solving new problems. Because the solutions are usually far reaching in their implications and because many disciplines are often affected by these solutions, the ecologist is under considerable pressure to act quickly and accurately from a limited base of information.

The key to successfully dispatching an ecologic problem usually is a good understanding of the interdigitation of the environmental stress in question and of all the natural and anthropogenic systems involved directly or indirectly with that stress. Such an understanding may be, and usually is, a combination of empirical and theoretical knowledge. Working to untangle, predict, or correct the nature and pathway of a factor (such as a pollutant) through an ecosystem requires that we understand the flux of that factor between the biotic and abiotic aspects of the ecosystem. Such knowledge involves three major areas of ecology upon which we have already touched. They are an understanding of (1) the *chemical disposition* of the factors involved (their chemical structure and characteristics); (2) the *physical disposition* of the factors (their pathway into and through the ecosystem, including its physical mixing characteristics); and (3) the *biologic disposition* of the factors (their metabolic implications and potential entry into food chains).

Each of these three fields of study is focused upon illuminating the pathways of various factors which pass through major ecosystems on the earth. When taken together, their combined study and integration results in attempting to understand *biogeochemical cycles*. In general, these cycles have evolved in such a way that life, taken as a whole, tends toward a stable, self-replicating balance of all those factors present. The principle ecologic means by which these factors remain in averaged balance is through *recycling*.

There are two general groups of cycling: (1) gaseous recycling, involving air and/or water; and (2) sedimentary recycling, which involves the crust of the earth. Of the 105 known chemical elements on the earth, 30 or 40 are clearly known to be *essential* or *limiting factors* to life and the study of their pathways through an ecosystem is called *nutrient cycling*. Carbon, hydrogen, oxygen, and nitrogen are needed for life processes in highest concentration, although the balance or imbalance of an ecosystem usually rests on the elements of least concentration, the limiting factors as discussed in the previous chapter.

The predictability and stability (i.e., balance) of cycles that are dominated by gaseous exchanges is usually great because the atmosphere provides a virtually infinite reservoir. It serves as a sink or as a source for many gaseous factors. Sedimentary-type cycles, in contrast, tend to be relatively unstable and widely different from one location to another because of the relatively finite limits and immobility of a terrestrial sink or source area. When a biogeochemical cycle is disturbed, which is usually the result of human induced perturbations, acyclic consequences result in too little of a factor or set of factors in one place and too much in another place. Although the ultimate outcome of such a problem is extremely variable and entirely dependent upon the factors involved and upon the nature of the subject ecosystem, general patterns can be defined. These can help in determining the critical factors or pathways involved in order to correct or lessen the problem being considered.

For example, nitrogen has both a gaseous phase and a sedimentary phase, and it represents a complex cycle (Figure 6.1). It is closely interwoven into the organic productivity of the ecosystem as an essential element in the synthesis of protoplasm. More generally, nitrogen is a very "malleable" element in that it is essentially ubiquitous; nitrogen gas comprises 79% of the atmosphere. It also occurs in a wide variety of compounds and it ranges in its chemical valence from $+5$ to -3. The most usual inorganic forms are the highly oxidized nitrite (NO_2) and ammonia (NH_4), nitrate (NO_3), and nitrogen gas (N_2). The organic forms consist primarily of amide and amino compounds, which form proteins. The assimilation of nitrates and ammonia by living organisms is the major pathway for conversion of the inorganic forms into organic compounds synthesized by a living system. Once incorporated into organic compounds, the nitrogen is eventually recycled by a series of interwoven steps (Figure 6.2). *Ammonification* breaks down amino acids, proteins, nucleic acids, and others into the equivalent amount of ammonia. This step may be followed by *nitrification,* which is the aerobic oxidation of ammonia and converts it to nitrite and then nitrate. Under anoxic (no oxygen) conditions, the sinks of nitrite and nitrate can be converted into molecular nitrogen (nitrogen gas).

Even though nitrogen is by far the greatest fraction of the atmosphere, it is regularly a limiting factor for plant growth. Nitrogen gas in the atmosphere is inert, chemically inactive, and unavailable for the required metabolic activities of most organisms. However, a few known genera of

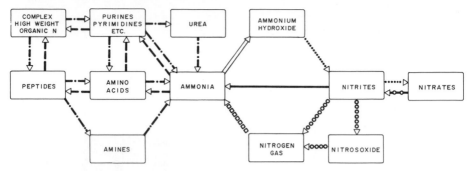

Figure 6.1. The organic and inorganic constituents of the simplified nitrogen cycle in the biosphere are shown above. These main transformations include:

———— Assimilation of nitrate	ooooooo Denitrification
— — — Assimilation of ammonia	□□□□□□ Nitrogen fixation
—·— Ammonification	═════ Simple oxydation
·········· Nitrification	

bacteria and several species of blue-green algae are capable of removing nitrogen gas from the environment and converting it into ammonia.

Metabolic energy is required for this conversion (called fixation) and is supplied by adenosine triphosphate (ATP). Such nitrogen-fixing organisms are therefore not limited by the scarcity of *reactive* nitrogen compounds because they can utilize the vast sink of atmospheric nitrogen gas. In fact, such organisms make nitrogen available to other species, such as the bacterial genus *Rhizobium,* which provides nitrates through its symbiotic relationship with the roots of legumes.

The natural fixation of nitrogen by bacteria is seldom of special concern to the applied ecologist dealing with problems associated with environmental impacts and pollution. It is not nearly as rare for him to deal with natural fixation of nitrogen by blue-green algae when they are associated with a bloom in a recreational lake or reservoir, causing both esthetic and ecologic water quality degradation. However, the most usual confrontation between applied ecologist and the nitrogen cycle involves the presence of nitrates, nitrites, and/or ammonia in aquatic environments resulting in accelerated *eutrophication.*

Eutrophication is a natural and inevitable ecologic process in every body of more or less standing water. It results from the enrichment of that water body by nutrients, such as nitrogen and phosphorus, which enhance the productivity of the plant biomass present. The acceleration of this natural aging process results when humans add more sediments, directly or indirectly. Causing an increase in erosion and sedimentation, reducing the flow of water to a system, or directly adding nitrates in the form of agricultural

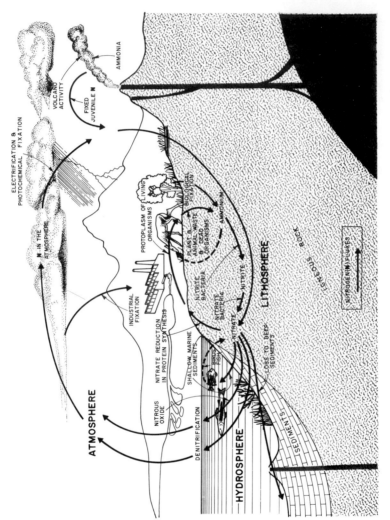

Figure 6.2. The nitrogen cycle involves both a gaseous and a sedimentary phase. Although the atmosphere contains 79% nitrogen gas, this vast reservoir is not usable by most organisms until the nitrogen is combined with oxygen or hydrogen and assimilated into plants. The nitrogen compounds may then be eaten and assimilated by animals.

fertilizer runoff or sewage to the water bodies augment the natural process. Because this artificial acceleration is invariably associated with stressing the natural environment, the process has generally been named *cultural eutrophication*. Eutrophication is a relative term and its mitigation is manageable only with the greatest difficulty.

One reason is because the nitrogen cycle is both complex and an integral part of the whole eutrophication issue. It appears that the nitrogen cycle is more complex in aquatic ecosystems than in terrestrial systems because of the interaction of sediments and the water column in combination with inputs to the system from the runoff. Any description of the eutrophication problem must take into account the retention and recycling of nitrogen

Figure 6.3. The nitrogen cycle in aquatic systems has been frequently studied with respect to its importance in eutrophication (ie. the enrichment of the water body with nutrients which leads to plant growth). Sediments serve as a reservoir of nitrogen compounds, among others, which are redistributed to the water column long after new additions of that nutrient may have ceased from other external sources.

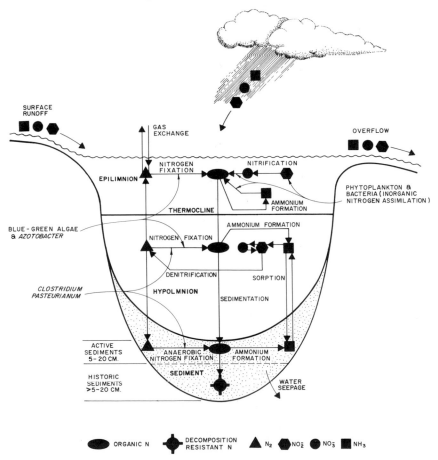

nutrients, which are initially removed from the water column, stored in the sediments, and subsequently redistributed to the water column. Because this process is slow and usually of a very site-specific nature, accurately determining the balance or imbalance that may be causing a problem is most challenging. The approach usually taken by the applied ecologist is to verify the general nature of the problem and to then remove or minimize the input of critical nutrients (Figure 6.3). If that approach produces some response in the system, further modifications are developed to accelerate, redirect, or correct the managed modification.

For example, if a community is developed on the shores of a lake and, after some time, it recognizes a eutrophication problem, an applied ecologist might proceed to correct the problem in the following three steps. First he would define the problem. Second, he would determine the most practical solution. Third, he would implement the solution and modify it as appropriate.

In order to define the problem, the ecologist must make a judgement of what pollutants and what pathways are the most likely primary cause of the problem. Field surveys and seasonal water quality and sediment analyses are necessary, which then allow the ecologist to construct an approximate model of the total input and output of the ecosystem in question. The mass of nutrients, biomass, sediments, and water that comprise the system are balanced as closely as possible to match the existing system. Such a *mass balance* or *stoichiometric* equation of the critical ecologic factors involved usually allows the ecologist to decide which facet of the problem should be corrected first.

Determining how to best achieve that correction usually requires close coordination with the appropriate public and private agencies in order to insure that solutions sought by the ecologist will be supported by the client and public and will be likely to achieve results acceptable to the client. Furthermore, the ecologist must determine that the long-range results of his work will serve the best ecologic interests of that client. In this example, if nutrients are determined to be the main problem subsequent to a mass balance preliminary determination, three basic solutions must be evaluated from a practical point of view: (1) excess nutrients can be removed at the source in order to prohibit their entry into the problem ecosystem: (2) accumulated nutrients (i.e., the sediments), which serve as endemic sources of enrichment, can be physically removed; or (3) some combination of both approaches can be sought.

Implementation of the final choice of solutions may require diversion of all sewage to a treatment plant that adequately treats the community wastes such that nutrients are not added to the receiving waterbody. Prohibition of the use of fertilizers that wash into the water body may be required. The staging and extent of dredging and the disposal of the dredge spoils must be planned to be effective with respect to the money and effort allocated to implementation. If the ecologic work plan is correct but cannot be effectively implemented, the results will not be cost effective. If only cosmetic

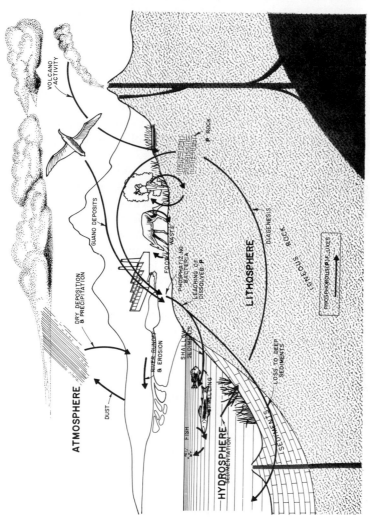

Figure 6.4. The phosphorus cycle does not involve a gaseous phase and is, therefore, a sedimentary cycle. It appears that this cycle is not balanced and that relatively large amounts of phosphorus are lost annually to the deep ocean sediments.

plans are implemented for lack of effort or funds, a temporary political solution may be achieved but the ecologic solution is assured of eventual failure under these circumstances.

Although the nitrogen cycle is essential to eutrophic systems and their control, usually it does not represent the limiting factor in fresh-water systems, although it may be limiting in coastal marine systems. As has been emphasized previously, phosphorus appears to generally be a critical limiting factor in terrestrial systems as well as many aquatic systems.

On the average, the concentration of nitrogen in natural waters is 23 times greater than phosphorus and, therefore, phosphorus is relatively rare. It is found in a relatively soluble mineral reservoir of certain rocks. These geologic deposits are eroded and thus release phosphate (PO_4^-) to the waters, which carry the element to intercepting ecosystems. Phosphorus is an essential constituent of protoplasm, participating structurally in the synthesis of RNA, DNA, phospholipids, and many other compounds. It is the primary vehicle for the distribution of metabolic energy through the functioning of ATP.

The biogeochemical cycle of phosphorus is considerably simpler than nitrogen because it is essentially a sedimentary cycle not involving any significant gaseous pathways (Figure 6.4). This fact of its physical chemistry also means that phosphorus is removed from the biosphere for exceedingly long periods, if not permanently. All phosphorus in an ecosystem will eventually be carried in water, sediments, or protoplasm to the sea, where it may be deposited in the shallow sediment or lost to the deep sediments. One estimate suggests that as much as 2 million tons of phosphorus are eroded into the sea each year and that about 60,000 tons are returned to the land from the sea each year in the form of guano, deposited on land by marine birds. The guano is derived from the fish upon which the birds feed.

Figure 6.5. Upwelling of coastal waters usually occurs in depths of about 200 meters. When climatic and oceanographic conditions coincide to move coastal waters off shore, deeper waters move up replacing the moved water. These vertical currents carry nutrient rich water into the euphotic zone where phytoplankton can grow and provide the base of a food web which will support the rich fisheries found in these regions.

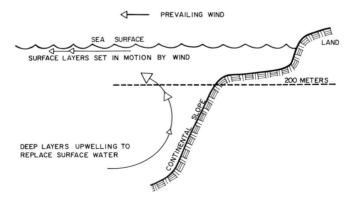

Phosphorus in the sea was incorporated at some point into the food chain of those fish.

Some of that phosphorus will be carried down into the deep sediments and, from there, some will be returned to the euphotic zone by *upwellings*. Offshore winds and currents move the surface waters of some coastlines offshore, which may cause the deeper water to rise from depths of a few hundred meters (Figure 6.5). These waters are usually relatively rich in nutrients, including phosphates, and upwellings are associated with rich fisheries (Figure 3.23). With respect to fresh-water systems in temperate climates, inorganic phosphorus (PO_4^-) is generally released to the hypolimnion over the winter. This is one important factor triggering the spring phytoplankton bloom when the thermocline breaks down, allowing convection currents to carry the nutrients from the hypolimnion into the euphotic zone (Figure 5.3).

The pathway of sulfur through the biosphere, like that of nitrogen, involves both gaseous and sedimentary cycles. However, it is simpler. Sulfur is a key constituent of several amino acids, the building blocks of proteins, and although relatively less sulfur is required by living organisms than nitrogen or phosphorus, it is still an essential element. Much as with nitrogen, an interrelationship of multiple sulfur cycles is required in order for a balance to be maintained between use and generation (Figure 6.6).

One particularly important aspect of the sulfur cycle in aquatic sediments is a consequence of its reaction with iron to form iron sulfide. The presence of that compound is required for insoluable phosphorus to be converted into soluble forms, which are then much more mobile and available for utilization by plants. Primary production is therefore directly dependent upon the sulfur cycle both for the availability of sulfate and for the release of phosphorus (in the form of phosphate) from the sediments.

Of course there are other essential "natural" biogeochemical cycles required for the balanced functioning of any stable ecosystem and many of these may be sought by the reader in the suggested readings listed in this book. The preceeding three examples provide examples of the general types encountered. However, it must be borne in mind that even in the absence of human intervention, many ecosystems do not necessarily contain balanced cycles. Furthermore, it is not *necessarily* good ecologic or management sense to create or maintain balanced ecosystems. What is important is that we understand how and why an ecosystem is unbalanced and that we acquire the skill to balance it if we decide to do so.

A number of both theoretical and applied studies have shown that ecosystems naturally evolve from one seral stage to another and that a progression is seen which eventually leads to a climax condition that is apparently stable. Until that final sere is attained, the pathway of nutrients or other factors is usually not balanced and may not even be stable. Logically, it appears that as ecosystems mature, once a resident biomass is established in the pioneer community of a new ecosystem in primary succession, they tend to conserve nutrients. Such a new community acts like a nutrient sponge, assimilating the nutrients that enter the system and

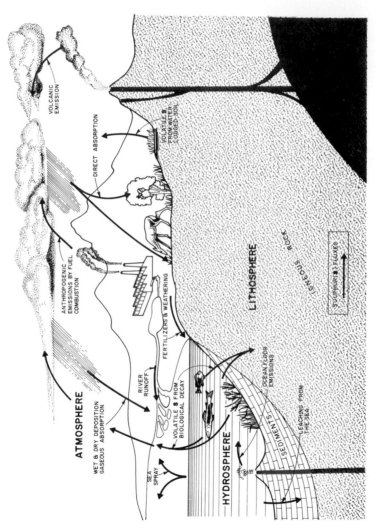

Figure 6.6. The sulfur cycle is considered to be a relatively simple gaseous/sedimentary cycle. Microorganisms in the soil are largely responsible for the conversion of hydrogen sulfide into sulfur and then into sulfate. The reverse reaction can also be present through the involvement of other bacteria. The addition of sulfur from anthropogenic sources increases relative to the amount of fossil fuels used. On a global basis, these additions may be insignificant when compared to natural volcanic emissions. Nevertheless, they are extremely significant with respect to regional problems of air pollution.

recycling them within the system. In that sense, the early successional seral stages are sinks. It appears that as long as this evolving ecosystem has an increasing *net* biomass, the capacity of the ecosystem as a nutrient sink dominates its capacity as a potential nutrient source.

In theory, when the climax sere is reached, the new production of biomass should equal the loss of old biomass and, therefore, the input to that ecosystem in the form of nutrients should be equal to the output of the system. In other words, production will equal respiration, and the recycling of nutrients will be balanced and stable (Figure 6.7). With respect to patterns of secondary succession, where there is a preexisting population that we can assume has been a climax sere, the pattern of nutrient pathways is initially quite different than in the pattern just described for primary succession.

In secondary succession, the soil has already been modified and will serve as a nutrient source when its biota is substantially changed, as with the devegetation of a forest stand. Clear cutting is a technique of harvesting timber that results in stripped land. In that state, losses of those elements limiting and/or essential to plant growth are lost from the system at a very much higher rate than is the case with an intermediate successional ecosystem. With respect to the retention of limiting and essential elements by intermediate seral stages as compared with a climax sere, a study of a forested ecosystem has shown that more than six and a half times as much nitrate leaves an old system than leaves a young one. That research suggests that ammonia carried into the ecosystem as precipitation is converted into nitrate and is utilized and retained by the younger system but passes through the older one. The same relative tendencies are demonstrated with the mass balance of other elements necessary to forest growth. Such nonessential elements as sodium (Na^+), in contrast, pass through both young and old ecosystems, unaffected by their different normal ecologic state of succession.

Figure 6.7. The assimilation of nutrients by the biomass of the seral succession of an ecosystem causes the early stages to act as a nutrient sponge. The magnitude of that sponge decreases as the climax seral stage is approached. At that point, the ecosystem does not have a net increase in biomass nor does it retain nutrients. That is, the same mass of nutrients leave the ecosystem as enter it.

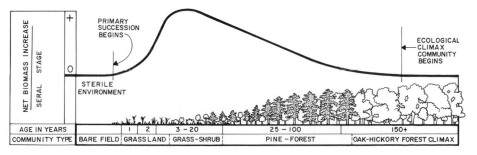

These characteristics are held in common by virtually every ecosystem. They lead to the important observation that the pathways of such factors as nutrients, through mature (i.e., climax) ecosystems are usually not retentive or conservative and that input into these systems is theoretically balanced by their output. In contrast, young and evolving ecosystems contain pathways that are retentive and conservative and their essential inputs are much higher than their outputs.

With respect to environmental management by the applied ecologist, these principles may be used to mitigate the impact of nutrient cycling. Maintenance of an intermediate seral stage ecosystem in the pathway of limiting or essential elements, such as nitrogen, phosphorus, potassium, magnesium, or calcium, will create a sink and remove a relatively large fraction of those elements from the environment. If the "ecologic youth" of the system is not maintained, it eventually becomes mature (i.e., climax) and then cease to serve as a sink. If its net productivity and/or biomass is reduced, it becomes a source of those elements that it initially had retained.

The most potentially solvable problems with which an applied ecologist may become involved usually concern the sorts of biodegradable stress and pathways outlined above. In them, the factors most directly linked with the problem are usually not foreign to the preexisting ecosystem. Instead, a desired state has been upset by their being present in too large or too small amounts. In those circumstances, the ecologist seeks either to return the mass balance to its previous state or to manipulate the new state in some acceptable way in order to redress and make the problem manageable or even productive. However, when the problem is caused by toxic and/or nonbiodegradable or ecologically "foreign" factors, the challenge is different and often frustrating and tragic.

One of the most prominent examples of this kind of applied ecology is found in the case of mercury poisoning of ecosystems. This modern environmental disorder has been traced to our technology. It became an issue in 1953 in the villages surrounding Minamata Bay on the Japanese island of Kyushu. Three years after the heavy metal-containing waste water of a chemical plant began being discharged into the relatively small, quiet bay, humans began to appear with what has come to be called Minamata's Disease. Its symptoms include numbness of lips and limbs; tunnel vision; lack of muscular coordination; loss of speech, hearing, and taste; emotional disturbance; loss of memory; convulsions; loss of sight; muscular contraction; elevated blood sugar; coma; and death.

The disease is caused by the ingestion of methyl mercury compounds. In Minamata Bay, the muds near the factory effluent channel contained more than 2000 ppm of mercury and tested fish and shellfish contained up to 50 ppm of mercury as compared with 1 ppb (part per billion) in the sea water. (The average concentration of mercury in the oceans in 0.03 ppb.) The humans involved showed elevated mercury levels in urine, blood, hair, brain, liver, and kidney.

One reason that no caution was exercised in disposing of mercury into the bay was simply that ignorance was bliss and we knew nothing of the danger. It was generally assumed that the mercury would not become mobile in the environment. However, recent investigations have shown that once in the sediments of an aquatic system, metallic mercury is acted upon by aerobic bacteria that convert the mercurous ion into either *mono-methyl mercury* (CH_3Hg) or *dimethyl mercury* (CH_3HgCH_3). These are *alkyl mercury compounds* which have physical characteristics critically different from metallic mercury (Figure 6.8). They are highly water soluble. They readily pass through cellular membranes. They are not easily degraded or purged from an organism. The half-life of methyl mercury in humans is approximately 70 days and some standards permit ingestion of 0.075 mg per day.

Mitigation of the negative ecologic effects from mercury contamination of a aquatic environment may include any one or all of at least six strategies, including the diversion of any new mercury from the receiving water body as the first step. Second, the sediments and their contained mercury may be removed for disposal elsewhere. Third, such inorganics as silica minerals may be mixed with the sedimenting mercury, creating an aerobic blanket. Here, ferric and manganese ions tend to bind the mercury, forming oxides that reduce the probability of mercury methylation. Fourth, the mercury may be converted into mercuric sulfide, with an associated low accessibility to biochemical methylation. Fifth, the pH of the environment can be increased, which favors biochemical methylation leading to the generation of volatile dimethyl mercury. This has a lower accumulation rate in fish than monomethyl mercury. Sixth, the sedimented mercury deposits may be covered and chemically sealed by use of mercury absorbents of inert materials that stop or decrease the release of alkyl mercury by the sediments to the water column.

However, no set of remedial methods can be universally promulgated at this relatively early stage in our understanding of the mercury passage through an ecosystem. The costs and benefits of a particular method must be evaluated for a particular site. The best and safest procedure, however, is to remove any new or continuing sources, while recognizing that in the absence of "sealing" or removing the historic sediments of mercury, they continue to release methylated mercury from that sink/source for 10 to 100 years at a minimum.

Such *persistence* of a toxic environmental contaminant is a characteristic of many of the other ecologic disturbances associated with our technologic advancement as well. In particular, dichlorodiphenyltrichloroethane (DDT) is a classic example of this circumstance. The benefits and some of the ecologic liabilities of this substance have been outlined in the preceding chapter and its possible carcinogenic and mutagenic characteristics are discussed here. Here, we can also consider certain crucial aspects of its dispersal in the environment and its entry into food chains.

DDT has been often applied to large areas of land by aerial spraying from aircraft. Mixed with inert material, spray rates of between 0.5 and 5 pounds

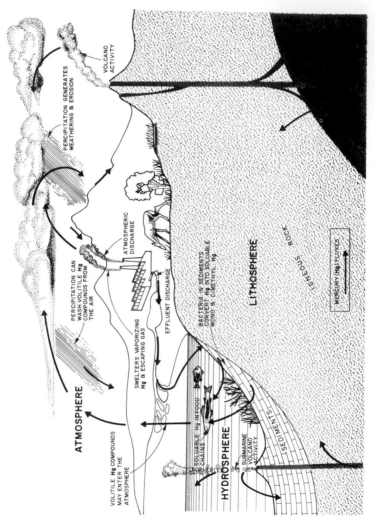

Figure 6.8. The mercury cycle in the biosphere centers upon the relative environmental mobility of certain compounds of the element. Mono and dimethyl mercury is evolved by the action of certain aerobic bacteria upon elemental (metallic) mercury. These compounds are soluable, become mobile, and enter the food chain where they tend to be concentrated.

of pure DDT per acre are used. Combined with all other forms of application, the worldwide distribution of DDT is ubiquitous and relatively high (Table 6.1). This is notable especially because DDT did not exist before 1874 and was not commercially available until a few years after 1939.

It is estimated that about 50% of the DDT applied to the land is volatilized and carried into the atmosphere, and this may in large part explain the remarkably broad dispersion of the chemical. In sufficiently high concentrations, this chlorinated hydrocarbon is directly toxic to organisms other than insects, such as several crustacean species. It affects many bird species by interferring with their steroid hormone production and the functioning of the carbonic anhydrase enzyme in the shell-forming epithelial lining of the uterus. That tissue normally builds the 98% calcium carbonate structure of the egg shell. The readily broken, thin shells laid by the poisoned birds cause unhatched young to die and the affected population to dwindle and be threatened as a result of this human synthesized, unintentional avian genocide.

An important characteristic of DDT was previously cited relative to its very low solubility in water, which assures its retention in the soil or on the vegetation to which it is applied. Its converse high solubility in fats brings with it many of the problems that parallel its pathway through an ecosystem. DDT is not readily metabolized and it tends to be stored and accumulated in the fat of those animals which ingest the chemical.

Concentrations of 9.9 ppm of DDT were found in the soil of a terrestrial environment that was sprayed with the insecticide to control the spread of Dutch Elm disease by its beetle vector. Earthworms that fed on the soil contained 141 ppm and robins that fed on the worms contained 444 ppm in their body fat. Even more dramatic accumulations have been discovered in the food webs of aquatic systems.

The general principle illustrated in this example of the concentration of a nonmetabolite (i.e., a nondegradable contaminant) in its passage through a food chain is called *bioaccumulation* or *biomagnification* (Figure 6.9). As the substance is passed from one trophic level to the next higher one, it is concentrated and may become a dominating and sinister stress at a considerable ecologic distance from the point or trophic level of its initial introduction into the biosphere.

Table 6.1. DDT Average Worldwide Dispersal

Soil	2–10 ppm
Ground water	0.001–0.2 ppm
Surface water	0.001–0.2 ppm
Rain	0.1–0.3 ppb
Trade winds	0.1–0.3 ppb
Human fat	6–12 ppm
Beef fat	0.5 ppm
Deep oceans	0.00 ppb

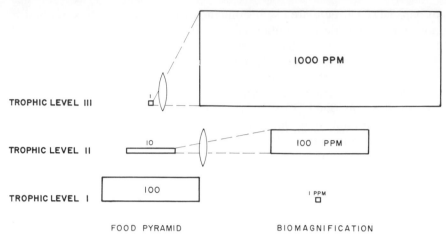

TROPHIC LEVEL III

TROPHIC LEVEL II

TROPHIC LEVEL I

FOOD PYRAMID   BIOMAGNIFICATION

Figure 6.9. The Eltonian (Food) Pyramid and Biomagnification are interrelated. Biomagnification is the biochemical process through which an element, or a compound like DDT, is concentrated in living tissue as it passes from the protoplasm of one trophic level to that of the next higher trophic level. When this occurs, the top carnivores of the pyramid will be exposed to relatively high concentrations of the metabolically incorporated pollutant even though that pollutant may have entered the biosphere at exceedingly low levels of contamination.

The best preventative ecologic medicine for this sort of problem is to cease, or at least control, all new releases to the environment and to understand the pathways of the problems. It may not be practically possible to correct the problems associated with previous activities, but knowing the ecologic processes involved and the nature and significance of what may be expected can help direct remedial methods and treatment effectively. Unfortunately, the nature of many of the pertinent pathways still remains outside our certain knowledge and we must deal with the most logical hypothesis available. That necessity is simultaneously the most frustrating and the most fascinating aspect of applied ecology. Interim decisions must be made and acted upon without the benefit of knowing with any great certainty that the basic information upon which that judgement is made is true, partly true, or not true. The applied ecologist seeks to be confident that this understanding is at least empirically true, even if the theory leading to the empirical truth is itself not true.

This fascination for relevant and important ecologic puzzles that have not yet been solved and the challenge of making new associations of information are traits shared by most scientists. Furthermore, the enigma of the piece of the puzzle that does not fit is usually the aspect of the challenge which spurs the greatest intensity of curiosity. For example, a recent short publication by Harvey and Steinhauer (in Nriagu, 1976) exemplified both the overview of a large ecologic problem and a challenge to test and modify an important fundamental principle relative to biomagnification. Their work also questioned the wisdom of making dire ecologic predictions based upon preliminary and usually sketchy knowledge of a particular situation.

As has been pointed out, neither PCB's nor DDT are naturally occuring chemicals. Relatively little information was available in 1970 when a number of "celebrities" forecasted the death of the seas as a result of environmental degradation from petroleum spillage, PCB's, and DDT contamination. However, as the years have passed examination of the marine food chain has revealed no consistency in the PCB's or DDT content of organisms taken from the various trophic levels, suggesting that in these marine studies, at least, biomagnification of PCB's and DDT does not exist.

Of particular interest was the observation that subsequent to 1972, there was a relatively dramatic drop in PCB's concentrations in the North Atlantic, coincident with the prohibition of many uses of the dieletric in both the United States (1970–1971) and Europe (1972–1973). In 1972, levels were as high as 88 ppt in the Sargasso Sea, 8 ppt in 1973 off of South American, and 7.7 ppt off the Lesser Antilles in 1974–1975. In the Sargasso Sea, it appeared that 2 mg of PCB's per square meter of sea surface were lost in the uppermost 100 m in 1 year. There were five possible explanations given:

1. The PCB's in the surface waters may have been diluted through mixing and diffusion. However, studies using radioactive tracers have shown that halving concentrations of strontium-90 take about 5 years, and the diminished concentration of PCB's had taken only 1 year.
2. The PCB's in the surface waters may have been absorbed on the surface of particulates descending to the deep ocean floor. However, studies have shown that about 96% of the materials present are recycled wholly within the euphotic zone and that whereas 500 g per square meter of falling solids per year are required to account for the decrease in PCB's concentration, only 13 g per square meter is actually present.
3. The PCB's in the surface waters may have gone through an evaporative co-distillation at low and midlatitudes, been carried in the atmosphere to the high latitudes with water vapor, where it may have been codistilled to be entrained (trapped or recirculated) in the cold, denser waters of the deep Atlantic.
4. The PCB's may have been biodegraded by marine microorganisms converting the tetrochlorobiphenyl component into water-soluble metabolites and CO_2.
5. The PCB's may have been incorrectly measured because an analytical methods change was incorporated in PCB's determinations which could result in lower PCB's measurements in 1973, and therefore, relatively higher measurements in 1972. However, quality control inspections did not support this circumstance as a possibility.

Taking all these possibilities into consideration we are able to judge that (1) PCB's is geochemically very stable, with little chemical alteration as it passes from surface waters to waters deeper than 5000 m; (2) PCB's is only very slowly biologically degraded; (3) less than 1% of the surface PCB's sinks, absorbed to sediments; and (4) only about 0.1% of the surface PCB's is found in the biomass present.

Therefore, it seems most probable, given our present knowledge, that once PCB's enter the ocean they tend to volatilize and move to higher latitudes, where they are washed from the air to accumulate in the cold waters. Their final degradation and/or accumulation is in the sediments and bottom waters of the polar oceans, unless upwelling brings them up to repeat the cycle.

To end this book on an "iffy" biogeochemical probability is most appropriate. Applied ecology often deals with probabilities and the ultimate skill of the professional applied ecologist is not to avoid decisions based on probabilities but to avoid dangerously premature and irretrievable decisions. When the fund of information is sure and complete, the decisions are ecologically clear and easy. Getting to that stage is the hard part. It is also the interesting part and requires constructing an approach to applied ecology that allows the student to make the correct judgements in a frontier of puzzling challenges.

Bibliography

Bormann, F. H., G. E. Likens, and J. M. Melillo. 1977. Nitrogen budget for an aggrading northern hardwood forest ecosystem. *Science,* **198,** 981–982.

Brill, W. J. 1977. Biological nitrogen fixation. *Sci. Am.* **236**(3), 68–81.

Carlson, D. A., K. D. Konyha, W. B. Wheeler, G. P. Marshall, and R. G. Zaylskie. 1976. Mirex in the environment: Its degradation to kepone and related compounds. *Science,* **194,** 939–941.

Chadwick, M. J., and G. T. Goodman. (eds.). 1975. *The Ecology of Resource Degradation and Renewal.* Wiley, New York, 480 pp.

Chapra, S. C., and A. Robertson. 1977. Great lakes eutrophication: the effect of point source control of total phosphorus. *Science,* **196,** 1448–1449.

Delwiche, C. C. 1970. The nitrogen cycle. *Sci. Am.,* **233**(3), 137–146.

Dillon, P. J., and F. H. Rigler. 1974. The phosphorus–chlorophyll relationship in lakes. *Limnol. Oceanogr.* **19**(5), 767–773.

D'Itri, P. A., and F. M. D'Itri. 1978. Mercury contamination: A human tragedy. *Environ. Man.,* **2**(1), 3—16.

Eisenbud, M. 1973. *Environmental Radioactivity.* (2nd ed.) Academic Press, New York, 542 pp.

Elder, D. L., and S. W. Fowler. 1977. Polychlorinated biphenyls: Penetration into the deep ocean by zooplankton fecal pellet transport. *Science,* **197,** 459–461.

Fagerstron, T., and B. A. Sell. 1973. Methyl mercury accumulation in the agnatic food chain. A model and some implications for research planning. *Ambio.,* **2,** 167–171.

Gilbert, R. G., R. C. Rice, H. Bouwer, C. P. Gerba, C. Wallis, and J. L. Melnick. 1976. Wastewater renovation and re-use: Virus removal by soil filtration. *Science,* **192,** 1004–1005.

Goldwater, L. J. 1971. Mercury in the environment. *Sci. Am.,* **224**(5), 15–21.

Harvey, G. R., and W. G. Steinhauer, 1974. Atmospheric transport of polychlorobiphenyls to the North Atlantic. *Atmos. Environ.* **8,** 287–290.

Hutchinson, G. E. 1973. Eutrophication. *Am. Scientist,* **61,** 269–279.

Hutchinson, G. L., and F. G. Viets, Jr. 1969. Nitrogen enrichment of surface water by absorption of ammonia volatilized from cattle feedlots. *Science,* **166,** 514–515.

Imboden, D. M. 1974. Phosphorus model of lake eutrophication. *Limnol. Oceanogr.,* **19**(2), 297–304.

Jensen, S. 1972. The PCB story. *Ambio.,* **1,** 123–131.

Jernelov, A., and H. Laum. 1973. Studies in Sweden on feasibility of some methods for restoration of mercury-contaminated bodies of water. *Environ. Sci. Technol.,* **7,** 712–718.

Keeney, D. R. 1973. The nitrogen cycle in sediment-water systems. *J. Environ. Qual..* **2**(1), 15–29.

Keeney, D. R., and R. L. Chen. 1974. The fate of nitrate in lake sediment columns. *Water Res. Bull.* **10**(6), 1162–1172.

Likens, G. E., and F. H. Bormann. 1974. Linkages between terrestrial and aquatic ecosystems. *BioScience,* **24**(8), 447–456.

Likens, G. E., F. H. Bormann, and N. M. Johnson. 1969. Nitrification: Importance to nutrient losses from a cut over forest ecosystem. *Science,* **163,** 1205–1206.

Likens, G. E., F. H. Bormann, R. S. Pierce, J. S. Eaton, and N. M. Johnson. 1977. *Biogeochemistry of a Forested Ecosystem.* Springer-Verlag, New York, 147 pp.

Lloyd-Jones, C. P. 1971. Evaporation of DDT. *Nature,* **229,** 65–66.

Middlebroods, E. J., D. H. Falkenborg, and T. E. Maloney. 1976. *Biostimulation and Nutrient Assessment.* Ann Arbor Science, Publ., Ann Arbor, 390 pp.

Mitchell, R. 1974. *Introduction to Environmental Microbiology.* Prentice-Hall, Englewood Cliffs, N.J., 385 pp.

Navratil, M., and J. Doblas. 1973. Development of pleural hyalinosis in long term studies of persons exposed to asbestos dust. *Environ. Res.,* **6,** 455–472.

Nriagu, J. O. (ed.). 1976. *Environmental Biogeochemistry.* Vol. 1. *Carbon, Nitrogen, Phosphorus, Sulfur, and Selenium Cycles.* Ann Arbor Science Publ., Ann Arbor, 423 pp.

O'Brien, W. J. 1972. Limiting factors in phytoplankton algae: Their meaning and measurement. *Science,* **178,** 616–617.

Parker, H. 1975. *Wastewater Systems Engineering.* Prentice-Hall, Englewood Cliffs, N.J. 412 pp.

Peakall, D. B. 1967. Pesticide induced enzyme breakdown of steroids in birds. *Nature,* **216,** 505–506.

Peakall, D. B. 1970. Pesticides and the reproduction of birds. *Sci. Am.,* **222**(4), 73–78.

Quraishi, M. S. 1977. *Biochemical Insect Control.* Wiley, New York; 280 pp.

Rodgers, J. L., Jr. 1976. *Environmental Impact Assessment, Growth Management, and the Comprehensive Plan.* Ballinger, Cambridge, Mass. 185 pp.

Rohl, A. N., A. M. Langer, and I. J. Sclikoff. 1977. Environmental asbestos pollution related to use of quarried serpentine rock. *Science,* **196,** 1319–1322.

Sawyer, C. N. 1947. Fertilization of lakes by agricultural and urban drainage. *J. New Engl. Water Works Assoc.,* **61,** 109–127.

Schindler, D. W. 1974. Eutrophication and recovery in experimental lakes: Implications for lake management. *Science,* **184,** 897–899.

Sebetich, M. J. 1975. Phosphorus kinetics of freshwater microcosms. *Ecology,* **56,** 1262–1280.

Sprangler, D. G., J. L. Spigarelli, J. M. Rose, and H. M. Miller, 1973. Methyl mercury: bacterial degradation in lake sediments. *Science,* **180,** 192–193.

Suffet, I. H. (ed.). 1977. *Fate of Pollutants in the Air and Water Environments.* Part 1. Wiley & Sons, New York, 484 pp.

Syers, J. K., R. F. Harris, and D. E. Armstrong. 1973. Phosphate chemistry in lake sediments. *J. Environ. Qual.,* **2**(1), 1–14.

True, W. P. (ed.). 1960. *The Smithsonian Treasury of science.* Vol. II. *Earth and Life.* Simon and Schuster, New York, 834 pp.

Vitousek, P. M., and W. A. Reiners. 1975. Ecosystems succession and nutrient retention: A hypothesis. *BioScience,* **25**(6), 376–381.

Vollenweider, R. A. 1968. *Scientific Fundamentals of the Eutrophication of Lakes and Flowing Waters, with Particular Reference to Nitrogen and Phosphorus as Factors in Eutrophication.* (DAS/CSI/68.27) Organization for Economic Cooperation and Development, Paris, 159 pp.

Vollenweider, R. A. 1975. Input–output models with special reference to the phosphorus loading concept in limnology. *Hydrologie,* **37**(1), 53–84.

Shelpdale, D. M. 1972. The contribution made by airborne pollutants to the pollution of large bodies of water. *Atmosphere,* **10**(1), 18–22.

Woodwell, G. M., P. P. Craig, and H. A. Johnson. 1971. DDT in the biosphere: Where does it go? *Science,* **174,** 1101–1107.

Woodwell, G. M., P. P. Craig, and H. A. Johnson. 1972. Atmospheric circulation of DDT. *Science,* **177,** 725.

Appendix. Conversion Factors

The symbols and units of measure that we use to describe and to study ecology change and seem to expand on virtually a daily basis. The following appendix of information is provided as an aid in translating these symbols and units. The first section provides a summary of the *Système Internationale* and those prefixes used, as well as a condensed conversion table of the most commonly used units. The second section provides a fairly comprehensive conversion table, including both common and some obscure units. The third section provides a metric conversion table pertaining most specifically to the dispersion of various forms of air pollutants but which is also useful in other areas, including the even distribution of a substance over a surface.

Section One

General SI (Système Internationale) Conversion Tables

SI Base Units

Quantity	Name of unit	Unit symbol
Length	metre	m
Mass	kilogram	kg
Time	second	s

Quantity	Name of unit	Unit symbol
Electric current	Ampere	A
Thermodynamic temperature	Kelvin	K
Luminous intensity	candela	cd
Amount of	mole	mol

Approved Numerical Prefixes

Exponential expression	Decimal equivalent	Prefix	Phonic	Symbol
10^{12}	1 000 000 000 000	tera	ter' a	T
10^{9}	1 000 000 000	giga	ji' ga	G
10^{6}	1 000 000	mega	meg' a	M*
10^{3}	1 000	kilo	kil' o	k*
10^{2}	100	hecto	hek' to	n
10	10	deka	dek' a	da
10^{-1}	0.1	deci	des' i	d
10^{-2}	0.01	centi	sen' ti	c*
10^{-3}	0.001	milli	mil' i	m*
10^{-6}	0.000 001	micro	mi' kro	μ*
10^{-9}	0.000 000 001	nano	nan' o	n
10^{-12}	0.000 000 000 001	pico	pe' ko	p
10^{-15}	0.000 000 000 000 001	femto	fem' to	f
10^{-18}	0.000 000 000 000 000 001	atto	at' to	a

*Most commonly used units.

Length

Units	Symbol	Meters
Angstrom	A	1×10^{-10}
Millimicron	mμ	1×10^{-9}
Micron	μ	.000001
Millimeter	mm	.001
Centimeter	cm	.01
Decimeter	dm	.10
Meter	m	1.0
Dekameter	dkm	10
Hectometer	hm	100
Kilometer	km	1000

Weight

Units	Symbol	Grams
Microgram	μg	.000001
Milligram	mg	.001
Centigram	cg	.01
Decigram	dg	.10
Gram	g	1.0
Dekagram	dkg	10
Hectogram	hg	100
Kilogram	kg	1000

Volume

Units	Symbol	Liters
Microliter	μl	.000001
Milliliter	ml	.001
Centiliter	cl	.01
Deciliter	dl	.10
Liter	l	1.0
Dekaliter	dkl	10.0
Hectoliter	hl	100.0
Kiloliter	kl	1000.0

Conversion Tables

Length

Unit	Centimeters	Meters	Kilometers	Inches	Feet
Centimeter	1.00	0.01	0.00001	0.39370	0.032808
Meter	100.00	1.00	0.001	39.370	3.2808
Kilometer	100000.	1000.00	1.00	39370.0	3280.8
Inch	2.54	0.02540	0.000025	1.00	0.08333
Foot	30.4801	0.304801	0.000305	12.00	1.00

1 angstrom = 0.0001 microns = 1×10^{-8} centimeters = 3.937×10^{-9} inches
1 micron = 1×10^4 angstroms = 1×10^{-4} centimeters = 3.397×10^{-5} inches
1 kilometer = 0.62137 mile 1 mile (U.S.) = 1.60935 kilometers

Mass and Weight

Unit	Grams	Grains avoir.	Drams avoir.	Ounces avoir.	Ounces Troy
Milligram	0.001	0.015432	0.000564	0.0000353	0.0000322
Gram	1.00	15.43236	0.564383	0.0352740	0.0321507
Kilogram	1000.0	15432.36	564.383	35.2740	32.1507
Dram (Av.)	1.771845	27.34375	1.00	0.0625	0.056966
Ounce (Av.)	28.3495	437.5	16.00	1.00	0.911458
Pound (Av.)	453.5924	7000.0	256.00	16.00	14.5833
Ounce (Tr.)	31.1035	480.00	17.55428	1.09714	1.00

1 cu. ft water at 60°F = 62.37 lbs av. 1 gal (U.S.) water at 62°F = 8.337 lbs av.

Volume and Capacity

Unit	Liters	Fl.ounce (U.S.)	Liq.pints (U.S.)	Cubic centimeters	Cubic inches
Milliliter	0.001	0.0338147	0.002113	1.00003	0.061025
Liter	1.00	33.8147	2.1134	1000.03	61.025
Cu. cm (cc)	0.001	0.0338147	0.002113	1.00	0.061023
Cu. inch	0.0163867	0.554113	0.034632	16.3872	1.00
Fl. dram (U.S.)	0.003696	0.125	0.007813	3.6967	0.225586
Fl. oz (U.S.)	0.0295729	1.00	0.0625	29.5737	1.80469
Liq. pint (U.S.)	0.473167	16.00	1.00	473.179	28.875
Liq. gal (U.S.)	3.7853	128.00	8.00	3785.43	231.00

1 gal (U.S.) = 0.83268 gals (British)   1 gal (British) = 1.20094 gals (U.S.)
1 minim = 0.0020833 Oz (U.S. fl.) = 0.016667 dram (U.S. fl.) = 0.061610 ml = 0.061612 cc

Area

Units	Square centimeters	Square inches
Sq. Cm.	1.00	.1550
Sq. Meter	10000.00	1550.00
Sq. Inch	6.451626	1.00
Sq. Foot	929.0341	144.00

Pressure

1 atmosphere = 14.696 lbs/sq. in.
1 atmosphere = 1.01325×10^6 dynes/sq. cm
1 atmosphere = 760.00 mm of mercury at 0°C
1 cm of mercury at 0°C = 0.013158 atmospheres

Section Two

To convert:	Into:	Multiply by:
	A	
Abcoulomb	statcoulombs	2.998×10^{10}
Acre (U.S.)	sq. chain (Gunters)	10.0
Acre	rods^2	160.0
Acre	square links (Gunters)	1×10^5
Acre	hectare or sq. hectometer	0.4047
Acres	sq. feet	43,560.0
Acres	sq. meters	4047.0
Acres	sq. miles	1.5625×10^{-3}
Acres	sq. yards	4840.0
Acres	cuerdas	1.0303
Acre-feet	cu. feet	43,560.0

To convert:	Into:	Multiply by:
Acre-feet	gallons	3.259×10^5
Amperes/sq. cm.	amps/sq. in.	6.452
Amperes/sq. cm.	amps/sq. meter	10^4
Amperes/sq. in.	amps/sq. cm	0.1550
Amperes/sq. in.	amps/sq. meter	1550.0
Amperes/sq. meter	amps/sq. cm	10^4
Amperes/sq. meter	amps/sq. in.	6.452×10^4
Ampere -hr	coulombs	3600.0
Ampere - hr	faradays	0.03731
Ampere/turns	gilberts	1.257
Ampere-turns/cm	amp-turns/in.	2.540
Ampere-turns/cm	amp-turns/meter	100.0
Ampere-turns/cm	gilberts/in	1.257
Ampere-turns/in	amp-turns/cm	0.3937
Ampere-turns/in.	amp-turns/meter	39.37
Ampere-turns/in.	gilberts/cm	0.4950
Ampere-turns/meter	amp-turns/cm	0.01
Ampere-turns/meter	amp-turns/in.	0.0254
Ampere-turns/meter	gilberts/cm	0.01257
Angstrom unit	inch	3937×10^{-9}
Angstrom unit	meter	1×10^{-10}
Angstrom unit	Micron (or μ)	1×10^{-4}
Anker (beer)	gallons	10.0
Are	acre (US)	0.02471
Ares	sq. yards	119.60
Ares	acres (British)	0.02471
Ares	sq. meters	100.0
Astronomical unit	kilometers	1.495×10^8
Atmospheres	ton sq. inch	0.007348
Atmospheres	cm mercury	76.0
Atmospheres	ft water (at 4°C)	33.90
Atmospheres	in. mercury (at 0°C)	29.92
Atmospheres	kg/sq. cm	1.0333
Atmospheres	kg/sp. meter	10,332
Atmospheres	pounds/sq. in.	14.70
Atmospheres	tons/sq ft	1.058

B

Bag (British)	Bushels	3
Barrels (U.S. dry)	cu. inches	7056.0
Barrels (U.S. dry)	quarts (dry)	105.0
Barrels (U.S. liquid)	gallons	31.5
Barrels (oil)	gallons (oil)	42.0
Bars	atmospheres	0.9869
Bars	dynes/sq. cm	10^6
Bars	kg/sq. meter	1.020×10^4
Bars	pounds/sq. ft	2089.0
Bars	pounds/sq. in.	14.50
Baryl	dyne/sq. cm	1.000
Bolt (U.S. cloth)	meters	36.576
BTU	liter-atmosphere	10.409
BTU	ergs	1.0550×10^{10}

To convert:	Into:	Multiply by:
BTU	foot-lbs	7.97
BTU	gram-calories	252.0
BTU	horsepower-hr	3.929×10^{-4}
BTU	joules	1054.8
BTU	kilogram-calories	0.2520
BTU	kilogram-meters	107.56
BTU	kilowatt-hr	2.928×10^{-4}
BTU/hr	foot-pounds/s	0.2162
BTU/hr	gram-cal/s	0.0700
BTU/hr	horsepower-hr	3.929×10^{-4}
BTU-hr	watts	0.2931
BTU/min	foot-lbs/s	12.96
BTU/min	horsepower	0.02358
BTU/min	kilowatts	0.01757
BTU/min	watts	17.58
BTU/sq. ft/min	watts/sq. in.	0.1221
Bucket (British dry)	cubic cm	1.818×10^4
Bushels (U.S. dry)	cu. ft	1.2444
Bushels	cu. in	2150.4
Bushels	cu. meters	0.03524
Bushels	liters	35.24
Bushels	pecks	4.0
Bushels	pints (dry)	64.0
Bushels	quarts (dry)	32.0
Butt (British, dry)	gallons	126

C

To convert:	Into:	Multiply by:
Calories, gram (mean)	BTU (mean)	3.9685×10^{-3}
Candle/sq cm	Lamberts	3.142
Candle/sq in	Lamberts	0.4870
Carat (metric)	milligrams	200.0
Centares (centiares)	sq. meters	1.0
Centigrade	Fahrenheit	$(°C \times 9/5) + 32$
Centigrams	grams	0.01
Centiliter	ounce fluid (US)	0.3382
Centiliter	cubic inch	0.6103
Centiliter	drams	2.705
Centiliters	liters	0.01
Centimeters	feet	3.281×10^{-2}
Centimeters	inches	0.3937
Centimeters	kilometers	10^{-5}
Centimeters	meters	0.01
Centimeters	miles	6.214×10^{-6}
Centimeters	millimeters	10.0
Centimeters	mils	393.7
Centimeters	yards	1.094×10^{-2}
Centimeter-dynes	cm-grams	1.020×10^{-3}
Cental	pounds	100
Cental	kilograms	45.359
Centimeter-dynes	meter-kg	1.020×10^{-8}
Centimeter-dynes	pound-feet	7.376×10^{-8}
Centimeter-grams	cm-dynes	980.7

To convert:	Into:	Multiply by:
Centimeter-grams	meter-kg	10^{-3}
Centimeter-grams	pound-feet	7.233×10^{-3}
Centimeters of mercury	atmospheres	0.01316
Centimeters of mercury	feet of water	0.4460
Centimers of mercury	kg/sq. meter	136.0
Centimeters of mercury	pounds/sq. ft	27.85
Centimeters of mercury	pounds/sq. in	0.1934
Centimeters/s	feet/min	1.1969
Centimeters/s	feet/s	0.03281
Centimeters/s	kilometers/hr	0.036
Centimeters/s	knots	0.1943
Centimeters/s	meters/min	0.6
Centimeters/s	miles/hr	0.02237
Centimeters/s	miles/min	3.728×10^{-4}
Centimeters/s/s	feet/s/s	0.03281
Centimeters/s/s	km/hr/s	0.036
Centimeters/s/s	meters/s/s	0.01
Centimeters/s/s	miles/hr/s	0.02237
Chain	inches	792.00
Chain	meters	20.12
Chains (surveyors' or Gunter's)	yards	22.00
Chaldron (U.S. dry)	meters³	1.2686
Chaldron (British dry)	meters³	1.1638
Circular mils	sq. cm	5.067×10^{-6}
Circular mils	sq. mils	0.7854
Circumference	radians	6.283
Circular mils	sq. inches	7.854×10^{-7}
Coomb (British dry)	meters³	0.145
Cords	cord feet	8
Cord feet	cu. feet	16
Coulomb	abcoulomb	0.1
Coulomb	statcoulombs	2.998×10^{9}
Coulombs	faradays	1.036×10^{-5}
Coulombs/sq. cm	coulombs/sq. in.	64.52
Coulombs/sq. cm	coulombs/sq. meter	10^{4}
Coulombs/sq. in.	coulombs/sq. cm	0.1550
Coulombs/sq. in.	coulombs/sq. meter	1550
Coulombs/sq. meter	coulombs/sq. cm	10^{-4}
Coulombs/sq. meter	coulombs/sq. in	6.452×10^{-4}
Cubic centimeters	cu. feet	3.531×10^{-5}
Cubic centimeters	cubic inches	0.06102
Cubic centimeters	cu. meters	10^{-6}
Cubic centimeters	cu. yards	1.308×10^{-6}
Cubic centimeters	gallons (U.S. liq.)	2.642×10^{-4}
Cubic centimeters	liters	0.001
Cubic centimeters	pints (U.S. liq.)	2.113×10^{-3}
Cubic centimeters	quarts (U.S. liq.)	1.057×10^{-3}
Cubic feet	bushels (dry)	0.8036
Cubic feet	cu. cm	28,317.0
Cubic feet	cu. inches	1728.0
Cubic feet	cu. meters	0.02832
Cubic feet	cu. yards	0.03704

To convert:	Into:	Multiply by:
Cubic feet	gallons (U.S. liq.)	7.48052
Cubit	inches	18.0
Cuerda	acres	0.97
Cuerda	meters^2	3930.39
Cubic feet	liters	28.32
Cubic feet	pints (U.S. liq.)	59.84
Cubic feet	quarts (U.S. liq.)	29.92
Cubic feet/min	cu. cm/s	472.0
Cubic feet/min	gallons/s	0.1247
Cubic feet/min	liters/s	0.4720
Cubic feet/min	pounds of water/min	62.43
Cubic feet/s	million gal/day	0.646317
Cubic feet/s	gallons/min	448.831
Cubic inches	cu. cm	16.39
Cubic inches	cu. feet	5.787×10^{-4}
Cubic inches	cu. meters	1.639×10^{-5}
Cubic inches	cu. yards	2.143×10^{-5}
Cubic inches	gallons	4.329×10^{-3}
Cubic inches	liters	0.01639
Cubic inches	mil-feet	1.061×10^{3}
Cubic inches	pints (U.S. liq.)	0.03464
Cubic inches	quarts (U.S. liq.)	0.01732
Cubic meters	bushels (dry)	28.38
Cubic meters	cu. cm	10^{6}
Cubic meters	cu. feet	35.31
Cubic meters	cu. inches	61,023.0
Cubic meters	cu. yards	1.308
Cubic meters	gallons (U.S. liq.)	264.2
Cubic meters	liters	1000.0
Cubic meters	pints (U.S. liq.)	2113.4
Cubic meters	quarts (U.S. liq.)	1057.0
Cubic yards	cu. cm	7.646×10^{5}
Cubic yards	cu. feet	27.0
Cubic yards	cu. inches	46,656.0
Cubic yards	cu. meters	0.7646
Cubic yards	gallons (U.S. liq.)	202.0
Cubic yards	liters	764.6
Cubic yards	pints (U.S. liq.)	1615.9
Cubic yards	quarts (U.S. liq.)	807.9
Cubic yards/min	cubic ft/s	0.45
Cubic yards/min	gallons/s	3.367
Cubic yards/min	liters/s	12.74

D

Dalton	gram	1.650×10^{-24}
Days (tropical, mean solar)	seconds	86,400.0
Days (sidereal, mean solar)	seconds	86,164.0
Decigrams	grams	0.1
Deciliters	liters	0.1
Decimeters	meters	0.1

To convert:	Into:	Multiply by:
Decistere	meters^3	1.0
Degrees (angle)	quandrants	0.01111
Degrees (angle)	radians	0.01745
Degrees (angle)	seconds	3600.0
Degrees/s	radians/s	0.01745
Degrees/s	revolutions/min	0.1667
Degrees/s	revolutions/s	2.778×10^{-3}
Dekagrams	grams	10.0
Dekaliters	liters	10.0
Dekameters	meters	10.0
Drachm	minims	60
Drams (apothecaries' or Troy)	ounces (avoirdupois)	0.1371429
Drams (apothecarie's or Troy)	ounces (Troy)	0.125
Drams (U.S. fluid or apoth.)	cubic cm	3.6967
Drams	grams	1.7718
Drams	grains	27.3437
Drams	ounces (avoir.)	0.0625
Dyne/cm	erg/sq. millimeter	0.01
Dyne/sq. cm	atmospheres	9.869×10^{-7}
Dyne/sq. cm	inch of mercury at 0°C	2.953×10^{-5}
Dyne/sq. cm	inch of water at 4°C	4.015×10^{-4}
Dynes	grams	1.020×10^{-3}
Dynes	joules/cm	10^{-7}
Dynes	joules/meter (newtons)	10^{-5}
Dynes	kilograms	1.020×10^{-6}
Dynes	poundals	7.233×10^{-5}
Dynes	pounds	2.248×10^{-6}
Dynes/sq. cm	bars	10^{-6}

E

To convert:	Into:	Multiply by:
Ell	cm	114.30
Ell	inches	0.45
Em, pica	inch	0.167
Em, pica	cm	0.4233
Erg/s	dyne-cm/s	1.000
Ergs	BTU	9.481×10^{-11}
Ergs	dyne-centimeters	1.0
Ergs	foot-pounds	7.376×10^{-8}
Ergs	gram-calories	0.2389×10^{-7}
Ergs	gram-cm	1.020×10^{-3}
Ergs	horsepower-hr	3.7250×10^{-14}
Ergs	joules	10^{-7}
Ergs	kg calories	2.389×10^{-11}
Ergs	kg meters	1.020×10^{-8}
Ergs	kilowatt hr	0.2778×10^{-13}
Ergs	watt hours	0.2778×10^{-16}
Ergs/s	BTU/min	5.688×10^{-9}
Ergs/s	ft lbs/min	4.425×10^{-6}
Ergs/s	ft lbs/s	7.3756×10^{-8}

To convert:	Into:	Multiply by:
Ergs/s	horsepower	1.341×10^{-10}
Ergs/s	kg calories/min	1.433×10^{-9}
Ergs/s	kilowatts	10^{-10}

F

Farads	microfarads	10^4
Faraday/s	ampere (absolute)	9.6500×10^4
Faradays	ampere-hours	26.80
Faradays	coulombs	9.649×10^4
Fathom	meter	1.828804
Fathoms	feet	6.0
Feet	centimeters	30.48
Feet	kilometers	3.048×10^{-4}
Feet	meters	0.3048
Feet	miles (naut.)	1.645×10^{-4}
Feet	miles (stat.)	1.894×10^{-4}
Feet	millimeters	304.8
Feet	mils	1.2×10^4
Feet of water	atmospheres	0.02950
Feet of water	in. of mercury	0.8826
Feet of water	kg/sq. cm	0.03048
Feet of water	kg/sq. meter	304.8
Feet of water	pounds/sq. ft	62.43
Feet of water	pounds/sq. in.	0.4335
Feet/min	cm/s	0.5080
Feet/min	feet/s	0.01667
Feet/min	km/hr	0.01829
Feet/min	meters/min	0.3048
Feet/min	miles/hr	0.01136
Feet/s	cm/s	30.48
Feet/s	km/hr	1.097
Feet/s	knots	0.5921
Feet/s	meters/min	18.29
Feet/s	miles/hr	0.6818
Feet/s	miles/min	0.01136
Feet/s/s	cm/s/s	30.48
Feet/s/s	km/hr/s	1.097
Feet/s/s	meters/s/s	0.3048
Feet/s/s	miles/hr/s	0.6818
Feet/100 feet	percent grade	1.0
Flagon (beer)	pint	2.0
Foot-candle	lumen/sq. meter	10.764
Foot-pounds	BTU	1.285×10^{-3}
Foot-pounds	ergs	1.356×10^7
Foot-pounds	gram-calories	0.3239
Foot-pounds	hp-hr	5.051×10^{-7}
Foot-pounds	joules	1.356
Foot-pounds	kg-calories	3.24×10^{-4}
Foot-pounds	kg-meters	0.1383
Foot-pounds	kilowatt-hr	3.766×10^{-7}
Foot-pounds/min	BTU/min	1.286×10^{-3}
Foot-pounds/min	foot-pounds/s	0.01667

To convert:	Into:	Multiply by:
Foot-pounds/min	horsepower	3.030×10^{-5}
Foot-pounds/min	kg-calories/min	3.24×10^{-4}
Foot-pounds/min	kilowatts	2.260×10^{-5}
Foot-pounds/s	BTU/hr	4.6272
Foot-pounds/s	BTU/min	0.07712
Foot-pounds/s	horsepower	1.818×10^{-3}
Foot-pounds/s	kg-calories/min	0.01943
Foot-pounds/s	kilowatts	1.356×10^{-3}
Furlongs	miles (U.S.)	0.125
Furlongs	rods	40.0
Furlongs	feet	660.0

G

To convert:	Into:	Multiply by:
Gallons (U.S.)	cu. cm	3,785.4
Gallons	cu. feet	0.1337
Gallons	cu. inches	231.0
Gallons	cu. meters	3.785×10^{-3}
Gallons	cu. yards	4.951×10^{-3}
Gallons	liters	3.785
Gallons (liq. Br. Imp.)	gallons (U.S. liq.)	1.20094
Gallons (U.S.)	gallons (Imp.)	0.83268
Gallons of water	pounds of water	8.3453
Gallons/min	cu. ft/s	2.228×10^{-3}
Gallons/min	liters/s	0.06308
Gallons/min	cu. ft/hr	8.0208
Gausses	lines/sq. in.	6.452
Gausses	webers/sq. cm	10^{-8}
Gausses	webers/sq. in.	6.452×10^{-8}
Gausses	webers/sq. meter	10^{-4}
Gilberts	ampere-turns	0.7958
Gilberts/cm	amp-turns/cm	0.7958
Gilberts/cm	amp-turns/in.	2.021
Gilberts/cm	amp-turns/meter	79.58
Gills (British)	cubic cm	142.07
Gills	liters	0.1183
Gills	pints (liq.)	0.25
Grade	radian	0.01571
Grains	drams (avoirdupois)	0.03657143
Grains (Troy)	grains (avoir.)	1.0
Grains (Troy)	grams	0.06480
Grains (Troy)	ounces (avoir.)	2.0833×10^{-3}
Grains (Troy)	pennyweight (Troy)	0.04167
Grains/U.S. gal	parts/million	17.118
Grains/U.S. gal	pounds/million gal	142.86
Grains/Imp. gal	parts/million	14.286
Grams	dynes	980.7
Grams	grains	15.43
Grams	joules/cm	9.807×10^{-5}
Grams	joules/meter (newtons)	9.807×10^{-3}
Grams	kilograms	0.001
Grams	milligrams	1000.0
Grams	ounces (avoir.)	0.03527

To convert:	Into:	Multiply by:
Grams	ounces (Troy	0.03215
Grams	poundals	0.07093
Grams	pounds (avoir.)	2.205×10^{-3}
Grams/cm	pounds/inch	5.600×10^{-2}
Grams/cu. cm	pounds/cu. ft	62.43
Grams/cu. cm	pounds/cu. in.	0.03613
Grams/cu. cm	pounds/mil-foot	3.405×10^{-7}
Grams/liter	grams/gal	58.417
Grams/liter	pounds/1000 gal	8.345
Grams/liter	pounds/cu. ft	0.062427
Grams/liter	parts/million	1000.0
Grams/sq. cm	pounds/sq. ft	2.0481
Gram-calories	BTU	3.9685×10^{-3}
Gram-calories	ergs	4.1868×10^{7}
Gram-calories	foot-pounds	3.074
Gram-calories	horsepower-hr	1.5593×10^{-6}
Gram-calories	kilowatt-hr	1.1630×10^{-4}
Gram-calories	watt-hr	1.1630×10^{-3}
Gram-calories/s	BTU/hr	14.286
Gram-centimeters	BTU	9.297×10^{-8}
Gram-centimeters	ergs	980.7
Gram-centimeters	joules	9.807×10^{-5}
Gram-centimeters	kg-cal	2.343×10^{-8}
Gram-centimeters	kg-meters	10^{-5}

H

To convert:	Into:	Multiply by:
Hand	cm	10.160
Hectares	acres	2.471
Hectares	sq. feet	1.076×10^{5}
Hectograms	grams	100.0
Hectoliters	liters	100.0
Hectometers	meters	100.0
Hectowatts	watts	100.0
Henries	millihenries	1,000.0
Hogsheads (British)	cubic feet	10.114
Hogsheads (U.S.)	cubic feet	8.42184
Hogsheads (U.S.)	gallons (U.S.)	63
Horsepower	BTU/min	42.418
Horsepower	foot-lbs/min	33,000.
Horsepower	foot-lbs/s	550.0
Horsepower (metric) (542.5 ft-lb/s)	horsepower (550 ft-lb/s)	0.9863
Horsepower (550 ft-lb/s)	horsepower (metric) (542.5 ft-lb/s)	1.014
Horsepower	kg-calories/min	10.69
Horsepower	kilowatts	0.7457
Horsepower	watts	745.7
Horsepower (boiler)	BTU/hr	33.479
Horsepower (boiler)	kilowatts	9.803
Horsepower-hr	BTU	2545
Horsepower-hr	ergs	2.6845×10^{13}
Horsepower-hr	foot-lbs	1.98×10^{6}
Horsepower-hr	gram-calories	641,190

To convert:	Into:	Multiply by:
Horsepower-hr	joules	2.6845×10^6
Horsepower-hr	kg-calories	641.3
Horsepower-hr	kg-meters	2.737×10^5
Horsepower-hr	kilowatt-hr	0.7457
Hours	days	4.167×10^{-2}
Hours	weeks	5.952×10^{-3}
Hundredweights (long)	pounds	112
Hundredweights (long)	tons (long)	0.05
Hundredweights (short)	ounces (avoir.)	1600
Hundredweights (short)	pounds	100
Hundredweights (short)	tons (metric)	0.0453592
Hundredweights (short)	tons (long)	0.0446429

I

Inches	centimeters	2.54
Inches	meters	2.54×10^{-2}
Inches	miles	1.578×10^{-5}
Inches	millimeters	25.40
Inches	mils	1000.0
Inches	yards	2.778×10^{-2}
Inches of mercury	atmospheres	0.03342
Inches of mercury	feet of water	1.133
Inches of mercury	kg/sq. cm	0.03453
Inches of mercury	kg/sq. meter	345.3
Inches of mercury	pounds/sq. ft	70.73
Inches of mercury	pounds/sq. in.	0.4912
Inches of water (at 4°C)	atmospheres	2.458×10^{-3}
Inches of water (at 4°C)	inches of mercury	0.07355
Inches of water (at 4°C)	kg/sq. cm	2.540×10^{-3}
Inches of water (at 4°C)	ounces/sq. in	0.5782
Inches of water (at 4°C)	pounds/sq. ft	5.2022
Inches of water (at 4°C)	pounds/sq. in.	0.03613
International ampere	ampere (absolute)	0.9998
International volt	volts (absolute)	1.0003
International volt	joules (absolute)	1.593×10^{-19}
International volt	joules	9.654×10^4

J

Joules	BTU	9.480×10^{-4}
Joules	ergs	10^7
Joules	foot-pounds	0.7376
Joules	kg-calories	2.389×10^{-4}
Joules	kg-meters	0.1020
Joules	watt-hr	2.778×10^{-4}
Joules/cm	grams	1.020×10^4
Joules/cm	dynes	10^7
Joules/cm	joules/meter (newtons)	100.0
Joules/cm	poundals	723.3
Joules/cm	pounds	22.48

To convert:	Into:	Multiply by:
	K	
Kilderkin (British)	meters^3	0.082
Kilograms	dynes	980,665
Kilograms	grams	1000.0
Kilograms	joules/cm	0.09807
Kilograms	joules/meter (newtons)	9.807
Kilograms	poundals	70.93
Kilograms	pounds	2.205
Kilograms	tons (long)	9.842×10^{-4}
Kilograms	tons (short)	1.102×10^{-3}
Kilograms/cu. meter	grams/cu. cm	0.001
Kilograms/cu. meter	pounds/cu. ft	0.06243
Kilograms/cu. meter	pounds/cu. in.	3.613×10^{-5}
Kilograms/cu meter	pounds/mil-foot	3.405×10^{-10}
Kilograms/meter	pounds/ft	0.6720
Kilograms/sq. cm	dynes	980,665
Kilograms/sq. cm	atmospheres	0.9678
Kilograms/sq. cm	feet of water	32.81
Kilograms/sq. cm	inches of mercury	28.96
Kilograms/sq. cm	pounds/sq. ft	2048
Kilograms/sq. cm	pounds/sq. in.	14.22
Kilograms/sq. meter	atmospheres	9.678×10^{-5}
Kilograms/sq. meter	bars	98.07×10^{-6}
Kilograms/sq. meter	feet of water	3.281×10^{-3}
Kilograms/sq. meter	inches of mercury	2.896×10^{-3}
Kilograms/sq. meter	pounds/sq. ft	0.2048
Kilograms/sq. meter	pounds/sq. in.	1.422×10^{-3}
Kilograms/sq. mm	kg/sq. meter	10^6
Kilogram-calories	BTU	3.969
Kilogram-calories	foot-pounds	307.4
Kilogram-calories	hp hr	1.560×10^{-3}
Kilogram-calories	joules	4186
Kilograms-calories	kg-meters	426.9
Kilogram-calories	kilojoules	4.186
Kilogram-calories	kilowatt hr	1.163×10^{-3}
Kilogram meters	BTU	9.297×10^{-3}
Kilogram meters	ergs	9.807×10^7
Kilogram meters	foot-pounds	7.233
Kilogram-meters	joules	9.807
Kilogram-meters	kg-calories	2.343×10^{-3}
Kilogram-meters	kilowatt-hr	2.724×10^{-6}
Kilolines	maxwells	1000.0
Kiloliters	liters	1000.0
Kilometers	centimeters	10^5
Kilometers	feet	3281
Kilometers	inches	3.937×10^4
Kilometers	meters	1000.0
Kilometers	miles	0.6214
Kilometers	millimeters	10^4
Kilometers	yards	1094
Kilometers/hr	cm/s	27.78
Kilometers/hr	feet/min	54.68
Kilometers/hr	feet/s	0.9113

To convert:	Into:	Multiply by:
Kilometers/hr	knots	0.5396
Kilometers/hr	meters/min	16.67
Kilometers/hr	miles/hr	0.6214
Kilometers/hr/s	cm/s/s	27.78
Kilometers/hr/s	ft/s/s	0.9113
Kilometers/hr/s	meters/s/s	0.2778
Kilometers/hr/s	miles/hr/s	0.6214
Kilowatts	BTU/min	56.90
Kilowatts	foot-lbs/min	4.425×10^4
Kilowatts	foot-lbs/s	737.6
Kilowatts	horsepower	1.341
Kilowatts	kg-calories/min	14.33
Kilowatts	watts	1000.0
Kilowatt-hr	BTU	3413
Kilowatt-hr	ergs	3.600×10^{13}
Kilowatt-hr	foot-lbs	2.655×10^6
Kilowatt hr	gram-calories	8.6001×10^5
Kilowatt-hr	horsepower-hr	1.341
Kilowatt-hr	joules	3.6×10^6
Kilowatt-hr	kg-calories	860.5
Kilowatt-hr	kg-meters	3.671×10^5
Kilowatt-hr	pounds of water evaporated from and at 212°F	3.53
Kilowatt-hr	pounds of water raised from 62° to 212°F	22.75
Knots	centimeters/s	51.48
Knots	feet/hr	6080
Knots	kilometers/hr	1.8532
Knots	nautical miles/hr	1.0
Knots	statute miles/hr	1.1516
Knots	yards/hr	2027
Knots	feet/s	1.689

L

To convert:	Into:	Multiply by:
Large (beer)	pints	1.0
League (statute)	miles	3.0
Light year	miles	5.9×10^{12}
Light year	kilometers	9.464×10^{12}
Lines/sq. cm	gausses	1.0
Lines/sq. in.	gausses	0.1550
Lines/sq. in.	webers/sq. cm	1.550×10^{-9}
Lines/sq. in.	webers/sq. in.	10^{-8}
Lines/sq. in.	webers/sq. meter	1.550×10^{-5}
Links (engineers')	inches	12.0
Links (surveyors')	inches	7.92
Liters	bushels (U.S. dry)	0.02838
Liters	cu. cm	1000.0
Liters	cu. feet	0.03531
Liters	cu. inches	61.03
Liters	cu. meters	0.001
Liters	cu. yards	1.308×10^{-3}
Liters	gallons (U.S. liq.)	0.2642
Liters	pints (U.S. liq.)	2.113

To convert:	Into:	Multiply by:
Liters	quarts (U.S. liq.)	1.057
Liters/min	cu. ft/s	5.885×10^{-4}
Liters/min	gal/s	4.403×10^{-3}
Lumens/sq. ft	foot-candles	1.0
Lumen	spherical candle power	0.07958
Lumen	watt	0.001496
Lumen/sq. ft	lumen/sq. meter	10.76
Lux	foot-candles	0.0929

M

Maxwells	kilolines	0.001
Maxwells	webers	10^{-8}
Megalines	maxwells	10^6
Megohms	microhms	10^{12}
Megohms	ohms	10^6
Meters	centimeters	100.0
Meters	feet	3.281
Meters	inches	39.37
Meters	kilometers	0.001
Meters	miles (naut.)	5.396×10^{-4}
Meters	miles (stat.)	6.214×10^{-4}
Meters	millimeters	1000.0
Meters	yards	1.094
Meters	varas	1.179
Meters/min	cm/s	1.667
Meters/min	feet/min	3.281
Meters/min	feet/s	0.05468
Meters/min	km/hr	0.06
Meters/min	knots	0.03238
Meters/min	miles/hr	0.03728
Meters/s	feet/min	196.8
Meters/s	feet/s	3.281
Meters/s	kilometers/hr	3.6
Meters/s	kilometers/min	0.06
Meters/s	miles/hr	2.237
Meters/s	miles/min	0.03728
Meters/s/s	cm/s/s	100.0
Meters/s/s	ft/s/s	3.281
Meters/s/s	km/hr/s	3.6
Meters/s/s	miles/hr/s	2.237
Meter-kilograms	cm-dynes	9.807×10^7
Meter-kilograms	cm-grams	10^5
Meter-kilograms	pound-feet	7.233
Microfarad	farads	10^{-4}
Micrograms	grams	10^{-6}
Microhms	megohms	10^{-12}
Microhms	ohms	10^{-6}
Microliters	liters	10^{-6}
Microns	centimeters	1×10^{-4}
Miles (naut.)	feet	6080.27
Miles (naut.)	kilometers	1.853
Miles (naut.)	meters	1853

To convert:	Into:	Multiply by:
Miles (naut.)	miles (statute)	1.1516
Miles (naut.)	yards	2027
Miles (statute)	centimeters	1.609×10^5
Miles (statute)	feet	5280
Miles (statute)	inches	6.336×10^4
Miles (statute)	kilometers	1.609
Miles (statute)	meters	1609
Miles (statute)	miles (naut.)	0.8684
Miles (statute)	yards	1760
Miles/hr	cm/s	44.70
Miles/hr	feet/min	88
Miles/hr	feet/s	1.467
Miles/hr	km/hr	1.609
Miles/hr	km/min	0.02682
Miles/hr	knots	0.8684
Miles/hr	meters/min	26.82
Miles/hr	miles/min	0.1667
Miles/hr/s	cm/s/s	44.70
Miles/hr/s	feet/s/s	1.467
Miles/hr/s	km/hr/s	1.609
Miles/hr/s	meters/s/s	0.4470
Miles/min	cm/s	2682
Miles/min	feet/s	88
Miles/min	km/min	1.609
Miles/min	knots/min	0.8684
Miles/min	miles/hr	60.0
Mil-feet	cu. inches	9.425×10^{-6}
Milliers	kilograms	1000.0
Milligrams	grains	0.01543236
Milligrams	grams	0.001
Milligrams/liter	parts/million	1.0
Millihenries	henries	0.001
Milliliters	liters	0.001
Millimeters	centimeters	0.1
Millimeters	feet	3.281×10^{-3}
Millimeters	inches	0.03937
Millimeters	kilometers	10^{-6}
Millimeters	meters	0.001
Millimeters	miles	6.214×10^{-7}
Millimeters	mils	39.37
Millimeters	yards	1.094×10^{-3}
Millimicrons	meters	1×10^{-9}
Million gals/day	cu. ft/s	1.54723
Mils	centimeters	2.540×10^{-2}
Mils	feet	8.33×10^{-5}
Mils	inches	0.001
Mils	kilometers	2.540×10^{-8}
Mils	yards	2.778×10^{-5}
Miner's inches	cu. ft/min	1.5
Minims (British)	cubic cm	0.059194
Minims (U.S. fluid)	cubic cm	0.061612
Minutes (angles)	degrees	0.01667

To convert:	Into:	Multiply by:
Minutes (angles)	quadrants	1.852×10^{-4}
Minutes (angles)	radians	2.909×10^{-4}
Minutes (angles)	seconds	60.0
Minutes (time)	week	9.9206×10^{-5}
Mynograms	kilograms	10.0
Myriameters	kilometer	10.0
Myriawatts	kilowatts	10.0
Myriameters	meters	10^4

N

Nail (British)	inch	2.25
Nepers	decibels	8.686
Newton	dynes	1×10^5
Nip (beer)	pints	0.25
Noggin (British)	gallon	0.0313

O

Ohm (international)	Ohm (absolute)	1.0005
Ohms	megohms	10^4
Ohms	microhms	10^4
Ounces (avoir.)	drams	16.0
Ounces (avoir.)	grains	437.5
Ounces (avoir.)	grams	28.349527
Ounces (avoir.)	pounds	0.0625
Ounces (avoir.)	ounces (Troy)	0.9115
Ounces (avoir.)	tons (long)	2.790×10^{-5}
Ounces (avoir.)	tons (metric)	2.835×10^{-5}
Ounces (fluid)	cu. inches	1.805
Ounces (fluid)	liters	0.02957
Ounces (Troy)	grains	480.0
Ounces (Troy)	grams	31.103481
Ounces (Troy)	ounces (avoir.)	1.09714
Ounces (Troy)	pennyweights (troy)	20.0
Ounces (Troy)	pounds (Troy)	0.08333
Ounces/sq in.	dynes/sq. cm	4309
Ounces/sq. in.	pounds/sq. in.	0.0625

P

Parsec	miles	19×10^{12}
Parsec	kilometers	3.084×10^{13}
Parts/million	grains/U.S. gal	0.0584
Parts/million	grains/Imp. gal	0.07016
Parts/million	pounds/million gal	8.345
Pecks (British)	cubic inches	554.6
Pecks (British)	liters	9.091901
Pecks (U.S.)	bushels	0.25
Pecks (U.S.)	cubic inches	537.605
Pecks (U.S.)	liters	8.80958
Pecks (U.S.)	quarts (dry)	8
Pennyweights (troy)	grains	24.0
Pennyweights (troy)	ounces (Troy)	0.05
Pennyweights (troy)	grams	1.55517

To convert:	Into:	Multiply by:
Pennyweights (troy)	pounds (Troy)	4.1667×10^{-3}
Pints (dry)	cu. inches	33.60
Pints (liq.)	cu. cm	473.2
Pints (liq.)	cu. feet	0.0167
Pints (liq.)	cu. inches	28.87
Pints (liq.)	cu. meters	4.732×10^{-4}
Pints (liq.)	cu yards	6.188×10^{-4}
Pints (liq.)	gallons	0.125
Pints (liq.)	liters	0.4732
Pints (liq.)	quarts (liq.)	0.5
Poise	gram/cm-s	1.00
Pounds (avoir.)	ounces (Troy)	14.5883
Poundals	dynes	13.825×10^{3}
Poundals	grams	14.10
Poundals	joules/cm	1.383×10^{-3}
Poundals	joules meter (newtons)	0.1383
Poundals	kilograms	0.01410
Poundals	pounds	0.031081
Pounds	drams (avoir.)	256
Pounds	dynes	4.4482×10^{5}
Pounds	grains	7000
Pounds	grams	453.5924
Pounds	joules/cm	0.04448
Pounds	joules/meter (newtons)	4.448
Pounds	kilograms	0.4536
Pounds	ounces	16.0
Pounds	ounces (Troy)	14.5833
Pounds	poundals	32.17
Pounds (avoir.)	pounds (Troy)	1.21528
Pounds	tons (short)	0.0005
Pounds (Troy)	grains	5760
Pounds (Troy)	grams	373.24177
Pounds (Troy)	ounces (avoir.)	13.1657
Pounds (Troy)	ounces (Troy)	12.0
Pounds (Troy)	pennyweight (Troy)	240.0
Pounds (Troy)	pounds (avoir.)	0.822857
Pounds (Troy)	tons (long)	3.6735×10^{-4}
Pounds (Troy)	tons (metric)	3.7324×10^{-4}
Pounds (Troy)	tons (short)	4.1143×10^{-4}
Pounds of water	cu. feet	0.01602
Pounds of water	cu. inches	27.68
Pounds of water	gallons	0.1198
Pounds of water/min	cu. ft/s	2.670×10^{-4}
Pound-feet	cm/dynes	1.356×10^{-2}
Pound-feet	cm/gram	13,825
Pound-feet	meter-kg	0.1383
Pounds/cu. ft	gram/cu. cm	0.01602
Pounds/cu. ft	kg/cu. meter	16.02
Pounds/cu. ft	pounds/cu. in.	5.787×10^{-4}
Pounds/cu. in.	gm/cu. cm	27.68
Pounds/cu. in.	kg/cu. meter	2.768×10^{4}
Pounds/cu. in.	pounds/cu. ft	1728
Pounds/cu. in.	pounds/mil-foot	9.425×10^{-4}

To convert:	Into:	Multiply by:
Pounds/ft	kg-meter	1.488
Pounds/in.	gm/cm	178.6
Pounds/sq. ft	atmospheres	4.725×10^{-4}
Pounds/sq. ft	feet of water	0.01602
Pounds/sq. ft	inches of mercury	0.01414
Pounds/sq. ft	kg/sq. meter	4.882
Pounds/sq. ft	pounds/sq. in.	6.945×10^{-3}
Pounds/sq. in.	atmospheres	0.06805
Pounds/sq. in.	feet of water	2.307
Pounds/sq. in.	inches of mercury	2.036
Pounds/sq. in.	kg/sq. meter	703.1
Pounds/sq. in.	pounds/sq. ft	144.0

Q

To convert:	Into:	Multiply by:
Quadrants (angle)	degrees	90.0
Quadrants (angle)	minutes	5400.0
Quadrants (angle)	radians	1.571
Quadrants (angle)	seconds	3.24×10^{5}
Quarts (U.S. dry)	cu. inches	67.20
Quarts (U.S. liq.)	cu. cm	946.4
Quarts (U.S. liq.)	cu. feet	0.03342
Quarts (U.S. liq.)	cu. inches	57.75
Quarts (U.S. liq.)	cu. meters	9.464×10^{-4}
Quarts (U.S. liq.)	cu. yards	1.238×10^{-3}
Quarts (U.S. liq.)	gallons	0.25
Quarts (U.S. liq.)	liters	0.9463
Quire	sheets	25

R

To convert:	Into:	Multiply by:
Radians	degrees	57.30
Radians	minutes	3438
Radians	quadrants	0.6366
Radians	seconds	2.063×10^{5}
Radians/s	degrees/s	57.30
Radians/s	revolutions/min	9.549
Radians/s	revolutions/s	0.1592
Radians/s/s	rev/min/min	573.0
Radians/s/s	rev/min/s	9.549
Radians/s/s	rev/s/s	0.1592
Revolutions	degrees	360.0
Revolutions	quadrants	4.0
Revolutions	radians	6.283
Revolutions/min	degrees/s	6.0
Revolutions/min	radians/s	0.1047
Revolutions/min	rev/s	0.01667
Revolutions/min/min	radians/s/s	1.745×10^{-3}
Revolutions/min/min	rev/min/s	0.01667
Revolutions/min/min	rev/s/s	2.778×10^{-4}
Revolutions/s	degrees/s	360.0
Revolutions/s	radians/s	6.283
Revolutions/s	rev/min	60.0
Revolutions/s/s	radians/s/s	6.283

To convert:	Into:	Multiply by:
Revolutions/s/s	rev/min/min	3600.0
Revolutions/s/s	rev/min/s	60.0
Rod	chain (Gunters)	0.25
Rod	meters	5.029
Rods (surveyors')	yards	5.5
Rods	feet	16.5
Rood (British)	acre	0.25
Rood (British)	perches^2	40

S

To convert:	Into:	Multiply by:
Scruples	grains	20
Seconds (angle)	degrees	2.778×10^{-4}
Seconds (angle)	minutes	0.01667
Seconds (angle)	quadrants	3.087×10^{-4}
Seconds (angle)	radians	4.848×10^{-6}
Skein	feet	360
Slug	kilogram	14.59
Slug	pounds	32.17
Small (beer)	pint	0.5
Sphere	steradians	12.57
Square centimeters	circular mils	1.974×10^5
Square centimeters	sq. feet	1.076×10^{-3}
Square centimeters	sq. inches	0.1550
Square centimeters	sq. meters	0.0001
Square centimeters	sq. miles	3.861×10^{-11}
Square centimeters	sq. millimeters	100.0
Square centimeters	sq. yard	1.196×10^{-4}
Square feet	acres	2.296×10^{-5}
Square feet	sq. cm	929.0
Square feet	sq. inches	144.0
Square feet	sq. meters	0.09290
Square feet	sq. miles	3.587×10^{-8}
Square feet	sq. millimeters	9.290×10^4
Square feet	sq. yards	0.1111
Square inches	sq. mils	1×10^6
Square inches	circular mils	1.273×10^6
Square inches	sq. cm	6.452
Square inches	sq. feet	6.944×10^{-3}
Square inches	sq. millimeters	645.2
Square inches	sq. yards	7.716×10^{-4}
Square kilometers	acres	247.1
Square kilometers	sq. cm	10^{10}
Square kilometers	sq. ft	1.076×10^6
Square kilometers	sq. inches	1.550×10^9
Square kilometers	sq. meters	10^6
Square kilometers	sq. miles	0.3861
Square kilometers	sq. yards	1.196×10^6
Square meters	acres	2.471×10^{-4}
Square meters	sq. cm	10^4
Square meters	sq. feet	10.76
Square meters	hectare	1×10^{-4}
Square meters	sq. inches	1550

To convert:	Into:	Multiply by:
Square meters	sq. miles	3.861×10^{-7}
Square meters	sq. millimeters	10^6
Square meters	sq. yards	1.196
Square miles	acres	640.0
Square miles	hectares	258.9998
Square miles	sq. feet	27.88×10^6
Square miles	sq. km	2.590
Square miles	sq. meters	2.590×10^6
Square miles	sq. yards	3.098×10^6
Square millimeters	circular mils	1973.5
Square millimeters	sq. cm	0.01
Square millimeters	sq. feet	1.076×10^{-3}
Square millimeters	sq. inches	1.550×10^{-3}
Square mils	circular mils	1.273
Square mils	sq. cm	6.452×10^{-6}
Square mils	sq. inches	10^{-6}
Square yards	acres	2.066×10^{-4}
Square yards	sq. cm	8361
Square yards	sq. feet	9.0
Square yards	sq. inches	1296
Square yards	sq. meters	0.8361
Square yards	sq. miles	3.228×10^{-7}
Square yards	sq. millimeters	8.361×10^5
Stone (British)	pounds (avoir.)	14

T

To convert:	Into:	Multiply by:
Temperature (Celsius) (°C) + 273.18	absolute temperature (°K)	1.0
Temperature (Celsius) (°C) + 17.78	temperature (°F)	1.8
Temperature (Fahrenheit) (°F) + 460	absolute temperature (°F)	1.0
Temperature (°F) − 32	temperature (°C)	5/9
Tons (long)	kilograms	1016
Tons (long)	pounds	2240
Tons (long)	tons (short)	1.120
Tons (metric)	kilograms	1.000
Tons (metric)	pounds	2.205
Tons (short)	kilograms	907.185
Tons (short)	ounces	32,000
Tons (short)	ounces (Troy)	29,166.66
Tons (short)	pounds	2000
Tons (short)	pounds (Troy)	2430.56
Tons (short)	tons (long)	0.89287
Tons (short)	tons (metric)	0.9072
Tons (short/sq. ft)	kg/sq. meter	9765
Tons (short)/sq. ft	pounds/sq. in.	13.889
Tons of water/24 hr	pounds of water/hr	83.333
Tons of water/24 hr	gallons/min	0.16643
Tons of water/24 hr	cu. ft/hr	1.3349
Tun (beer)	gallons	216
Tun (liquid)	gallons	252

To convert:	Into:	Multiply by:
V		
Volt/inch	volt/cm	0.39370
Volt (absolute)	statvolts	0.003336
W		
Watts	BTU/hr	3.4138
Watts	BTU/min	0.05696
Watts	ergs/s	10^7
Watts	foot-lbs/min	44.254
Watts	foot-lbs/s	0.7378
Watts	horsepower	1.341×10^{-3}
Watts	horsepower (metric)	1.596×10^{-3}
Watts	kg-calories/min	0.01433
Watts	kilowatts	0.001
Watts (abs.)	BTU (mean)/min	0.05696
Watts (abs.)	joules/s	1
Watt-hours	BTU	3.413
Watt-hours	foot-pounds	2655.3
Watt-hours	gram-calories	860.01
Watt-hours	horsepower/hr	1.341×10^{-3}
Watt-hours	kilogram-calories	0.86001
Watt-hours	kilogram-meters	367.10
Watt-hours	kilowatt-hr	0.001
Watt (international)	watt (absolute)	1.0002
Webers	maxwells	10^8
Webers	kilolines	10^5
Webers/sq. in.	gausses	1.550×10^7
Webers/sq. in.	lines/sq. in.	10^4
Webers/sq. in.	webers/sq. cm	0.1550
Webers/sq. in.	webers/sq. meter	1440
Webers/sq. meter	gausses	10^4
Webers/sq. meter	lines/sq. in.	6.452×10^4
Webers/sq. meter	webers/sq. cm	10^{-4}
Weber/sq. meter	weber/sq. in.	6.452×10^{-4}
Y		
Yards	centimeters	91.44
Yards	kilometers	9.144×10^{-4}
Yards	meters	0.9144
Yards	miles (naut.)	4.934×10^{-4}
Yards	miles (stat.)	5.682×10^{-4}
Yards	millimeters	914.4
Year (leap)	days	366
Year (tropical, mean solar)	days (mean solar)	365.2422
Year (tropical, mean solar)	hours (mean solar)	8765.8128
Year (sidereal)	hours (mean solar)	8766.144

Section Three

Common Metric Conversions for Consideration of Air Pollution

Air Quality Data (25°C; 760 mm Hg)

Ppm SO_2	× 2620	= $\mu g/m^3$ SO_2	(sulfur dioxide)
Ppm CO	× 1150	= $\mu g/m^3$ CO	(carbon monoxide)
Ppm CO_x	× 1.15	= mg/m^3 CO	(carbon monoxide)
Ppm CO_2	× 1800	= $\mu g/m^3$ CO_2	(carbon dioxide)
Ppm CO_2	× 1.8	= mg/m^3 CO_2	(carbon dioxide)
Ppm NO	× 1230	= $\mu g/m^3$ NO	(nitrogen oxide)
Ppm NO_2	× 1880	= $\mu g/m^3$ NO_2	(nitrogen dioxide)
Ppm O_3	× 1960	= $\mu g/m^3$ O_3	(ozone)
Ppm CH_4	× 655	= $\mu g/m^3$ CH_4	(methane)
Ppm CH_4	× 0.655	= mg/m^3 CH_4	(methane)
Ppm CH_3SH	× 2000	= $\mu g/m^3$ CH_3SH	(methyl mercaptan)
Ppm C_3H_8	× 1800	= $\mu g/m^3$ C_3H_8	(propane)
Ppm C_3H_8	× 1.8	= mg/m^3 C_3H_8	(propane)
Ppm F^-	× 790	= $\mu g/m^3$ F^-	(fluoride)
Ppm H_2S	× 1400	= $\mu g/m^3$ H_2S	(hydrogen sulfide)
Ppm NH_3	× 696	= $\mu g/m^3$ NH_3	(ammonia)
Ppm HCHO	× 1230	= $\mu g/m^3$ HCHO	(formaldehyde)

For other elements and/or compounds:

$$ppm \times \frac{M}{0.02445} = \mu g/m^3, \text{ where } M = \text{molecular weight}$$

Other Useful Conversion Factors

To convert from:	To:	Multiply by[a]:
Ppm by volume (20°C)	milligrams/cu. m	$\dfrac{M}{24.04}$
	micrograms/cu. m	$\dfrac{M}{0.02404}$
	micrograms/liter	$\dfrac{M}{24.04}$
	ppm by weight	$\dfrac{M}{28.8}$
	pounds/cu. ft	$\dfrac{M}{385.1 \times 10^6}$
Ppm by weight	milligrams/cu. m	1.198
	micrograms/cu. m	1.198×10^3
	micrograms/liter	1.198
	ppm by volume (20°C)	$\dfrac{28.8}{M}$
	pounds/cu. ft	7.48×10^{-6}
Pounds/cu. ft	milligrams/cu. m	16.018×10^6
	micrograms/cu. m	16.018×10^9
	micrograms/liter	16.018×10^6
	ppm by volume (20°C)	$\dfrac{385.1 \times 10^6}{M}$
	ppm by weight	133.7×10^3

Particle count

To convert from:	To:	Multiply by:
No./cu. m	no./liter	0.001
	no./cu. cm	1.0×10^{-6}
	no./cu. ft	28.317×10^{-3}
No./liter	no./cu. m	1000.0
	no./cu. cm	0.001
	no./cu. ft	28.316
No./cu. cm	no./cu. m	1.0×10^6
	no./liter	1000.0
	no./cu. ft	28.316×10^3
No./cu. ft	no./cu. m	35.314
	no./liter	35.315×10^{-3}
	no./cu. cm	35.314×10^{-6}

Dustfall

To convert from:	To:	Multiply by:
Tons/sq. mile	pounds/acre	3.125
	pounds/1000 sq. ft	0.07174
	grams/sq. m	0.3503
	kilograms/sq. km	350.3
	milligrams/sq. m	350.3
	milligrams/sq. cm	0.03503
	grams/sq. ft	0.03254
Pounds/acre	5 tons/sq. mile	0.32
	pounds/1000 sq. ft	0.023

To convert from:	To:	Multiply by[a]:
	grams/sq. m	0.1121
	kilograms/sq. km	112.1
	milligrams/sq. m	112.1
	milligrams/sq. cm	0.01121
	grams/sq. ft	0.0104
Pounds/1000 sq. ft	tons/sq. mile	13.94
	pounds/acre	43.56
	grams/sq. m	4.882
	kilograms/sq. km	4882.4
	milligrams/sq. m	4882.4
	milligrams/sq. cm	0.4882
	grams/sq. ft	0.4536
Grams/sq. m	tons/sq. mile	2.855
	pounds/acre	8.921
	pounds/1000 sq. ft	0.2048
	kilograms/sq. km	1000
	milligrams/sq. m	1000
	milligrams/sq. cm	0.1
	grams/sq. ft	0.0929
Grams/cu. m	milligrams/cu. m	1000.0
	grams/cu. ft	0.02832
	micrograms/cu. m	1.0×10^6
	micrograms/cu. ft	28.317×10^3
	pounds/1000 cu. ft	0.06243
Micrograms/cu. m	milligrams/cu. m	0.001
	grams/cu. ft	28.317×10^{-9}
	grams/cu. m	1.0×10^{-6}
	micrograms/cu. ft	0.02832
	pounds/1000 cu. ft	62.43×10^{-9}
Micrograms/cu. ft	milligrams/cu. m	35.314×10^{-3}
	grams/cu. ft	1.0×10^{-6}
	grams/cu. m	35.314×10^{-6}
	micrograms/cu. m	35.314
	pounds/1000 cu. ft	2.2046×10^{-6}
Pounds/1000 cu. ft	milligrams/cu. m	16.018×10^3
	grams/cu. ft	0.35314
	micrograms/cu. m	16.018×10^6
	grams/cu. m	16.018
	micrograms/cu. ft	353.14×10^3

Atmospheric gases

Milligrams/cu. m	micrograms/cu. m	1000.0
	micrograms/liter	1.0
	ppm by volume (20°C)	$\dfrac{24.04}{{}^aM}$
	ppm by weight	0.8347
	pounds/cu. ft	62.43×10^{-9}
Micrograms/cu. m	milligrams/cu. m	0.001
	micrograms/liter	0.001
	ppm by volume (20°C)	$\dfrac{0.02404}{{}^aM}$

To convert from:	To:	Multiply by[a]:
	ppm by weight	834.7×10^{-6}
	pounds/cu. ft	62.43×10^{-12}
Micrograms/liter	milligrams/cu. m	1.0
	micrograms/cu. m	1000.0
	ppm by volume (20°C)	$\dfrac{24.04}{{}^a M}$
	ppm by weight	0.8347
	pounds/cu. ft	63.43×10^{-9}
Kilograms/sq. km	tons/sq. mile	2.855×10^{-3}
	pounds/acre	8.921×10^{-3}
	pounds/1000 sq. ft	204.8×10^{-6}
	grams/sq. m	0.001
	milligrams/sq. m	1.0
	milligrams/sq. cm	0.0001
	grams/sq. ft	92.9×10^{-6}
Milligrams/sq. m	tons/sq. mile	2.855×10^{-3}
	pounds/acre	8.921×10^{-3}
	pounds/1000 sq. ft	204.8×10^{-6}
	grams/sq. m	0.001
	kilograms/sq. km	1.0
	milligrams/sq. cm	0.0001
	grams/sq. ft	92.9×10^{-6}
Milligrams/sq. cm	tons/sq. mile	28.55
	pounds/acre	89.21
	pounds/1000 sq. ft	2.048
	grams/sq. m	10.0
	kilograms/sq. km	10.0×10^3
	milligrams/sq. m	10.0×10^3
	grams/sq. ft	0.929
Grams/sq. ft	tons/sq. mile	30.73
	pounds/acre	96.154
	pounds/1000 sq. ft	2.204
	grams/sq. m	10.764
	kilograms/sq. km	10.764×10^3
	milligrams/sq. m	10.764×10^3
	milligrams/sq. cm	1.0764
Airborne particulate matter		
Milligrams/cu. m	grams/cu. ft	283.2×10^{-6}
	grams/cu. m	0.001
	micrograms/cu. m	1000.0
	micrograms/cu. ft	28.52
	pounds/1000 cu. ft	62.43×10^{-6}
Grams/cu. ft	milligrams/cu. m	35.3145×10^3
	grams/cu. m	35.314
	micrograms/cu. m	35.314×10^6
	micrograms/cu. ft	1.0×10^6
	pounds/1000 cu. ft	2.2046

${}^a M$ = molecular weight.

Glossary

Abatement: the method of reducing the degree or intensity of pollution; also the use of such a method.

Abbe refractometer: an instrument used for determining the refractive index of liquids, minerals, and gemstones. Its operation is based on the measurement of the critical angle.

Abicoen: the nonbiotic elements within a habitat.

Abiotic: nonliving, as opposed to biotic; abiotic factors controlling biologic activity include pH, temperature, moisture, and chemicals.

Ablation: the combined processes by which a glacier wastes and melts away.

Absolute temperature: temperature measured in degrees Celsius from absolute zero, $-273.18°C$. Absolute temperatures are given either as "degrees absolute" (e.g., $150°A$) or as "degrees Kelvin" (e.g., $150°K$).

Absolute zero: the temperature at which all thermal motion of atoms and molecules ceases: $-273.18°C$.

Absorption: the penetration of a substance into or through another. For example, in air pollution control absorption is the dissolving of a soluble gas, present in an emission, in a liquid that can be extracted; the taking up of one substance into the body of another.

Abundance: the number of individuals of a species in an area, population, or community.

Abyssal: referring to that part of the ocean between a depth of about 2000 and 6000 m.

Abyssal hills: small irregular hills, rising to a height of 30 to 1000 m that cover large areas of the ocean floor. They are especially common in the Pacific Ocean. Relatively small topographic features of the deep ocean floor ranging to 610–915 m high and a few miles wide.

Abyssal plain: a very flat portion of the ocean floor underlain by sediments. The slope of this feature is less than 1 : 1000; flat, nearly level areas that occupy the deepest portions of many ocean basins.

Accelerate: to cause to develop or progress more quickly.

Accelerator: in radiology, a device for imparting high velocity to charged particles, such as electrons or protons. These fast particles can penetrate matter and are known as radiation.

Acclimation: the physiologic and behavioral adjustments of an organism to changes in its immediate environment.

Acclimatization: the acclimation or adaptation of a particular species over several generations to a marked change in the environment.

Acclimatory response: reversible change in the morphology or physiology of an organism in response to environmental change.

Accretion: (1) The gradual addition of new land to old by the deposition of sediment carried by the water of a stream. (2) The process by which inorganic bodies grow larger, by the addition of fresh particles to the outside. (3) A theory of continental growth by the addition of successive geosynclines. (4) In soils, the process of alluviation is usually one of the addition of minerals by accretion.

Acid: (1) A substance containing hydrogen which dissociates to form hydrogen ions when dissolved in water (or which reacts with water to form hydromium ions). (2) A substance capable of donating protons to other substances. (3) A term applied to igneous rocks having a higher percentage of silica than orthoclase, the limiting figure commonly adopted being 66%.

Acid mine drainage: drainage from certain mines of water containing minerals and having a low pH. The low pH is commonly caused by oxidation of iron sulfide to sulfuric acid. Mine water usually contains a high concentration of iron.

Acid soil: a soil with an acid reaction; for practical purposes with a pH of less than 6.6.

Acidic solution: A liquid whose hydrogen ion concentration is greater than its hydroxyl ion concentration, or whose pH is less than 7.0.

Actinomycetes: moldlike bacteria involved in the decomposition of organic matter; may be responsible for tastes and odors in water supplies.

Activated carbon: a highly absorbent form of carbon, used to remove odors and toxic substances from gaseous emissions. In advanced waste treatment, activated carbon is used to remove dissolved organic matter from waste water. A form of powdered carbon used to absorb recalcitrant organic compounds from waste waters as well as tastes and odors from drinking water.

Acquired character: character not inherited but acquired by an individual organism during its lifetime as a result of use or disuse according to its mode of life or the conditions under which it has lived.

Acre: a measure of surficial land area, containing in the United States and England 43,560 ft^2.

Acre-foot: the volume of liquid or solid required to cover 1 acre to a depth of 1 ft.

Acre-inch: the quantity of water, soil or other material that will cover 1 acre 1 in. deep.

Actium: a plant–animal community on a rocky seashore.

Activated sludge: sludge that has been aerated and subjected to bacterial action, used to remove organic matter from sewage; the mixed microbial population in the activated sludge treatment process for sewage treatment.

Activated sludge process: the process of using biologically active sewage sludge to hasten breakdown of organic matter in raw sewage during secondary waste treatment.

Active fault: a fault along which there is recurrent movement, usually indicated by small, periodic displacements or seismic activity.

Acute toxicity: any poisonous effect produced within a short period of time, usually up to 24–96 hr, resulting in severe biologic harm and often death.

Adaptability: capacity for evolutionary changes. Adaptability may depend on the phenotype's tolerance of environmental change as well as on the genetic variability of the population.

Adaptation: a change in structure or habit or an organism that produces better adjustment to the environment; a genetically determined characteristic that enhances the ability of an organism to cope with its environment.

Adaptive radiation: the adaptation, through natural selection, of the descendants of a taxon to a multitude of environments and habits by differentiating into different species and even higher taxonomic categories; evolutionary diversification of species derived from a common ancestor into a variety of ecologic roles.

Adaptive zone: a unit of environment occupied by a single kind of organism; a particular environmental opportunity requiring similar adaptations for diverse species. Species in different adaptive zones usually differ by major morphologic or physiologic characteristics.

Adenosine triphosphate (ATP): a molecule composed of adenosine and three phosphate groups. The phosphates are bound by high-energy linkages and are associated with energy transfer in living cells.

Adiabatic: a condition in which heat is neither gained nor dissipated.

Adiabatic thermodyne: the relationship of pressure and volume when a gas or other fluid is compressed or expanded without either giving out or receiving heat. See **isothermal**.

Adiabatic gradient: a temperature gradient in a column of material such that essentially no heat enters or leaves the system upon such processes as convection.

Adulterants: chemicals or substances that by law do not belong in a food, plant, animal, or pesticide formulation.

"A" horizon: zone of alluviation in alluviated soils; the uppermost zone in the soil profile, from which soluble salts and colloids have been leached, and in which organic matter has accumulated.

Adsorption: the adhesion of a substance to the surface of a solid or liquid. Adsorption is often used to extract pollutants by causing them to be attached to such adsorbents as activated carbon or silica gel. Hydrophobic, or water-repulsing, adsorbents are used to extract oil from waterways in oil spills. The adherence of a gas, liquid, or dissolved material on the surface of a solid. (1) Adhesion of molecules of gases, or of ions or molecules in solutions, to the surface of solid bodies with which they are in contact. (2) The behavior of a multicomponent fluid system that results in a dissolved material becoming more concentrated, or less concentrated, at an interface than in the body of the solution.

Advanced waste treatment: waste water treatment beyond the secondary or biologic stage; it includes removal of such nutrients as phosphorus and nitrogen and a high percentage of suspended solids. Advanced waste treatment, known as tertiary treatment, is the "polishing stage" of waste water treatment and produces a high-quality effluent.

Aeolian: refers to the wind, or to a soil that has been moved by the wind or that is subject to such movement.

Aerate: to expose to the action of the air; supply or charge with air.

Aeration: the process of being supplied or impregnated with air. Aeration is used in waste water treatment to foster biologic and chemical purification.

Aerial: relating to the air or atmosphere. "Subaerial" is applied to phenomena occurring under the atmosphere, "subaqueous" to phenomena occurring under water.

Aerial photograph: a photograph of the earth's surface taken from the air. It is usually one of a series taken from an aircraft moving in a systematic pattern at a given altitude in order to obtain a mosaic for mapping land divisions, geology, soil, vegetation, topography, etc.

Aerobe: an oxygen-requiring organism.

Aerobic: refers to life or processes than can occur only in the presence of oxygen.

Aerobic organism: an organism that thrives in the presence of oxygen.

Aeroplankton: microorganisms found floating in the air.

Aerosol: a suspension of liquid or solid particles in the air.

Afterburner: an air pollution abatement device that removes undesirable organic gases through incineration.

Aggregate: (1) To bring together; to collect or unite into a mass. (2) Composed of a mixture of substances separable by mechanical means. (3) The mineral material, such as sand, gravel, shells, slag, broken stone, or combinations thereof, with which cement or bituminous material is mixed to form a mortar or concrete. Fine aggregate may be considered as the material that passes a 6.35 mm screen, and coarse aggregate as the material that does not pass a 6.35 mm screen.

Aggressive mimicry: resemblance of predators or parasites to harmless species to deceive potential prey or hosts into ignoring them.

Agonistic: competitive, combative; specifically, in aggressive encounters between individuals.

Agricultural pollution: the liquid and solid wastes from all types of farming, including runoff from pesticides, fertilizers, and feedlots; erosion and dust from plowing, animal manure and carcasses, and crop residues and debris. It has been estimated that agricultural pollution in the United States has amounted to more than 2½ billion tons per year.

Air curtain: a method for mechanical containment of oil spills. Air is bubbled through a perforated pipe, causing an upward water flow that retards the spreading of oil. Air curtains are also used as barriers to prevent fish from entering a polluted body of water.

Air mass: a widespread body of air with properties that have been established while the air has been situated over a particular region of the earth's surface and that undergoes specific modifications while in transit away from that region.

Air monitoring: see **monitoring**.

Air pollution: the presence of contaminants in the air in concentrations that prevent the normal dispersive ability of the air and that interfere directly or indirectly with human health, safety, or comfort or with the full use and enjoyment of our property.

Air pollution episode: the occurrence of abnormally high concentrations of air pollutants usually caused by low winds and temperature inversion and accompanied by an increase in illness and death. See **inversion**.

Air quality control region: an area designated by the Federal government where two or more communities—either in the same or different states—share a common air pollution problem.

Air quality criteria: the levels of pollution and lengths of exposure at which adverse effects on health and welfare occur.

Air quality standards: the prescribed level of pollutants in the outside air that cannot be exceeded legally during a specified time in a specified geographic area.

Algae (algal cells): a group of plants variously one celled, colonial, or filamentous, containing chlorophyll and other pigments and having no true root, stem, or leaf.

Algae are found in water or damp places and include seaweeds and pond scum. Class Thallophytes, includes single-celled plants and common seaweeds.

Algal bloom: a proliferation of living algae on the surface of lakes, streams, or ponds. Algal blooms are stimulated by phosphate enrichment.

Algicide: a specific chemical highly toxic to algae. Algicides are often applied to water to control nuisance algal blooms.

Algology: the study of algae

Algorithm: a set of step by step instructions for performing a numerical or algebraic operation, as in digital computer applications; a mathematical formula.

Alkali: (1) sodium carbonate or potassium carbonate, or more generally any bitter-tasting salt found at or near the surface in arid and semiarid regions. (2) A strong base, e.g., NaOH or KOH. (3) An alkali metal.

Alkali soil: a soil with a strong basic reaction. For practical purposes with a pH of 8.5 or higher.

Alkaline solution: a liquid whose hydroxyl ion concentration is greater than its hydrogen ion concentration, or whose pH is greater than 7.0. Sea water is slightly alkaline, having a pH between 7.5 and 8.4.

Alkalinity: the capacity of a water to accept protons, i.e., hydrogen ions. It is usually expressed as millequivalents per liter.

Alkalosis: excessive alkalinity (high pH) of the blood caused by rapid removal of carbon dioxide as by hyperventilation.

Allochronic speciation: separation of a population into two or more evolutionary units as a result of reproductive isolation because two subpopulations in the same area mate at different times.

Allochthonous: (1) a term applied to rocks of which the dominant constituents have not been formed in situ. See **Autochthonous**. (2) Usually refers to material resources that have originated outside the division of an ecosystem under consideration.

Allopatric: occurring in different places; usually refers to geographically separated populations.

Allopatric speciation: separation of a population into two or more evolutionary units as a result of reproductive isolation caused by geographic separation of two subpopulations.

Allopolyploidy: increase in the number of chromosome sets in a fertilized egg that receives two sets of chromosomes because of hybrid parentage.

Alluvial fan: a cone-shaped deposit of alluvium made by a stream where it runs out onto a level plain or meets a slower stream. The fans generally form where streams issue from mountains upon the lowland.

Alluvium: (1) a general term for all detrital deposits resulting from the operations of modern rivers, thus including the sediments laid down in river beds, flood plains, lakes, fans at the foot of mountain slopes, and estuaries. (2) The rather consistent usage of the term throughout its history makes it quite clear that alluvium is intended to apply to stream deposits of comparatively recent time, that the subaqueous deposits of seas and lakes are not intended to be included, and that permanent submergence is not a criterion. Alluvium may become lithified, as has happened frequently in the past, and may then be termed ancient alluvium.

Alpha particle: a positively charged particle emitted by certain radioactive materials. It is the least penetrating of the three common types of radiation (alpha, beta, and gamma) and is usually not dangerous to plants, animals, or humans. A high-energy helium nucleus (two protons, two neutrons) emitted by some heavy radioactive nuclei.

Alpine: of, pertaining to, or like the alps or any lofty mountain; resembling a great mountain range of southern Europe called the Alps. Implies high elevation, particularly above tree line, and cold climate.

Artricial: see **nidicolous**.

Altruism: in an evolutionary sense, enhancing the fitness of an unrelated individual by acts that reduce the evolutionary fitness of the altruistic individual.

Ambient air: any unconfined portion of the atmosphere; the outside air.

Ammate: a compound, ammonium sulfamate, used as a relatively short-lived herbicide.

Ammocolous: an organism that grows in sand.

Ammonia stripping: a process for the removal of ammonia from waste water.

Amoeba: a genus of unicellular protozoan organisms of microscopic size, existing in nature in large numbers; many live as parasites and some species are pathogenic to humans.

Amorphous: without form. Applied to rocks and minerals having no definite crystalline structure.

Amplify: to increase the output of one flow of energy by interaction with a second flow.

Amplitude: the elevation of the crest of a wave or ripple above the adjacent troughs. Hydrodyn: One-half the wave height.

Anabolism: the biochemical processes in which energy is consumed by the cell. At the same time biosynthesis occurs.

Anadromous: type of fish that ascends rivers from the sea to spawn.

Anaerobe: an organism that does not require atmospheric oxygen.

Anerobic: refers to life or processes that occur in the absence of oxygen; a condition where oxygen is absent. The Black Sea is an example.

Anaerobic organism: a microorganism that thrives best, or only, when deprived of oxygen.

Angle of repose: the maximum slope or angle at which a material, such as soil or loose rock, remains stable. When exceeded, mass movement by slipping as well as by water erosion may be expected. Synonym: critical slope.

Angstrom unit: (often anglicized to Angstrom: abbreviated A or Å) A unit of length, 10^{-8} cm, commonly used in structural crystallography.

Anion: A negatively charged ion. Examples are chlorine, Cl^-, and oxygen, $O^=$.

Annelida: the phylum of invetebrate animals which includes the segmented worms.

Annual flood: The highest flow a river normally reaches during the year. This point is not usually a flood in the ordinary meaning, because there is no overflow.

Annual ring: the layer of xylem (wood) formed by a 1-year growth of cambium.

Anoxic: an environment devoid of oxygen.

Antagonism: destruction or prevention of growth of one organism by another.

Anthropogenic: that which is derived from the presence of humans; conditions resulting from human activity.

Antibiotic: an organic chemical produced by one microorganism that kills or inhibits another microorganism.

Anticoagulant: a chemical that interferes with blood clotting, often used as a rodenticide.

Antidegration clause: a provision in air quality and water quality laws that prohibits deterioration of air or water quality in areas where the pollution levels are presently below those allowed.

Apatetic: the coloration of an animal that causes it to resemble physical features of the habitat.

Aphotic zone: that part of the ocean where not enough light is present for photosynthesis by plants.

Apomixis: reproduction without sexual union, as in parthenogenesis (which see).

Aposematism: conspicuous appearance of an organism warning that it is noxious or distasteful; warning coloration.

Apostatic selection: selective predation on the most abundant forms in a population, regardless of their appearance, leading to balanced polymorphism; the stable occurrence of more than one form in a population.

Aquaculture: farming of the ocean, whereby such organisms as fish, algae, and shellfish are grown under controlled conditions. At present this technique is only used in near-shore areas.

Aquaculture project: a controlled discharge of pollutants to enhance growth or propagation of harvestable fresh-water, estuarine, or marine life plant or animal species.

Aquatic: growing or living in or upon water.

Aquatic plants: plants that grow in water either floating on the surface, growing up from the bottom of the body of water, or growing under the surface of water.

Aquatic agriculture: the management of natural and/or artificial means to increase the production of fish, oysters, or other animals or plants in fresh or salt waters.

Aquifer: a layer of rock, sand, or gravel through which water can pass (see **permeability**); an underground bed or stratum of earth, gravel or porous stone that contains water; the place in the ground where groundwater is naturally stored.

Aquifuge: a rock that contains no interconnected openings or interstices and therefore neither absorbs nor transmits water.

Arable: a soil that will satisfactorily produce cultivated crops; not to be confused with tillable.

Archipelago: any sea or broad sheet of water interspersed with many islands or with a group of islands; also, such a group of islands.

Arctic: (1) the region within the Arctic Circle (66°30′ N). (2) Geography: Lands north of the 10°C July isotherm (or that of whichever month is warmest) provided the mean temperature for the coldest month is not higher than 0°C.

Area source: in the air pollution, any small individual fuel combustion source, including any transportation sources. This is a general definition; area source is legally and precisely defined in Federal regulations. See **point source**.

Areal map: a geologic map showing the horizontal area or extent of rock units exposed at the surface.

Arid: said of a climate characterized by dryness, variously defined as rainfall insufficient for plant life or for crops without irrigation; less than 10 in. of annual rainfall; or a higher evaporation rate than precipitation rate. Synonym: dry.

Arrhenotoky: a sex determination system in which males develop from unfertilized, haploid eggs and thus have only one chromosome complement.

Arroyo: (1) the channel of an ephermeral or intermittent stream, usually with vertical banks of unconsolidated material 2 ft or more high.

Artesian: describes underground water trapped under pressure between layers of impermeable rock. An artesian well is one that taps artesian water. Refers to ground water under sufficient hydrostatic head to rise to above the aquifer containing it.

Artesian water: ground water that is under sufficient pressure to rise above the level at which it is encountered by a well but that does not necessarily rise to or above the surface of the ground.

Artesian well: one in which the water level rises above the top of the aquifer, whether or not the water flows at the land surface.

Artifacts: manmade objects of prehistoric age, such as weapons or tools of flint.

Artificial selection: intentional manipulation by humans of the fitness of individuals in a population to produce a desired evolutionary response.

Artificial substrate: a device placed in the water for a period extending to a few weeks that provides living spaces for a multiplicity of drifting and natural-born organisms that would not otherwise be at the particular spot because of limiting physical habitat. Examples of artificial substrates include glass slides, tiles, bricks, wooden shingles, concrete blocks, multiplate-plate samplers, and brush boxes.

Asbestos: (1) a mineral fiber with countless industrial uses; a hazardous air pollutant when inhaled; its effects as a water pollutant are under scrutiny. (2) White, gray, green-gray, or blue-gray fibrous variety of amphibole, usually tremolite or actinolite, or of chrysotile. Blue asbestos is crocidolite.

A-Scale sound level: the measurement of sound approximating the auditory sensitivity of the human ear. The A-Scale sound level is used to measure the relative noisiness or annoyance of common sounds.

Ascospores: the spores produced by *Ascomycetes*.

Aseptic: procedure that maintains sterility.

Asexual: refers to any type of reproduction that does not involve the union of sex cells (gametes).

Aspect: the compass direction toward which a slope faces.

Aspect diversity: variations in the outward appearance of species that live in the same habitat and that are eaten by visually hunting predators.

Asphalt: a brown to black solid or semisolid bituminous substance occurring in nature but also obtained as a residue from the refining of certain petroleum. Asphalt melts between 65° and 95°C and is soluble in carbon disulfide.

Asset: anything that can depreciate and requires work to be maintained.

Assimilation: incorporation of any material into the tissues, cells, and fluids of an organism.

Assimilation: conversion or incorporation of absorbed nutrients into protoplasm. Also refers to the ability of a body of water to purify itself of organic pollution. (1) The incorporation into a magma of material originally present in the wall rock. The term does not specify the exact mechanism of results; the "assimilated" material may be present as crystals, including wall rock elements, or as a true solution in the liquid phase of the magma. The resulting rock is called hybrid. Also termed magmatic assimilation. (2) The uptake of food material for production of new biomass.

Assimilation efficiency: a percentage expressing the proportion of energy in ingested food that is assimilated into the bloodstream of an organism.

Association: a group of species occurring in the same place.

Asymmetrical: (1) without proper proportion of parts; unsymmetrical. (2) Crystallography: having no center, plane, or axis of symmetry.

Atlantic-type coastline: a discordant coastline, especially one as developed in many areas around the Atlantic Ocean; e.g., the SW coastline of Ireland and the NW coastlines of France and Spain. Antonym: Pacific-type coastline.

Atmosphere: the layer of air surrounding the earth. (1) The gaseous envelope surrounding the earth. The atmosphere is odorless, colorless, and tasteless; very mobile, flowing readily under even a slight pressure gradient; elastic, compressible, capable of unlimited expansion, a poor conductor of heat, but able to transmit vibrations with considerable velocity. Its weight has been calculated as 5.9×10^{15} tons. One-half the mass of the atmosphere lies below 3.46 miles high. The ordinary term for the mixture of gases comprising the atmosphere is air, which also includes water vapor and solid and liquid particles. (2) A unit of pressure: a normal atmosphere is equal to the pressure exerted by a vertical column of mercury 760 mm in height, at 0°C, and with gravity taken at 980.665 cm/s^2, equal to about 14.7 pounds per square inch.

Atoll: a circular coral reef surrounding a lagoon.

Atom: the smallest component of an element that has all the properties of the element. An atom consists of protons, electrons, and neutrons.

Atomic bond: attraction exerted between atoms and ions. Four types are: metallic, ionic or polar, homopolar or coordinant, and residual or van der Waals. Bonding may be intermediate between these types.

Atomic pile: a nuclear reactor.

Atomic weight: the relative weight of an atom on the basis that the oxygen atom has a weight of 16. The atomic weight is essentially equal to the number of protons in the atom. Average relative weight of the atoms of an element referred to an arbitrary standard of 16.0000 for the atomic weight of oxygen. The atomic weight scale used by chemists takes 16.0000 as the average atomic weight of oxygen atoms as they occur in nature; the scale used by physicists takes 16.00435 as the atomic weight of the most abundant oxygen isotope. Division by a factor of 1.000272 converts an atomic weight on the physicists' scale to the weight on the chemists' scale.

Attractant: a chemical or agent that lures insects or other pests by olfactory stimulation.

Audiometer: an instrument for measuring hearing sensitivity.

Aufwuchs: periphyton; organisms attached to submerged surfaces above the bottom.

Autecology: the study of single individuals in relation to ecologic processes, in contrast to synecology, which is the study of whole communities in relation to their environment. The study of the individual organisms or species rather than the community. Life history and behavior, rather than adaptation to environment, are usually emphasized.

Anthigenic deposits: deposits that have formed in place before the sediment is buried. They usually have precipitated directly from the sea water.

Autocatalytic: self-catalyzing. Many chemical reactions are autocatalytic.

Autoclave: a high-pressure steam sterilizer.

Autochthonous: Materials which have originated within the division of an ecosystem under consideration.

Autolysis: self-destruction of cells by the action of autolytic or intracellular enzymes.

Autotrophic bacteria: bacteria that produce their own food from inorganic compounds.

Autotrophic organisms: those organisms that utilize light energy or the oxidation of inorganic chemicals as their sole energy source.

Autotrophy: fixation of light energy from the sun or use of inorganic compounds for food, as by plants and some bacteria (see **Heterotrophy**).

Average fitness of a population: (Noted as ^{r}). The average fitness of the genotypes in a population weighted by their frequencies.

Axenic: growth of a microorganism alone in a medium without other organisms being present.

Bacillus: a rod-shaped bacterium with specific physiologic characteristics.

Backfill: the material used to refill a ditch or other excavation, or the process of doing so.

Background level: with respect to air pollution, amounts of pollutants present in the ambient air from natural sources.

Background radiation: normal radiation present in the lower atmosphere from cosmic rays and from earth sources.

Backwash: return flow of water on a beach after the advance of a wave.

Bacteria: single-celled microorganisms that lack chlorophyll. Some bacteria are capable of causing human, animal, or plant diseases; others are essential in pollution control because they break down organic matter in the air and in the water.

Bactericidal: able to kill bacteria.

Bacteriophage: a virus that infects bacteria.

Bacteriostatic: capable of preventing bacterial growth but not of killing the bacteria.

Baffle: any deflector device used to change the direction of flow or the velocity of water, sewage, or products of combustion, such as fly ash or coarse particulate matter. Also used in deadening sound.

Baghouse: an air pollution abatement device used to trap particulates by filtering gas streams through large fabric bags, usually made of glass fibers.

Baguio: a tropical cyclone in the Philippine Islands.

Bajada: an outwash slope with long straight longitudinal profiles occuring in the southwestern United States.

Balanced polymorphism: maintenance of more than one allele in a population by the selective superiority of the heterozygote over both types of homozygotes.

Baling: a means of reducing the volume of solid waste by compaction.

Ballistic separator: a machine that separates inorganic from organic matter in a composting process.

Band application: with respect to pesticides, the application of the chemical over or next to each row of plants in a field.

Bar screen: in waste water treatment, a screen that removes large floating and suspended solids.

Barophile: a microorganism that lives at high pressures, usually in the depths of the ocean.

Barrens: an area relatively barren of vegetation in comparison with adjacent areas because of adverse soil or climatic conditions, wind, or other adverse environmental factors; for example, sand barrens or rock barrens.

Barrier beach: offshore bar. This term refers to a single elongate sand ridge rising slightly above the high-tide level and extending generally parallel with the coast, but separated from it by a lagoon. The term should apply to islands and spits.

Barrier reef: a coral reef that is separated from the coast by a lagoon that is too deep for coral growth. Generally, barrier reefs follow the coasts for long distances, often with short interruptions, termed passes.

Basal application: with respect to pesticides, the application of the pesticide formulation on stems or trunks of plants just above the soil line.

Basalt: an igneous rock, commonly found on the sea floor, mainly composed of feldspar and pyrozene minerals. Basalt rocks are thought to underlie most of the ocean basin.

Base map: a map on which information may be placed for purposes of comparison or geographic correlation.

Basidiospore: a sexual spore produced by the *Basidiomycetes*.

Basidium: the specialized part of the mycelium of *Basidiomycetes* on which basidiospores are formed.

Basin: See river basin.

Batesian mimicry: resemblance of an edible species (mimic) to an unpalatable species (model) to deceive predators.

Bathyal: referring to that part of the ocean between depths of about 200 and 2000 m.

Bathymetry: the measuring of the depth of the ocean.

Bathythermograph: an instrument used to measure temperature in the ocean.

Bay: (1) a recess in the shore or an inlet of a sea or lake between two capes or headlands, not as a cove. Bight, embayment. (2) A swampy area, usually oval-shaped and covered with brush; local on South Atlantic Coast. (3) Mining: An open space for waste between two packs in a longwall working.

Bayou: a lake, or small sluggish secondary stream, often in an abandoned channel or river delta; local on Gulf Coast of U.S. One of the half-closed channels of a river delta; local on Mississippi Delta.

Beach: the gently sloping shore of a body of water which is washed by waves or tides, especially the parts covered by sand or pebbles.

Bearing: the direction of a line with reference to the cardinal points of the compass. True bearing: the horizontal angle between a ground line and a geographic meridian. A bearing may be referred to either the south or the north point (N 30°E, or S 30°W). Magnetic bearing: the horizontal angle between a ground line and the magnetic meridian. A magnetic bearing differs from a true bearing by the exact angle of magnetic declination of the locality.

Beaufort wind scale: a system of estimating wind velocities, originally based (1806) by its inventor, Admiral Sir Francis Beaufort of the British Navy, on the effects of various wind speeds on the amount of canvas that a full-rigged frigate of the early nineteenth century could carry; since modified and widely used in international meteorology.

Bedrock: (1) the solid rock underlying auriferous gravel, sand, clay, etc., and upon which the alluvial gold rests. (2) Any solid rock exposed at the surface of the earth or overlain by unconsolidated material.

Bench mark: a relatively permanent material object, natural or artificial, bearing a marked point whose elevation above or below an adopted datum (such as sea level) is known. The usual designation is BM or PBM (permanent bench mark). A temporary or supplemental bench mark (TBM) is of a less permanent nature and the elevation may be less precise.

Benthic: pertaining to aquatic bottom or sediment habitats.

Benthic or benthonic: the area of the ocean bottom inhabited by marine organisms.

Benthic region: the bottom of a body of water. This region supports the benthos, a type of life that not only lives upon, but contributes to the character of the bottom. The bottom of all waters; the substratum that supports the benthos.

Benthos: bottom-dwelling organisms. The benthos includes (1) sessile animals, such as the sponges, barnacles, mussels, and oysters, some of the worms, and many attached algae; (2) creeping forms, such as snails and flatworms; and (3) burrowing forms which include most clams and worms. Organisms that live on or in close contact (such as certain fish) with the ocean bottom. The shallow-water bottom organisms of the sea. The plant and animal life whose habitat is the bottom of a sea, lake, or river.

Benthos: aquatic organisms attached to or resting on the bottom or living in the bottom sediments.

3,4-Benzpyrene: a strongly carcinogenic aromatic hydrocarbon.

Berm: (1) terraces that originate from the interruption of an erosion cycle. (2) A nearly horizontal portion of the beach or backshore formed by the deposit of material by wave action. (3) A mound of earth forming a wall-like barrier against water movement. Also a noise barrier of this nature constructed to mitigate vehicular noise in residential neighborhoods.

Beryllium: a metal that when airborne has adverse effects on human health; it has been declared a hazardous air pollutant. It is primarily discharged by such operations as machine shops and ceramic and propellant plants and foundries.

Beta particle: (1) an elementary particle emitted by radioactive decay that may cause skin burns. It is easily stopped by a thin sheet of metal. (2) A high-energy electron emitted by radioactive nucleus.

B horizon: alluvial horizon in soils that are illustrated in Fig. 3.11. The lower soil zone that is enriched by the deposition or precipitation (acretion) of material from the overlying zone or A horizon.

Between habitat specialization: restriction of the distribution of a population to a narrow range of habitats.

Binary fission: division of a cell into two daughter cells; the process by which bacteria reproduce.

Binomial system: system by which organisms are known by first a generic name and second a specific or trivial name; subspecies or varieties receive a third name.

Bioassay: the employment of living organisms to determine the biologic effect of some substance, factor, or condition.

Biochemical oxygen demand (BOD): a measure of the quantity of oxygen used in the biochemical oxidation of organic matter in a specified time, at a specified temperature, and under specified conditions. In general, a high BOD indicates the presence of a large amount of organic material. It is normal to find a BOD of 2–3 mg/liter in river waters receiving natural drainage. The amount of oxygen required to degrade organic matter present in a sample, usually held in the dark at 20° for 5 days. A measure of the amount of oxygen consumed in the biologic processes that break down organic matter in water. Large amounts of organic waste use up large amounts of dissolved oxygen; the greater the degree of pollution, therefore, the greater the BOD, the amount of oxygen required by the biologic population of a water sample to oxidize the organic matter in that water. It is usually determined over a 5-day period under standardized laboratory conditions and hence may not represent actual field conditions.

Biocoenose: an assemblage of organisms that live together as an interrelated community; A natural ecologic unit.

Biodegradable: the process of decomposing quickly as a result of the action of microorganisms.

Biodegradation: the destruction or mineralization of either natural or synthetic organic materials by the microorganisms of soils, waters, or waste water treatment systems.

Biogeochemical cycling: the cycling of chemical constituents through a biologic system.

Biogeochemical cycle: the cycling of such chemical elements as nitrogen, carbon, and sulfur from the environment into organic substances and back into the environment.

Biogeography: the study of the geographic distribution of plants and animals and the reasons for their distribution.

Biologic control: a method of controlling pests by means of introduced or naturally occurring predatory organisms, sterilization, or the use of inhibiting hormones, etc., instead of by mechanical or chemical means; the direct use of living organisms or their products to control other organisms that have become pests.

Biologic magnification: the concentration of certain substances up a food chain; a very important mechanism in concentrating pesticides and heavy metals in such organisms as fish.

Biologic oxidation: the process by which bacterial and other microorganisms feed on complex organic materials and decompose them. The self-purification of waterways and activated sludge and trickling filter waste water treatment processes depend on this principle. The process is also called biochemical oxidation.

Bioluminescence: the production of light by living organisms as a result of a chemical reaction; light energy produced by luminescent organisms.

Biomass: weight of living material expressed as a dry weight, in all or part of an organism, population, or community. Commonly expressed as weight per unit area, a biomass density.

Biome: a major terrestrial climax community composed of plants and animals; a major ecologic zone or region corresponding to a climatic zone or region; a major community of plants and animals associated with a stable environmental life zone or region (e.g., northern coniferous forest, Great Plains, tundra).

Biomonitoring: the use of living organisms to test the suitability of effluent for discharge into receiving waters and to test the quality of such waters downstream from a discharge.

Biophile: an element that is required by or is found in the bodies of living organisms. The list of such elements includes C, H, O, N, I, S, Cl, I, Br, Ca, Mg, K, Na V, Fe, Mn, and Cu. All may belong also to the chalcophile or lithophile groups.

Biosphere: the portion of the earth and its atmosphere capable of supporting life; a collective term for the area of habitat of the organisms of the earth; the portion of the earth that is inhabited by living organisms; the zone of air, land, and water at the surface of the earth that is occupied by living organisms. (1) Zone at and adjacent to the earth's surface where all life exists. (2) All living organisms of the earth.

Biostablizer: a machine used to convert solid waste into compost by grinding and aeration.

Biota: all the species of plants and animals occurring within a certain area; the animal and plant life of a region; flora and fauna collectively.

Biotic environment: biologic components of an organism's surroundings that interact with it, including competitors, predators, parasites, and prey. Interaction within a population are subclassified as the social, sexual, and parent–offspring environment.

Biotic succession: the natural replacement of one or more groups of organisms occupying a specific habitat by new groups. The preceding groups in some ways prepare or favorably modify the habitat for succeeding groups.

Biotape: an area where the principal habitat conditions and the living forms adapted to the conditions are uniform.

Black fly larvae (Simuliidae): aquatic larvae that produce a silklike thread with which they anchor themselves to objects in swift waters. With a pair of fan-shaped structures, a larva of this type produces a current of water toward its mouth and from this water ingests smaller organisms. The adults are terrestrial; females feed on the blood of higher animals.

Black mud: a mud formed in lagoons, sounds, or bays, in which there is a poor circulation or weak tides. The color is generally black because of black sulfides of iron and organic matter.

Bloodworms (Chironomidae): cylindrical elongated midge larvae with pairs of prolegs on both the first thoracic and last abdominal segments. Although many species are blood-red in color, some are pale yellowish, yellowish red, brownish, pale greenish yellow, and green. Most feed on diatoms, algae, tissues of aquatic plants, decaying organic matter, and plankton. Some are associated with rich organic deposits. Midge larvae are important as food for fishes.

Bloom: a proliferation of living algae and/or other aquatic plants on the surface of lakes or ponds. Blooms are frequently stimulated by phosphate enrichment. A readily visible concentrated growth or aggregation of plankton (plant and animal).

Blowout: in drilling a well by the rotary method an unexpected volume of gas under pressure sometimes "blows" the mud-laden drilling fluid from the hole, thus putting an end to drilling until controlled. The term is also used in standard-tool drilling when the flow of gas is sufficient to interfere with drilling operation.

BOD: See **biochemical oxygen demand**.

BOD$_5$: the amount of dissolved oxygen consumed in 5 days by biologic processes breaking down organic matter in an effluent. See **biochemical oxygen demand**.

Bog: wet, spongy land usually poorly drained, highly acid, and rich in plant residue. Morass; swamp. (1) Common name in Scotland and Ireland for a wet spongy morass, chiefly composed of decaying vegetable matter or peat. (2) A swamp or tract of wetland, covered in many cases with peat.

Boiling point: the temperature at which a liquid starts to boil. For pure water this temperature is 100°C or 212°F at normal pressure.

Boom: a floating device that is used to contain oil on a body of water.

Boreal: Northern. Often refers to the coniferous forest regions that stretch across Canada, northern Europe, and Asia.

Boron: powerful absorber of neutrons used—usually in alloy steel—for reactor control rods, etc.

Borrow pits: pits resulting from the excavation of mineral deposits, i.e., quarrys.

Botanical pesticide: a plant-produced chemical used to control pests; for example, nicotine, strychnine, or pyrethrun.

Brackish water: a mixture of fresh and salt water.

Breeder: a nuclear reactor that produces more fuel than it consumes.

Breeder reactor: reactor that produces more fissile nuclei than it consumes.

Broadcast application: with respect to pesticides, the application of a chemical over an entire field, lawn, or other area.

Broken stick model: a model of relative abundance of species obtained by random division of a line representing the resources of an environment into segments.

Buffer: a chemical used to prevent changes in pH.

Burial ground (graveyard): a place for burying unwanted radioactive materials to prevent radiation escape, the earth or water acting as a shield. Such materials must be placed in water-tight, noncorrodible containers so the radioactive material cannot leach out and invade underground water supplies.

Burner reactor: reactor that consumes more fissile nuclei than it produces.

Cadmium: See **heavy metals**.

Calcareous: containing calcium carbonate.

Calcareous algae: seaweed that remove carbon dioxide from the shallow water in which they live and as a consequence secrete or deposit a more or less solid calcareous structure.

Calorie: a unit of heat; the amount of heat needed to raise the temperature of 1 g of water from 3.5° to 4.5°C. Various prefixes are used if the degree of temperature change is at different magnitudes.

Canopy: the uppermost layer consisting of crowns of trees or shrubs in a forest or woodland.

Capillarity: the force that causes water to rise in a constricted space through molecular attraction, often against the pull of gravity; the attractive force between two unlike molecules, illustrated by the rising of water in capillary tubes of hairlike diameters or the drawing up of water in small interstices, as those between the grains of a rock.

Capsule: a layer of mucoid material surrounding a bacterial cell.

Carbon dioxide (CO_2): a colorless, odorless, nonpoisonous gas that is a normal part of the ambient air. Carbon dioxide is a product of fossil fuel combustion, and some researchers have theorized that excess CO_2 raises atmospheric temperatures.

Carbon-14: a radioactive isotope that can be used for dating. This isotope is especially useful in dating material that was once alive, because all living matter contains carbon. The half-life of carbon-14 is 5570 ± 30 years. A radioactive isotope of carbon with atomic weight 14, produced by collisions between neutrons and atmospheric nitrogen. Useful in determining the age of carbonaceous material younger than 30,000 years.

Carbon monoxide (CO): a colorless, odorless, highly toxic gas that is a normal byproduct of incomplete fossil fuel combustion. Carbon monoxide, one of the major air pollutants, can be harmful in small amounts if breathed over a certain period of time.

Carcinogenic: cancer producing.

Carcinogenic agent: a material that causes cancer. Compare mutagenic, causing mutations, and teratagenic, causing birth defects.

Carnivore: an organism that consumes mostly flesh.

Carrying capacity: amount of animal life, human life, or industry that can be supported indefinitely on available resources; Number of individuals that the resources of a habitat can support.

Catabolism: the biochemical processes of a cell that generate energy for use by the cell. At the same time substrates are degraded.

Catalytic converter: an air pollution abatement device that removes organic contaminants by oxidizing them into carbon dioxide and water through chemical reaction. Can be used to reduce nitrogen oxide emissions from motor vehicles.

Catastrophic waves: large waves, resulting from intense storms or submarine slumping, that can cause immense damage and loss of life.

Cation: a positively charged ion. Examples are hydrogen, H^+, and sodium, Na^+.

Caustic soda: sodium hydroxide (NaOH), a strongly alkaline, caustic substance used as the cleaning agent in some detergents.

Cells: with respect to solid waste disposal, earthen compartments in which solid wastes are dumped, compacted, and covered over daily with layers of earth.

Celsius scale: a thermometric scale, proposed in 1742 by Anders Celsius. Today defining 100°C as the boiling point of water and 0°C as the melting point of ice.

Centigrade: temperature units (°C).

Centrifugal collector: any of several mechanical systems using centrifugal force to remove aerosols from a gas stream.

Centrifugal force: a force caused by rotation which causes motion away or out from the rotating object.

Cesium: particularly cesium-137; fission product, biologically hazardous beta emitter.

cfs: cubic feet per second, a measure of the amount of water passing a given point.

Channelization: the straightening and deepening of streams to permit water to move faster, to reduce flooding, or to drain marshy acreage for farming. However, channelization reduces the organic waste assimilation capacity of the stream and may disturb fish breeding and destroy the stream's natural beauty.

Character convergence: evolution of similar appearance or behavior in unrelated species for the purpose of facilitating direct interaction between individuals. Also called social mimicry.

Character displacement: divergence in the characteristics of two otherwise similar species where their ranges overlap, caused by selective effects of competition between the species in the area of overlap.

Character divergence: evolution of differences between similar species occurring in the same areas, caused by the selective effects of competition.

Check dam: a small, low dam in a watercourse to decrease stream velocity and to promote sedimentation of eroded material.

Chemical oxygen demand (COD): a measure of the amount of oxygen required to oxidize organic and oxidizable inorganic compounds in water. The COD test, like the BOD test, is used to determine the degrees of pollution in an effluent.

Chemoautotroph: an organism that utilizes oxidation of inorganic chemicals for its energy and growth; an organism that oxidizes inorganic compounds (often hydrogen sulfide) to obtain energy for synthesis of organic compounds, e.g. sulfur bacteria.

Chemostat: an apparatus used to grow bacteria continuously in a specific growth phase.

Chemosterilant: a pesticide chemical that controls pests by destroying their ability to reproduce.

Chemotaxis: movement of a cell in response to a chemical attractant or repellent.

Chemotherapy: control of disease by treatment with chemicals.

Chilling effect: the lowering of the earth's temperature because of the increase of atmospheric particulates that inhibit penetration of the sun's energy.

China syndrome: a Western expression; possibly consequence of core meltdown, when a molten mass of intensely radioactive material plummets through vessel and containment and into the earth beneath, in the direction of China.

Chlorides: considered a major anion, especially in ocean waters where average concentrations are as high as 19 g per 1000 ml of seawater.

Chlorinated hydrocarbons: a class of generally long-lasting, broad-spectrum insecticides of which the best known is DDT, first used for insect control during World War II. Other similar compounds include aldrin, dieldrin, heptachlor, chlordane, lindane, endrin, mirex, benzene hexachloride (BHC), and toxaphene. The qualities of persistence and effectiveness against a wide variety of insect pests were long regarded as highly desirable in agriculture, public health and home uses. However, later research has revealed that these same qualities may represent a potential hazard through accumulation in the food chain and persistence in the environment.

Chlorination: the application of chlorine to drinking water, sewage, or industrial waste for disinfection or oxidation of undesirable compounds.

Chlorinator: a device for adding a chlorine-containing gas or liquid to drinking or waste water.

Chlorine-contact chamber: in a waste treatment plant, a chamber in which effluent is disinfected by chlorine before it is discharged to the receiving waters.

Chlorophyll: a green pigment in photosynthetic organisms that acts as an electron donor in the photosynthetic process; a group of green pigments, found in plants and essential for photosynthesis.

Chloroplast: the portion of eucaryotic and plant cells where photosynthesis occurs.

Chlorosis: yellowing or whitening of normally green plant parts. It can be caused by disease organisms, by lack of oxygen or nutrients in the soil, or by various air pollutants.

Chromium: See **heavy metals**.

Chromosome: any of several small bodies, found in the nucleus of the cell, which bear the genetic material.

C horizon: the layer of weathered bedrock at the base of a soil. It has undergone little alteration by organisms and is presumed to be similar in chemical, physical, and mineralogic composition to the material from which at least a portion of the overlying solum has developed.

Chronic: marked by long duration or frequent recurrence, as a disease.

Cilia: hairs on certain organisms that are responsible for motility.

Clarification: in waste water treatment, the removal of turbidity and suspended solids by settling, often aided by centrifugal action and chemically induced coagulation.

Clarifier: in waste water treatment, a settling tank that mechanically removes settled solids from wastes.

Classification of communities: attempts to understand biotic communities by assigning them to defined categories or "community types" based on shared characteristics. Classification criteria may include formation type, dominant overstory or understory species, presence of certain "indicator" species, etc. All communities would be assigned to the various class types and viewed as possessing the characteristics of the class defintion.

Clay: (1) a size term denoting particles, regardless of mineral composition, with diameter less than :1-/256 mm (4 μm). (2) A group of hydrous aluminosilicates (clay minerals). (3) A sediment of soft plastic consistency composed primarily of fine-grained minerals. (4) In engineering, any surficial material that is unconsolidated.

Climate, oceanic: the type of climate characteristic of land areas near oceans which contribute to the humidity and at the same time have a moderating influence on temperature and range of temperature variation. Synonym: marine climate.

Climax: an ecosystem that is constant or repeating in pattern and that replaces itself; an ecosystem at a steady state. The terminal community of sere which is in dynamic equilibrium with the prevailing climate. The major world climaxes are equivalent to formations and biomes. The term is also used in connection with any subdivision, such as climax association.

Climax or mature community: the theoretical endpoint of ecologic succession. Usually the oldest and most stable communities, a climax is defined as a community where gross productivity equals respiration, and therefore net productivity equals zero. Many ecologists argue that the concept of a fixed climax has outlived its usefulness, and one should rather speak of communities in terms of relative maturity. End of a successional sequence; the last stage of the sere (which see). A community that has reached stability under a particular set of environmental conditions.

Climograph: A graphic representation of the relationship between rainfall and temperature.

Cline: change in population characteristics over a geographic area, usually related to a corresponding environmental change.

Clone: a population of cells formed from one cell.

Coadaptation: evolution of characteristics of two or more species to their mutual advantage.

Coagulation: the clumping of particles in order to settle out impurities; often induced by such chemicals as lime or alum. An aggregation or flocculation of cells.

Coast: a strip of land of indefinite width (may be several miles) that extends from the seashore inland to the first major change in terrain features.

Coastal zone: coastal waters and adjacent lands that exert a measurable influence on the uses of the sea and its ecology.

Coccus: a spherical bacterium

COD: see **chemical oxygen demand**. This factor gives an approximation of the quantities of oxidizable materials present in waters and wastes.

Coefficient of haze (COH): a measurement of visibility interference in the atmosphere.

Coencytic organisms: organisms that consist of protoplasm containing many nuclei. Slime molds are coenocytic at one stage in their development.

Coevolution: development of genetically determined traits in two species to facilitate some interaction, usually mutually beneficial. See **counterevolution**.

Coexistence: occurrence of two or more species in the same habitat; usually applied to potentially competing species.

Coffin: a thick-walled container (usually lead) used for transporting radioactive materials.

COH: See **coefficient of haze**.

Coliform bacteria: a group of bacteria predominantly inhabiting the intestines of man or animals, but also occasionally found elsewhere; fecal coliform bacteria or those organisms associated with the intestine of warm-blooded animals that are used commonly to indicate the presence of fecal material and the potential presence of organisms capable of causing disease in man.

Coliform, fecal: coliforms originating in the intestines of warm-blooded animals, i.e., *Escherichia coli.*

Coliform index: an index of the purity of water based on a count of its coliform bacteria.

Coliform organism: any of a number of organisms common to the intestinal tract of man and animals whose presence in waste water is an indicator of pollution and of potentially dangerous bacterial contamination.

Coliform, total: all bacterial organisms of the coliform group, including *Excherichia coli* and *Aerobacter aerogenes.* These organisms are classified as being aerobic and facultative anaerboic, gram negative, nonsporulating bacilli that produce acid and gas from the fermentation of lactose.

Coliphage: a virus that infects *E. coli.*

Colligative properties: those properties that vary with the number of chemical elements in a solution and not with the composition of the elements. In sea water with increasing salinity, boiling point and osmotic pressure increase, and freezing point and vapor pressure decrease.

Colloidal particles: very small particles, usually smaller than 0.00024 mm.

Colluvium: a general term applied to loose and incoherent deposits, usually at the foot of a slope or cliff and brought there chiefly by gravity. Talus and cliff debris are included in such deposits.

Colony: a visible growth of a protist on a solid nutrient medium.

Combined sewers: a sewerage system that carries both sanitary sewage and storm water runoff. During dry weather, combined sewers carry all waste water to the treatment plant. During a storm only part of the flow is intercepted because of plant overloading; the remainder goes untreated to the receiving stream.

Combustion: burning. Technically, a rapid oxidation accompanied by the release of energy in the form of heat and light. It is one of the three basic contributing factors to air pollution; the others are attrition and vaporization.

Commensalism: an association between two organisms that benefits one or both organisms.

Comminution: mechanical shredding or pulverizing of waste; a process that converts it into a homogeneous and more manageable material. Used in solid waste management and in the primary stage of waste water treatment.

Comminutor: a device that grinds solids to make them easier to treat.

Community: a group of populations living closely together as an association; an integrated group of organisms inhabiting a common area. These organisms may be dependent on each other or possibly upon the environment. The community may be defined by its habitat or by the composition of the organisms. An association of interacting populations, usually delimited by their interactions or by spatial occurrence.

Compaction: reducing the bulk of solid waste by rolling and tamping.

Compensation depth: the depth at which the oxygen produced by a plant during photosynthesis equals the amount the plant needs for respiration (during a 24-hr period).

Compensation point: stage at which respiration and photosynthesis balance each other; usually refers to the lower limit of the euphotic zone. Also, the level of oxygen in the environment at which an organism's oxygen intake just balances its oxygen consumption.

Competition: use or defense of a resource by one individual that reduces the availability of that resource to other individuals.

Competitive exclusion principle: the hypothesis, based on theoretical considerations and laboratory experiments, that two or more species cannot coexist on a single resource that is scarce relative to demand for it.

Compost: relatively stable decomposed organic material.

Composting: a controlled process of degrading organic matter by microorganisms. (1) Mechanical: a method in which the compost is continuously and mechanically mixed and aerated. (2) Ventilated cell: compost is mixed and aerated by being dropped through a vertical series of ventilated cells. (3) windrow: an open-air method in which compostable material is placed in windrows, piles, or ventilated bins or pits and occasionally turned or mixed. The process may be anaerobic or aerobic. A process for decomposing solid wastes to a stable product for use in agriculture.

Condensation: the transformation of water from a vapor to a liquid, such as occurs when vapor in the atmosphere is changed to droplets of rain.

Cone of depression: a conical dimple in the water table surrounding a well, caused by pumping. The faster water is pumped, the deeper and steeper the cone becomes.

Congeneric: belonging to the same genus.

Conjugation: the combination of two sexual forms of an organism to yield a new individual.

Conservative elements: elements in sea water whose ratio to other conservative elements remains constant. Examples are chlorine, sodium, and magnesium.

Conspecific: belonging to the same species.

Consumer: an organism that grows and obtains its energy by utilizing the organic materials of other organisms, living or dead. Contrast producers, which use light or inorganic energy.

Consumer: organism, human being, or industry that maintains itself by transforming a high-quality energy source.

Consumptive use: the use of water, especially in irrigation, in such a way that it is converted to vapor and returned to the atmosphere. Therefore, it can no longer be directly returned to the stream or underground source from which it has originated.

Contact pesticide: a chemical that kills pests on contact with the body, rather than by ingestion (stomach poison).

Containment: structure within a reactor building—or the building itself—that acts as a barrier to contain any radioactivity which may escape from the reactor itself.

Contaminate: to introduce external microorganisms to a sample that is either sterile or under controlled growth conditions.

Contamination: radioactivity where it should not be.

Continent: large landmass rising more or less abruptly above the deep ocean floor; includes marginal areas that are shallowly submerged. At present, continents constitute about one-third of the earth's surface.

Continental drift: the concept that the continents can drift on the surface of the earth because of the weakness of the suboceanic crust, much as ice can drift through water. As proposed by the German meteorologist Alfred Wegener in 1912, the theory that the continents were once joined into a landmass that broke apart into several landmasses which then drifted, their shapes changing somewhat, and eventually arrived at their present positions as they continue to move.

Continental island: (1) an island that is near to and geologically related to a continent, as are the British Isles. (2) Continental islands are merely detached fragments of the continent near which they stand and from which they are separated, in almost all cases, by shoal water. The limit between deep and shallow water is drawn at the 100-fathom line (183 m), and nearly all continental islands rest upon submarine platforms that are under water less than 100 fathoms (183 m) deep and run into the submerged continental shelf.

Continental margin: that portion of the ocean adjacent to the continent and separating it from the deep sea. The continental margin includes the continental shelf, continental slope, and continental rise.

Continental rise: an area of gentle slope (usually less than half a degree or 1 : 100) at the base of the continental slope; submarine surface beyond the base of the continental slope, generally with gradient less than 1:1000, occurring at depths from about 1373 to 5185 m and leading down to abyssal plains.

Continental shelf: the shallow part of the sea floor immediately adjacent to the continent. It generally has a smooth seaward slope and terminates seaward at an abrupt change in slope, beginning the continental slope. Gently sloping, shallowly submerged marginal zone of the continents extending from the shore to an abrupt increase in bottom inclination; greatest average depth less than 183 m, slope generally less than 1 : 1000, local relief less than 18.3 m, width ranging from very narrow to more than 321.8 km.

Continental slope: a declivity averaging about 4° that extends from the seaward edge of the continental shelf down to the continental rise or deep sea floor. Continuously sloping portion of the continental margin with gradient of more than 1 : 40, beginning at the outer edge of the continental shelf and bounded on the outside by a rather abrupt contour.

Contour: (1) outline on an object. (2) Line connecting points of equal value on a map or diagram, most commonly points of equal elevation on a map.

Contour interval: the difference in value between two adjacent contour lines. Generally, it refers to the difference in elevation between two adjacent contour lines.

Contrails: long, narrow clouds caused by the disturbance of the atmosphere during passage of high-flying jets.

Convection currents: motion within a fluid caused by differences in density or temperature; transfer of material caused by differences in density, generally brought about by heating. Characteristic of the atmosphere and bodies of water.

Convergent evolution: (1) the process whereby phylogenetic stocks, which are not closely related genetically, produce similar appearing forms. (2) The evolution toward a common adaptation (when it occurs in forms that have independently developed similar adaptive features) even though they are far removed from each other genetically; development of characteristics with similar functions in species that live in the same kind of environment but in different places.

Coolant: (1) a substance, usually liquid or gas, used for cooling any part of a reactor in which heat is generated, including the core, the reflector, shield, and other elements that may be heated by absorption of radiation. (2) Liquid (water, molten metal) or gas (carbon dioxide, helium, air) pumped through reactor core to remove heat generated in the core.

Cooling pond: deep tank of water into which irradiated fuel is discharged upon removal from a reactor, there to remain until shipped for reprocessing.

Cooling tower: a device to remove excess heat from water used in industrial operations, notably in electric power generation.

Core: (1) the heart of a nuclear reactor where energy is released. (2) The region of a reactor containing fuel (and moderator, if any) within which the fission reaction is occurring.

Coriolis force: an apparent force created by the earth's rotation. This force causes moving objects to turn to the right in the northern hemisphere and to the left in the southern hemisphere. The apparent force caused by the earth's rotation which serves to deflect a moving body on the surface of the earth to the right in the northern hemisphere, but to the left in the southern hemisphere.

Counteradaptation: evolution of characteristics of two or more species to their mutual disadvantage.

Counterevolution: development of traits in population in response to exploitation, competition, or other detrimental interaction with another population.

Courtship: any behavioral interaction between individuals of opposite sexes that facilitates mating.

Covalent bond: the bond or linkage between two atoms in a molecule, formed by the sharing of electrons.

Cover material: soil that is used to cover compacted solid waste in a sanitary landfill.

Critical: refers to a chain reaction in which the total number of neutrons in one "generation" of a chain reaction is the same as the total number of neutrons in the next "generation" of the chain; that is, a system in which the neutron density is neither increasing or decreasing.

Critical mass: the minimum amount of a fissionable material, such as uranium-235 or plutonium-239, that is required to sustain fission in a nuclear reactor. For ^{239}Pu, the critical mass is about 5 kg.

Cross-bedding: (1) the arrangement of laminations of strata transverse or oblique to the main planes of stratification of the strata concerned; inclined, often lenticular, beds between the main bedding planes, found only in granular sediments. (2) The term should be applied to inclined bedding found only in profiles at right angles to the current direction.

Crossed Nicols: two Nicol prisms placed so that their vibration planes are mutually at right angles. Optical minerology: an anisotropic crystal is interposed between the Nicol prisms in order to observe optical interference effects. The petrographic microscope is normally used with nicol prisms (or equivalent polarizing devices) in the crossed position.

Crypsis: an aspect of the appearance of organisms whereby they avoid detection by others; usually applied to the prey of visually hunting predators.

Cultural change: any modification in characteristics specific to a population that is transmitted by teaching and learning, rather than by genetic mechanisms.

Cultural eutrophication: acceleration by man of the natural aging process of bodies of water.

Culture: (1) a defined microbial population growing in a nutrient medium. (2) Ways of life, language, social interaction, government, religion, etc., of a group of people.

Curie: (1) amount of radioactive material giving off 37,000 million radioactive emissions per second; radioactivity of 1 g of radium. (2) a measure of radioactivity.

Cutie-pie: a portable instrument equipped with a direct reading meter used to determine the level of radiation in an area.

Cybernetics: the study of control. Increasingly utilized in ecology.

Cyclone collector: a device used to collect large sized particulates from polluted air by centrifugal force.

Cytochrome: essential electron carriers utilized in respiratory processes.

Damselfly nymph (Odonata): the immature damselfly. This aquatic insect nymph has an enormous grasping lower jaw and three flat leaflike gill plates that project from the posterior of the abdomen. Nymphs live most of their lives searching for food among submerged plants in still water; a few cling to plants near the current's edge and a very few cling to rocks in flowing water. The carnivorous adults capture lesser insects on the wing.

Datum elevation: a level reference elevation used in mapping, the time datum used is near the surface.

Diameter Breast Height (DBH): a height of 1.3 m (4.5 ft) above average ground surface or above the root collar to measure the diameter of standing trees.

DDT: the first of the modern chlorinated by hydrocarbon insecticides: chemical name, 1,1,1-tricholoro-2,2-*bis*(*p*-chloriphenyl)ethane. It has a half-life of 15 years, and it residues can become concentrated in the fatty tissues of certain organisms, especially fish. Because of its persistence in the environment and its ability to accumulate and magnify in the food chain, the EPA has banned the registration and interstate side of DDT for nearly all uses in the United States effective December 31, 1972.

Dead and down fuel: a flammable fuel that is both dead and located on the forest floor. It is also possible to have live and standing fuels (living plants) and dead and standing fuels (grasses, forbs, or shrubs which are dessicated but still standing). Dead and down fuels are divided among size classes based on the average time lag of their fuel moisture content when relative humidity changes. A 1-hr time lag fuel would require about 1 hr to increase its fuel moisture after a humidity increase: a 10-hr fuel would require about 10 hr, etc.

Dead reckoning: a type of navigation that uses only the speed and direction of the ship to estimate its position.

Death rate: the percentage of newborns dying during a specified interval (see **mortality**).

Decay: radioactive transformation.

Decibar: a measure of pressure equal to 1/10 normal atmospheric pressure and approximately equal to the pressure change of 1 m depth in sea water.

Decibel (dB): a unit of measurement 10 × the logarithm of the ratio between two levels of power. In most cases the denominator is a threshold level.

Decline: to lessen in force, value, function, etc.; used of storage, assets, energy.

Decomposers: consumers, especially microbial consumers, that change their organic food into mineral nutrients; usually microconsumer organisms that break down organic matter and thus aid in recycling nutrients.

Decomposition: reduction of the net energy level and change in chemical composition of organic matter because of the actions of aerobic or anerobic microorganisms. Metabolic breakdown of organic materials; the by-products are released energy and simple organic and inorganic compounds (see **respiration**).

Decontamination: transfer of unwanted radioactivity to a less undesirable location.

Deep-scattering layer: a sound-reflecting layer caused by the presence of certain organisms in the water. The layer, or layers, which may be 100-m thick, usually rises toward the surface at night and descends when the sun rises.

Deep sea drilling project: a large-scale scientific project whose main aim is to drill numerous deep holes into the sediments on the ocean floor.

Delta: a deposit of sand or sediment, usually triangular, formed at the mouth of rivers. (1) An alluvial deposit, usually triangular, at the mouth of a river. (2) A tidal delta is a similar deposit at the mouth of a tidal inlet, put there by tidal currents. (3) A wave delta is a deposit made by large waves that run over the top of a spit or bar beach and down the landward side. (4) A deposit of sediment formed at the mouth of a river, in either the ocean or a lake, which results in progradation of the shore line. The name delta (Greek capital letter Δ) has been applied because of the triangular shape assumed by the salients formed by deltas built out from the straight coast.

Demand (economic): the desire for a commodity together with the ability to pay for it; also, the amount of some commodity that people are ready and able to buy at a certain price.

Denitrification: enzymatic reduction of nitrates by bacteria to nitrogen gas.

Density: (1) the mass or quantity of a substance in grams per cubic centimeter. (2) The quality of being dense, close, or compact. (3) The quantity of electricity per unit of volume at a point in space, or the quantity of electricity per unit of area at a point on a surface. (4) In an ecologic sense density is usually subdivided into two categories: (a) crude density, the number of biomass per unit of total area; and (b) specific density, the number of biomass per unit of habitable (colonizable) area. Weight per unit volume.

Density dependent: having influence on individuals in a population that varies with the number of individuals per unit area in the population.

Density independent: having influence on individuals in a population that does not vary with the number of individuals per unit area in the population.

Deoxyribonucleic acid (DNA): a nucleic acid found in most organisms.

Depreciation: a decrease in value of assets through deterioration.

Depuration: purifying.

Dermal toxicity: the ability of a pesticide chemical to poison an animal or human by skin absorption.

Dermatitis: any inflammation of the skin. One type may be caused by the penetration beneath the skin of cercaria found in water; this form of dermatitis is commonly called "swimmer's itch."

Desalination: the process of removing salt from water. Most often it is a distillation process in which salt water is evaporated and then condensed, leaving the salt as residue. A variety of processes whereby the salts are removed from sea water, resulting in water that can be used for human consumption; salt removal from sea or brackish water.

Desert: a region so devoid of vegetation as to be incapable of supporting any considerable population. Four kinds of deserts, may be distinguished: (1) the polar ice and snow deserts, marked by perpetual snow cover and intense cold; (2) the middle latitude deserts, in the basinlike interiors of the continents, such as the Gobi, characterized by scant rainfall and high summer temperatures; (3) the tradewind deserts, notably the Sahara, the distinguishing features of which are negligible precipitation and large daily temperature range; and (4) coastal deserts, where there is a cold current on the western coast of a large land mass, such as occurs in Peru.

Desiccant: a chemical agent that may be used to remove moisture from plants or insects, causing them to wither and die.

Desiccate: to dry up, to deprive or exhaust of moisture; to preserve by drying.

Detergent: synthetic washing agent that, like soap, lowers the surface tension of water, emulsifies oils, and holds dirt in suspension. Environmentalists have critized detergents because most contain large amounts of phosphorus-containing compounds that contribute to the eutrophication of waterways. Cleaning agents. Detergents may be soft, easily biodegraded, or hard, resistant to biodegradation.

Detrital deposits: sedimentary deposits resulting from the erosion and weathering of rocks.

Detritus: dead organic material. (1) Material produced by the disintegration and weathering of rocks that has been moved from its site of origin. (2) A deposit of such material. (3) Any fine particulate debris, usually of organic origin but sometimes defined as organic and inorganic debris. (4) Freshly dead or partially decomposed organic matter.

Deuterium: hydrogen-2, heavy hydrogen; its nucleus consists of one proton plus one neutron, rather than the one proton only of ordinary hydrogen.

Deuterium oxide, heavy water: water in which the hydrogen atoms are heavy hydrogen.

Developmental response: acquisition of one of several alternative forms by an organism, depending on the environmental conditions under which it grows.

Diapause: temporary interruption in the growth of embryos and larvae of insects, usually associated with a dormant period.

Dialysis: use of a semipermeable membrane. to separate water from dissolved materials. Separation of a colloid from ions and molecules in true solution, e.g., permitting ions to diffuse through a membrane that is impervious to the colloidal particles.

Diatomaceous earth (diatomite): a fine siliceous material, resembling chalk, used in waste water treatment plants to filter sewage effluent to remove solids. May also be used as inactive ingredients in pesticide formulations applied as dust or powder.

Diatometer: an apparatus that holds microscopic slides in the water. It is held in place by means of floats and an anchor. Living diatoms, by means of their thin gelatinous coating, become attached to the glass slides. The slides are removed from the diatometer at intervals generally of 14 days, dried, and shipped to the laboratory for study, identification, and enumeration.

Diatoms: organisms closely associated with algae that are characterized by the presence of silica in the cell walls, which are sculptured with striae and other markings, and by the presence of a brown pigment associated with the chlorophyll.

Dichlorodiphenyltrichloroethane: insecticide known to accumulate in animal tissues, DDT.

Diel: refers to the 24-hr period of day and night.

Diffused air: a type of sewage aeration. Air is pumped into the sewage through a perforated pipe.

Diffusion: random movements of molecules or heat from regions of concentration to regions of scarcity.

Digester: in a waste water treatment plant, a closed tank that decreases the volume of solids and stabilizes raw sludge by bacterial action.

Digestion: the biochemical decomposition of organic matter. Digestion of sewage sludge takes place in tanks where the sludge decomposes, resulting in partial gasification, liquefaction, and mineralization of pollutants. Chemical breakdown of food into a form that can be assimilated by an organism.

Dilution ratio: the ratio of the volume of water of a stream to the volume of incoming waste. The capacity of a stream to assimilate waste is partially dependent upon the dilution ratio.

Diluvium: name given to all coarse superficial accumulations, which were formerly supposed to have resulted from a general deluge; now employed as a general term for all the glacial and fluvioglacial deposits of the ice age.

Dimorphism: occurrences of two forms of individuals in a population.

Dioecy: in plants, the occurrence of reproductive organs of both sexes on different individual plants. See **monoecy.**

Diploid: pertaining to cells or organisms having two sets of chromosomes. See **ploidy.**

Direct competition: exclusion of individuals from resources by aggressive behavior or use of toxins by other individuals.

Discharge: the rate of flow of surface or underground water, generally expressed in cubic feet per second. It also refers to the emptying of a river into a lake or ocean.

Disinfectant: a chemical that kills microorganisms.

Disinfection: effective killing by chemical or physical processes of all organisms capable of causing infectious disease. Chlorination is the disinfection method commonly employed in sewage treatment processes.

Dispersal: movement of organisms away from the place of birth or from centers of population density.

Dispersant: a chemical agent used to break up concentrations of organic material. In cleaning oil spills, dispersants are used to disperse oil from the water surface.

Dispersion: pattern of spacing of individuals in a population.

Display: behavior and accompanying morphologic modifications that have evolved expressly for the purpose of communication.

Disruptive selection: evolutionary disadvantage of individuals in a population that have intermediate traits, which leads to the divergence of distinct subpopulations with different extreme traits.

Dissolved oxygen (DO): the oxygen dissolved in water or sewage. Adequate dissolved oxygen is necessary for the life of fish and other aquatic organisms and for the prevention of offensive odors. Low dissolved oxygen concentrations generally are caused by the discharge of excessive organic solids having high BODs, the result of inadequate waste treatment.

Dissolved solids: the total amount of dissolved material, organic and inorganic, contained in water or wastes. Excessive dissolved solids make water unpalatable for drinking and unsuitable for industrial uses.

Distillation: the removal of impurities from liquids by boiling. The steam, condensed back into liquid, is almost pure water; the pollutants remain in the concentrated residue.

Diurnal: referring to tides, one low and one high tide within one lunar day (about 24 hr and 50 min).

Divergence: the flow of water in different directions away from a particular area or zone; often associated with areas of upwelling.

Diversity: the number of species that lives together in an ecosystem; a measure of the variety of species in a community that takes into account the relative abundance of each species. Variety; number of differences, as number of different species.

DNA: deoxyribonucleic acid, concentrated mainly in the nuclear structures of organisms.

Dominance (genetic): ability of a genetic trait (allele) to mask the expression of an alternative form of the same gene when both are present in the same cell (that is, in a heterozygote).

Dominance hierarchy: orderly ranking of individuals in a group, based on the outcome of aggressive encounters.

Dormancy: a period of reduced biologic activity; for example, hibernation, estivation, and the period before seed germination.

Dose: in radiology, the quantity of energy or radiation absorbed. Amount of energy delivered to a unit mass of a material by radiation travelling through it.

Dose rate: time rate at which radiation delivers energy to a unit mass of a material through which it is traveling.

Dosimeter (dosemeter): an instrument used to measure the amount of radiation a person has received.

Dragonfly nymph (Odonata): the immature dragonfly. This aquatic insect nymph has gills on the inner walls of its rectal respiratory chamber. It has an enormous grasping lower jaw that it can extend forward to a distance several times the length of its head. Although many of these nymphs climb among aquatic plants, most sprawl in the mud, where they lie in ambush to await their prey.

Drainage basin: the geographic area within which all surface water tends to flow into a single river or stream via its tributaries.

Dredging: a method for deepening streams, swamps, or coastal waters by scraping and removing solids from the bottom. The resulting mud may be deposited in marshes in a process calling filling. Dredging and filling can disturb natural ecologic cycles. For example, dredging can destroy oyster beds and other aquatic life; filling can destroy the feeding and breeding grounds for fish species.

Drift: see **genetic drift.**

Drowned coast: the presence of certain long, narrow channels that are *largely* free from islands suggests that a subsidence of the coast has transformed the lower portions of the old river valleys into tidal estuaries, thus changing a hilly land surface into an archipelago of small islands. A shoreline exhibiting these characteristics is termed a "drowned coast."

Dry limestone process: a method of controlling air pollution caused by sulfur oxides. The polluted gases are exposed to limestone, which combines with oxides of sulfur to form manageable residues.

Dump: a land site where solid waste is disposed of in a manner that does not protect the environment.

Dust: fine-grained particulate matter that is capable of being suspended in air.

Dustfall jar: an open-mouthed container used to collect large particles that fall out of the air. The particles are measured and analyzed.

Dystrophic lakes: lakes between the eutrophic and swamp stages of aging. Such lakes are shallow and have high humus content, high organic matter content, low nutrient availability, and high BOD.

Earth tide: the rising and falling of the surface of the solid earth in response to the same forces that produce the tides of the sea. Semidaily earth tides fluctuate between 7 and 15 cm.

Earthquake: a sudden motion of the earth caused by faulting or volcanic activity. Earthquakes can occur in the near surface rocks or down to as deep as 700 km below the surface. The actual area of the earthquake is calied the focus; the point on the earth's surface above the focus is called the epicenter.

Ebb tide: a nontechnical term referring to that period of tide between a high water and the succeeding low water; falling tide.

Echo sounder: a survey instrument that determines the depth of water by measuring the time required for a sound signal to travel to the bottom and return.

Echo sounding: a method of determining the depth of the ocean by measuring the time interval between the emission of an acoustic signal and its return or echo from the sea floor. The returning signal is usually printed to give a visual picture of the topography of the sea floor. The instrument used in this method is called an echo sounder.

Ecologic: pertaining to the living environment.

Ecologic counterparts: species of different phylogenetic origin that occur in different areas and that have converged phenotypically to fulfill similar ecologic roles.

Ecologic efficiency: percentage of energy in biomass produced by one trophic level (which see) that is incorporated into biomass by the next highest trophic level.

Ecological impact: the total effect of an environmental change, either natural or manmade.

Ecological release: expansion of habitat and food preferences by populations in regions of low species diversity, resulting from reduced interspecific competition.

Ecology: the branch of science dealing with the relationship between organisms and their environment; the interrelationships of living things to one another and to their environment or the study of such interrelationships; the branch of biology that deals with the mutual relations of living organisms and their environments, and the relations of organisms to each other; the study of environmental systems, such as forests, lakes, seas, and urban areas, especially the living aspects; the study of mutual relationships between organisms and their environment, including the rock substrate.

Economic geology: deals with geologic materials of practical utility and the application of geology to engineering. Generally used today in the narrow sense to include only the application of geology to economically important mineral materials.

Economic poisons: those chemicals used to control insects, rodents, plant diseases, weeds, and other pests and also to defoliate economic crops such as cotton.

Ecosphere: See **biosphere.** The earth considered as one giant ecosystem.

Ecosystem: the interacting system of a biologic community and its nonliving environment; the functioning together of the biological community and the nonliving environment; a community together with its environment; an ecologic system; that system composed of interacting organisms and their environments; the result of interaction between biologic, geochemical, and geophysical systems. The geobiocoenose (which see). Ecologic system. Examples: forest, coral reef, old field, pond, aquarium.

Ecotone: a habitat created by the juxtaposition of distinctly different habitats; an edge habitat.

Ecotope: physical environment of a habitat. See also **biotope.** a genetically differentiated subpopulation that is restricted to a specific habitat.

Edaphic: a term referring for the soil conditions or types as ecologic factors. Pertaining to, or influenced by, soil conditions.

Eddy current: a circular movement of water of comparatively limited area formed on the side of a main current.

Edge water: the water surrounding or bordering oil or gas in a pool. Edge water usually encroaches on a field after much of the oil and gas has been recovered and the pressure has become greatly reduced.

Effective permeability: the observed permeability of a porous medium to one fluid phase under conditions of physical interaction between this phase and other fluid phases present.

Effective porosity: (1) Hydrology: often used in same sense as specific yield. It is the ratio of the volume of water, oil, or other liquid, which after being saturated with that liquid, it will yield under any specified hydraulic conditions to its own volume. (2) The property of rock or soil containing intercommunicating interstices, expressed as a percent of bulk volume occupied by such interstices.

Effective size: soil mechanics: the maximum diameter of the smallest 10% of the particles of a sediment. Equals the grain diameter at the nintieth percentile in sedimentary petrography.

Effective size of grain: Hydrology: the diameter of the grains in an assumed rock or soil that would transmit water at the same rate as the rock or soil under consideration but that is composed of spherical grains of equal size and arranged in a specified manner.

Effluent: a discharge of pollutants into the environment, partially or completely treated or in its natural state. Generally used in regard to discharges into waters. (1) Flowing forth out of. Geology: Flowing out, as lava through fissures in the side of a volcano, as a river from a lake. (2) Anything that flows forth. Geography: a stream flowing out of another or forming the outlet of a lake. (3) A discharge of pollutants into the environment, partially or completely treated or in its natural state. Generally used in regard to discharges into waters.

Egestion: elimination of undigested food material.

Electric conductivity: a measure of the ability that a material has for conducting electricity.

Electrodialysis: a process that uses electric current and an arrangement of permeable membranes to separate soluble minerals from water. Often used to desalinize salt or brackish water. The removal of ions from water by transfer across a membrane using electric current. Particularly useful in desalination.

Electron: negatively charged particle; much lighter than proton or neutron.

Electrostatic precipitator: an air pollution control device that removes particulate matter by imparting an electrical charge to particles in a gas stream for mechanical collection on an electrode.

Elevation: (1) a particular height or altitude above a general level, as, the height of a locality above the level of the sea, of a building, etc.; above the level of the ground. (2) In the United States, term generally refers to height in feet above mean sea level.

Elution: purification by washing and pouring off the lighter matter suspended in water, leaving the heavier portions behind.

Eluviated horizon: the uppermost, leached horizon in a soil. If there is leaching, the A horizon is leached first, but some soils are leached throughout and others are not leached.

Eluvium: atmospheric accumulation *in situ,* or at least only shifted by wind, in distinction to alluvium, which requires the action of water.

Embayment: (1) An indentation in a shoreline forming an open bay. (2) The formation of a bay. (3) The term describing a continental border area that has sagged concurrently with deposition so that an unusually thick section of sediment results. An embayment is similar to a basin of sedimentation or a geosyncline, and some embankments may be one flank of a larger subsiding feature. (4) Used in structural sense to designate a reentrant of sedimentary rocks into a crystalline massif.

Emergency episode: See **air pollution episode.**

Emission: See **effluent.** Generally used in regard to discharges into air.

Emission factor: the average amount of a pollutant emitted from each type of polluting source in relation to a specific amount of material processed. For example, an emission factor for a blast furnace (used to make iron) would be a number of pounds of particulates per ton of raw materials.

Emission inventory: a list of air pollutants emitted into a community's atmosphere, in amounts (usually tons) per day, by type of source. The emission inventory is basic to the establishment of emission standards.

Emission standard: the maximum amount of a pollutant legally permitted to be discharged from a single source, either mobile or stationary.

Emulsion: a colloidal dispersion of one liquid in another.

Encroachment: the advancement of water, replacing withdrawn oil or gas in a reservoir.

Endemic: an organism that is common to the ecosystem at any time. Particularly applicable to pathogens.

Endogenous: produced from within; originating from or due to internal causes. Synonyms: Endogenetic, endogenic. Contrasted with exogenous.

Endotoxin: a poison produced by a microbial cell that is released following cell lysis.

Energy: quantity that accompanies all processes and is measured by the amount of heat it becomes; a quantity necessary for useful work.

Energy budget: the balance between energy inputs and energy demands in an ecosystem.

Energy chain: a sequence of energy-transforming systems each of which recieves energy from the preceding system and supplies it to the next.

Energy conservation law: energy flowing into a system is equal to energy stored in the system plus energy flowing out of it (all measured in heat equivalents).

Energy degradation law: whether stored or being used, concentrations of energy spontaneously disperse, losing their potential for doing work.

Energy resource: the natural supply of energy available for use. Several major sources are the earth's internal heat (geothermal energy), fossil fuels (principally coal, natural gas, and oil), hydropower (rivers, ocean currents, tides, and waves) nuclear energy, solar energy, and wind.

Enrichment: the addition of nitrogen, phosphorus, and carbon compounds or other nutrients into a lake or other waterway that greatly increases the growth potential for algae and other aquatic plants. Most frequently, enrichment results from the inflow of sewage effluent or from agricultural runoff.

Enteric: dealing with human intestines.

Entropy: a measure of disorder, depreciation. (1) A measure of the unavailable energy in a system, i.e., energy that cannot be converted into another form of energy. (2) A measure of the mixing of different kinds of sediment; high entropy is approached when a single class of sediment is pure (i.e., unmixed).

Environment: the sum of all external conditons and influences affecting the life, development, and ultimately the survival of an organism; the abiotic and biotic factors controlling the life of an organism; surroundings, a zone of the biosphere occupied by ecologic systems or human activities; the sum total of all the external conditions that may act upon an organism or community, to influence its development or existence. For example, the surrounding air, light, moisture, temperature, wind, solid, and other organisms are all parts of the environment. Surroundings of an organism, including the plants and animals with which it interacts.

Environmental geochemistry: the distribution and interrelationship of the chemical elements and radioactivity at the earth's surface, including rocks, water, air, and biota.

Environmental geology: the application of geologic principles and knowledge to problems created by human occupancy and exploitation of the physical environment.

Environmental gradient: a continuum of conditions ranging between extremes, as the gradation from hot to cold environments.

Environmental impact statement: a document prepared by a Federal agency on the environmental impact of its proposals for legislation and other major actions significantly affecting the quality of the human environment. Environmental impact statements are used as tools for decision making.

Enzyme: an organic catalyst; an organic compound in living cells that accelerates specific biochemical transformations without itself being affected.

Eolian: (Aeolian, Obs.) (1) Applied to deposits arranged by the wind, as the sands and other loose materials along shores, etc. (from Eolus, the Greek god of winds). Subaerial is often used in much the same sense. (2) Applied to the erosive action of the wind, and to deposits that are made by the transporting action of the wind.

Ephemeral stream: a stream or portion of a stream that flows only in direct response to precipitation.

Epicenter: See **earthquake.** The point on the earth's surface directly above the focus of an earthquake.

Epidemic: a large incidence of a disease in a population.

Epidemiology: the study of diseases as they affect populations.

Epilimnion: that region of a body of water that extends from the surface to the thermocline and does not have a permanent temperature stratification. The warm circulating upper level of a large body of water. The warm, oxygen-rich surface layers of a lake or other body of water that lie above the thermocline (which see). The depth of the epilimnion is usually determined by the depth to which vertical mixing of water occurs.

Epipelagic: the upper portion of the oceanic province extending from the surface to a depth of about 185 m; also the animals that occupy it.

Epiphyte: a plant that lives wholly, but nonparasitically, on other plants, usually above the ground.

Epiplankton: those organisms which habitually live upon a floating object to which they are attached or on which they move freely.

Epithelial layer: cellular tissue covering all the free body surfaces, cutaneous, mucous, and serous, including the glands and other structures derived therefrom.

Equilibrium: equilibrium exists in any system when the phases of the system do not undergo any change of properties with the passage of time and provided the phases have the same properties when the same conditions, with respect to variants, are again reached by a different procedure.

Equilibrium isocline: a line on the population graph designating combinations of competing populations, or predator and prey populations, for which the growth rate of one of the populations is zero.

Equilibrium model: any hypothetical or actual representation of a system in which two balancing forces act to maintain stability in the system.

Equivalent radius: a measure of sedimentary particle size equal to the radius of a spherical particle of the same density that would have the same settling rate as the sedimentary particles.

Erosion: the wearing away of the land surface by wind or water. Erosion occurs naturally from weather or runoff but is often intensified by human land-clearing practices. The wearing down of the earth's surface by water; a term that describes the physical and chemical breakdown of a rock and the movement of these broken or dissolved particles from one place to another; the group of processes whereby earthy or rock material is loosened or dissolved and removed from any part of the earth's surface. It includes the processes of weathering, solution, corrasion, and transportation. The mechanical wear and transportation are affected by running water, waves, moving ice, or winds, which use rock fragments to pound or grind other rocks to powder or sand.

Erratic: (1) large water-worn and ice-borne blocks (boulders) which are scattered so generally over the higher and middle latitudes of the northern hemisphere. (2) A transported rock fragment different from the bedrock on which it lies, either free or as part of a sediment. The term is generally applied to fragments transported by glacier ice or by floating ice.

Escarpment: (1) To cut steeply; a steep face terminating high lands abruptly. (2) A slope; a steep descent; a declivity. Geology: The steep face frequently presented by the abrupt termination of stratified rocks.

Esthetic insults: a degradation of beauty through arrogant mistreatment.

Estivation: reduction of biologic activity by an organism during the summer; more generally, during periods that are hot, or dry, or both.

Estuarine: of, pertaining to, or formed in an estuary.

Estuary: an arm of the sea containing aquatic life, especially the wide mouth of a river, where the tide meets the current. Drainage channel adjacent to the sea in which the tide ebbs and flows. Some estuaries are the lower courses of rivers or smaller streams, others are no more than drainage ways that lead sea water into and out of coastal swamps.

Ethic: (as work ethic) idea or action which is considered right or moral.

Eucaryote: a protistan cell having a distinct nucleus.

Euphotic zone: that part of the water which receives sufficient sunlight for plants to be able to photosynthesize. The zone in water where there is sufficient light for algal productivity. Surface layer of water to the depth of light penetration at which photosynthesis of organisms just balances their respiration (see **compensation point**). The layer of water which receives ample sunlight for photosynthesis. It varies with the light extinction coefficient. Usually no deeper than about 60 m.

Euryhaline: tolerant of considerable difference in salinity; generally refers to marine organisms.

Eutrophic: term used to refer to a lake, pond, etc.: productive, rich in plant nutrients, minerals, and organisms, but often with variable conditions; referring to a body of water with abundant nutrients and high productivity.

Eutrophic lakes: shallow lakes, weed choked at the edges and very rich in nutrients. The water is characterized by large amounts of algae, low water transparency, low dissolved oxygen, and high BOD. A shallow lake that has good primary productivity, usually abundant littoral vegetation, and dense plankton population. Blooms occur when nutrient and physical conditions are right. During later stages of eutrophication the bottom water tends to become depleted of oxygen during the summer.

Eutrophic waters: waters with a good supply of nutrients; they may support rich organic production, such as algal blooms.

Eutrophication: the normally slow aging process by which a lake evolves into a bog or marsh and ultimately assumes a completely terrestrial state and disappears. During eutrophication the lake becomes so rich in nutritive compounds, especially nitrogen and phosphorus, that algae and other microscopic plant life becomes superabundant, thereby "choking" the lake and causing it eventually to dry up. Eutrophication may be accelerated by many human activities. The enrichment of a body of water with nutrients; the intentional or unintentional enrichment of water; Enrichment of bodies of water, primarily caused by sewage and runoff from fertilized agricultural land.

Evaporation: the transformation of water into a vapor. This physical change occurs when heat is absorbed by the liquid, as when boiling water turns to steam.

Evaporation ponds: shallow, artificial ponds where sewage sludge is pumped, permitted to dry, and either removed or buried by more sludge.

Evapotranspiration: the process by which water, evaporated from the earth and given off by plants and animals, is returned to the atmosphere as vapor. The sum of transpiration by plants and evaporation from water surfaces and the soil. Potential evapotranspiration is the amount of evapotranspiration that would occur if water were superabundant and is a function largely of temperature and humidity. A term embracing that portion of the precipitation returned to the air through direct evaporation or by transpiration of vegetation, no attempt being made to distinguish between the two.

Everglade: a tract of swampy land covered mostly with tall grass; a swamp or inundated tract of low land; local in the southern United States.

Evolution: the theory that life on earth has developed gradually, generally by change and branching of and within species, and generally from simple to complex.

Evolutionary opportunism: the principle that adaptations are based on whatever genetic variation is available in a population.

Excited: having an excess of energy.

Excretion: waste products produced by a cell.

Exothermic: designating, or pertaining to, a reaction that occurs with a liberation of heat. Antonym: andothermic.

Exotoxin: a poison excreted by a microbial cell.

Exploit: (1) to make complete use of; to utilize. (2) To use in research or experiment; to explore.

Exploitation: removal of individuals or biomass (which see) from a population by predators or parasites.

Exploitation equilibrium: stability between predator and prey populations in which counterevolutionary responses of one just balance those of the other.

Exponential growth: the phase in the growth curve, usually of bacteria, in which the statistical population is growing at an exponential rate.

Exponential rate of increase (r). Rate at which a population is growing at a particular instant, expressed as a proportional increase per unit of time.

Exudate: metabolic products excreted by microorganisms.

Fabric filters: a device for removing dust and particulate matter from industrial emissions, much like a home vacuum cleaner bag. The most common use of fabric filters is the baghouse.

Facilitate: to make easy or easier.

Facultative anaerobe: a microorganism capable of growing either aerobically or anaerobically.

Fall overturn: a physical phenomenon that may take place in a body of water during the early autumn. The sequence of events leading to fall overturn include: (1) cooling of surface waters, (2) density change in surface waters producing convection currents from top to bottom, (3) circulation of the total water volume by wind action, and (4) vertical temperature equality, 4°C. The overturn results in a uniformity of the physical and chemical properties of the water.

Fallout: radioactive fission products created by nuclear explosions, which descend from the atmosphere onto the surface of the earth.

Fallow land: land left unfarmed for one or more growing seasons, in order to kill weeds, make the soil richer, etc.

Fast breeder reactor: a nuclear reactor that, while utilizing ^{235}U, alters ^{238}U to a fissionable fuel and therefore produces more nuclear fuel than it consumes.

Fat clay: clay of relatively high plasticity. Opposite: lean clay.

Fathom: a common unit measure of depth equal to 6 ft (1.83 m).

Fault: a break in the earth's crust, causing a slippage and displacement of layers of the crust. The fault may raise a layer of rock anywhere from a few inches to thousands of feet higher than the adjoining layers. This affects both the location and flow of the underground water, often causing springs to appear. A fracture of rock along which the opposite sides have been relatively displaced; a fracture or fracture zone along which there has been displacement of the sides relative to one another parallel to the fracture.

Fault line: the intersection of a fault surface with the surface of the earth or with any artificial surface of reference.

Fault zone: a fault, instead of being a single clean fracture, may be a zone hundreds or thousands of feet wide; the fault zone consists of numerous interlacing small faults or a confused zone of gouge, breccis, or mylonite.

Fauna: the animals of any place and time that lived in mutual association. The limitations of any fauna are relative. They may be interpreted broadly, as of a fauna of a continent or of a geologic period, or restrictively, as of the fauna of a small area of the sea bottom during a single season. A paleontologic fauna consists only of those animals the remains of which are preserved as fossils.

Fecal coliform bacteria: a group of organisms common to the intestinal tracts of humans and of animals. The presence of fecal coliform bacteria in water is an indicator of pollution and of potentially dangerous bacterial contamination.

Fecal pellet: excreta, mainly of invertebrates, present especially in modern marine deposits but also as fossils in sedimentary rocks. Most are of simple ovoid form and 1 mm or less in size. More rarely they are rod shaped, with either longitudinal or transverse sculpturing. Coporlites (which see) are of similar origin but much larger. Synonym: faecal pellet.

Fecal streptococci: another species of bacteria found to be normal inhabitants of the large intestine of man and other animals.

Fecundity: rate at which an individual produces offspring, usually expressed only for females.

Feedback: a flow from the products of an action back to interact with the action.

Feedback inhibition: inhibition of an enzyme caused by a product of that enzyme's activity on a substrate.

Feedlot: a relatively small, confined land area for raising cattle. Although an economic method of fattening beef, feedlots concentrate a large amount of animal wastes in a small area. This excrement cannot be handled by the soil as it could be if the cattle were scattered on open range. In addition, runoff from feedlots contributes excessive quantities of nitrogen, phosphorus, and potassium to nearby waterways, thus contributing to eutrophication.

Fellfield: a topgraphic feature and community type which is caused by frequent snow (avalanche), mud, or rock slides. These communities are usually very wet (hydrophytic) for at least part of the year, seldom contains a tree stratum, and often exhibit characteristic species adapted to survive such extreme conditions.

Fen: a low-lying land area partly covered by water.

Fermentation: anaerobic degradation of a substrate.

Fetch: the distance over the sea surface that the wind blows, in the area where seas are generated. (1) In wave forecasting, the continuous area of water over which the wind blows in essentially a constant direction. Sometimes used synonymously with fetch length. Also generating area. (2) In wind setup phenomena, for inclosed bodies of water, the distance between the points of maximum and minimum water surface elevations. This would usually coincide with the longest axis in the general wind direction.

Fiber: slender, threadlike tissues formed by plants or animals and used for paper, clothing, and building materials.

Filling: the process of depositing dirt and mud in marshy areas to create more land for real estate development. Filling can disturb natural ecologic cycles. See **dredging.**

Film badge: a piece of masked photographic film worn like a badge by nuclear workers to monitor an exposure to radiation. Nuclear radiation darkens the film.

Filter sand: sand suitable for use in filtering the suspended matter from water.

Filtration: in waste water treatment, the mechanical process that removes particulate matter by separating water from solid material usually by passing it through sand.

Fines: (1) the fine fraction of a sediment or the product of rock crushing, particularly that which passes through a grading sieve. (2) The fraction of sand and gravel finer than 0.074 mm in particle diameter.

Fingerling: a small fish about the length of a finger, or a young fish up to the end of its first year.

Fireline: the advancing flaming front of a fire; also a natural or manmade strip cleared of flammable fuels which will prevent a fire from crossing the strip.

Fish-protein concentrate: an odorless, tasteless, protein concentrate that can be made from almost any kind of fish. This product is thought by many to be a partial solution to some of the world's food problems.

Fission: rupture of a nucleus into two lighter fragments (fission products) plus free neutrons—either spontaneously or as a consequence of absorption of a neutron. (1) The splitting of an atomic nucleus into at least two parts (other atomic nuclei) plus the ejection of two or more neutrons (and occasionally other particles) plus the emission of energy. (2) Separation of calyx of a coral by cleavage. (3) Reproduction of the asexual type in which one individual divides to form two new ones, as in coelenterates.

Fitness: genetic contribution by an individual's descendants to future generations of a population.

Flagellum: a threadlike portion of many motile microbial cells that is responsible for the cell motility.

Floc: a clump of solids formed in sewage by biologic or chemical action. A small, light, loose mass, as of a fine precipitate.

Flocculation: in waste water treatment, the process of separating suspended solids by chemical creation of clumps or flocs. Aggregation of cells in liquid.

Floodplain: a strip of land bordering a stream or river, consisting of sediment laid down over the centuries by the river when it moves laterally within the valley walls and is subject to overflow high floods. That portion of a river valley, adjacent to the river channel, which is built of sediments during the present regimen of the stream and which is covered with water when the river overflows its banks at flood stages.

Flood tide: (1) the flow, or rising toward the shore, is called flood tide, and the falling away, ebb tide. (2) A nontechnical term referring to that period of tide between low water and the succeeding high water; a rising tide.

Flowmeter: in waste water treatment, a meter that indicates the rate at which waste water flows through the plant.

Flue gas: a mixture of gases resulting from combustion and emerging from a chimney. Flue gas includes nitrogen oxides, carbon oxides, water vapor, and often sulfur oxides or particulates.

Fluidization: process in which gas passes through loose, fine-grained material, mixes with it, and causes it to flow like a liquid; may occur at the time of volcanic eruption as in a glowing avalanche.

Fluorides: gaseous, solid, or dissolved compounds containing fluorine, emitted into the air or water from a number of industrial processes. Fluorides in the air are a cause of vegetation damage and, indirectly, of livestock damage.

Flume: a channel, either natural or manmade, which carries water.

Fluvial: of, or pertaining to, rivers; growing or living in streams or ponds; produced by river action, as a fluvial plain.

Fly ash: all solids, including ash, charred paper, cinders, dust, soot, or other partially incinerated matter, that are carried in a gas stream.

Focus: See **earthquake.**

Fog: liquid particles formed by condensation of vaporized liquids.

Fogging: the application of a pesticide by rapidly heating the liquid chemical, thus forming very fine droplets with the appearance of smoke. Fogging is often used to destroy mosquitos and blackflies.

Food chain: a complex system that involves many different organisms, each of which is the food for an organism higher up in the chain or sequence; the sequence of organisms in which each is food for a higher member of the sequence.

Food chain or food web: the sequence of organisms that feed on each other, starting at the phytoplankton and leading to very large organisms; an abstract representation of the passage of energy through populations in the community; the dependence of organisms upon others in a series for food. The chain begins with plants or scavenging organisms and ends with the largest carnivores.

Food waste: animal and vegetable waste resulting from the handling, storage, sale, preparation, cooking, and serving of foods; commonly called garbage.

Foot: the bottom of a slope, grade, or declivity. A term for the lower part of any elevated landform.

Fossil fuels: coal, oil, and natural gas; so called because they are derived from the remains of ancient plant and animal life.

Founder principle: the principle that a population started by a small number of colonists contains only a small fraction of the genetic variation of the parent population.

Fractionation: the separation or division into different components that occurs with some isotopes in the marine environment.

Fracture zone: a large linear and irregular area of the sea floor, characterized by ridges and seamounts. These features are commonly associated with the median ridge common to most ocean basins.

Freezing point: The temperature at which a liquid freezes or solidifies at normal pressure. For fresh water this temperature is 0°C or 32°F; for normal sea water (salinity about 35 $^o/_{oo}$) the freezing temperature is -1.9°C or about 28°F.

Fresh water: water with less than 0.2% salinity, e.g., 2000 ppm not the same as potable (= drinkable) water.

Frequency: the number of cycles or events in a given period of time.

Fringing reef: the coral reefs around other lands or islands rest on the bottom along the shores. They are either fringing or barrier reefs, according to their position. Fringing reefs are attached directly to the shore, whereas barrier reefs, like artificial moles, are separated from the shore by a body of water usually termed a lagoon.

Frost line: the maximum depth to which the ground becomes frozen, it may be given for a particular winter, for the average of several winters, or for the extreme depth ever reached.

Fuel assembly, fuel element: single unit of fuel plus cladding which can be individually inserted into or removed from reactor core.

Fume: tiny solid particles commonly formed by the condensation of vapors of solid matter.

Fumigant: a pesticide that is burned or evaporated to form a gas or vapor that destroys pests. Fumigants are often used in buildings or greenhouses.

Functional response: change in the rate of exploitation of prey by an individual predator as a result of a change in prey density (see **numerical response).**

Fungi: small, often microscopic plants without chlorophyll. Some fungi infect and cause disease in plants or animals; other fungi are useful in stabilizing sewage or in breaking down wastes for compost.

Fungicide: a pesticide chemical that kills fungi or prevents them from causing diseases, usually on plants of economic importance. See **pesticide.**

Fusiform: (adj.) spindle shaped; narrowed both ways from a swollen middle.

Fusion: the combination of two light nuclei to form a single heavier nucleus. (1) Act or operation of melting or rendering liquid by heat. (2) State of being melted or dissolved by heat. (3) Union or blending of things as if melted together. (4) The combination of certain light nuclei, such as deuterium and tritium, forming a heavier nucleus and releasing energy.

Game fish: those species of fish sought by sports fishermen; for example, salmon, trout, black bass, striped bass, etc. Game fish are usually more sensitive to environmental changes and water quality degradation than "rough" fish.

Gamete: a haploid cell that fuses with another haploid cell of opposite sex during fertilization to form the zygote. In animals the male gamete is called the sperm and female gamete the egg, or ovum. A sexual cell that joins with another cell of opposite sex to form a zygote and ultimately a new organism.

Gamma ray: high-energy electromagnetic radiation of great penetrating power emitted by nucleus; rays or quanta of energy emitted by radioactive substances corresponding to x rays and visible light but having a much shorter wavelength than light; waves of radiant nuclear energy. Gamma rays are the most penetrating of the three types of radiation and are best stopped by dense materials, such as lead.

Garbage: See **food waste.**

Garbage grinding: a method of grinding food waste by a household disposal, for example, and washing it into the sewer system. Ground garbage then must be disposed of as sewage sludge.

Geiger counter: an electrical device that detects the presence of radioactivity.

Gelatinous matrix: Jellylike intercellular substance of a tissue; a semisolid material surrounding the cell wall of some algae.

Gemorphology: the study of earth's form and its evolution, both of which owe much to the action of water in rivers and glaciers.

Gene: a unit of genetic information. In biochemistry, gene refers to the part of the DNA molecule that encodes a single enzyme or structural protein unit.

Gene flow: exchange of genetic traits between populations by movement of individuals, gametes, or spores.

Gene locus: segment of a chromosome on which a gene resides.

Generalist: a species with broad food or habitat preferences, or both.

Generation time: average age (T_c) at which a female gives birth to her offspring, or average time (T) for a population to increase by a factor equal to the net reproductive rate.

Generator: a device that converts mechanical energy into electrical energy.

Genetic drift: change in allele frequency from random occurrences, variations in fecundity and mortality, in a population; gradual change with time in the genetic composition of a continuing population resulting from the random selection or elimination of genetic features. The effect is unrelated to the benefits or detriments of the genes involved.

Genetic feedback: evolutionary response of a population to the adaptations of competitors, predators, or prey.

Genetic load: selective details sustained by a population because of geotypes that deviate from the genotype with the maximum fitness.

Genome: the entire genetic complement of the individual.

Genotype: all the genetic characteristics that determine the structure and functioning of an organism; often applied to a single gene locus to distinguish one allele, or combination of alleles, from another. The genetic constitution of an organism or a species in contrast to its observable characteristics (the latter is the phenotype).

Genus: a group of organisms containing closely related species. A group of species believed to have descended from a common direct ancestor that are similar enough to constitute a useful unit at this level of taxonomy.

Geochemical cycle: the sequence of stages in the migration of elements during geologic changes. A major cycle, proceeds from magma to igneous rocks to sediments to sedimentary rocks to metamorphic rocks and possibly through magmatites back to magma. A minor or exogenic cycle, proceeding from sediments to sedimentary rocks to weathered material and back to sediments again.

Geochemistry: the study of (a) the relative and absolute abundances of the elements and of the atomic species (isotopes) in the earth, and (b) the distribution and migration of the individual elements in the various parts of the earth (atmosphere, hydrosphere, crust, etc.) and in minerals and rocks, with the object of discovering principles governing this distribution and migration. Geochemistry may be defined very broadly to include all parts of geology that involve chemical changes or may be focused more narrowly on the distribution of the elements.

Geodesic line: a line of shortest distance between any two points on any mathematically defined surface. Also termed a geodesic.

Geodetic datum: a datum consisting of five quantities: the latitude and the longitude of an initial point, the azimuth of a line from this point, and two constants necessary to define the terrestrial spheroid. It forms the basis for the computation of horizontal control surveys in which the curvature of the earth is considered.

Geographic center: the geographic center of an area on the earth has been defined as that point on which the area would balance if it were a plate of uniform thickness. In other words, it is the center of gravity of that plate.

Geologic horizon: a term used to denote a particular level in the rocks or a particular thickness of rocks at a particular level.

Geologic hazard: a geologic condition, either natural or manmade, that poses a potential danger to life and property. Examples are earthquakes, landslides, flooding, faulting, beach erosion, land subsidence, pollution, waste disposal, and foundation and footing failures.

Geologic map: a map showing surface distribution of rock varieties, age relationships, and structural features.

Geometric rate of increase (λ) Factor by which the size of a population changes over a specified period (see **exponential rate of increase**).

Geomorphic: of, or pertaining to, the figure of the earth or the form of its surface; resembling the earth.

Geomorphology: the study of the shape of the earth's surface, and the processes that control and modify these features.

Geophysical: the form geophysical is used to the exclusion of any other that might be derived from the term geophysics.

Geostrophic current: a current where the horizontal pressure gradient is balanced by the Coriolis force. These currents can be calculated by careful measurement of temperature and salinity at closely spaced localities.

Geosyncline: a long linear basin that contains large (a thousand or more meters) accumulations of sediments. These basins may be uplifted, folded, and faulted, eventually forming a mountain range.

Germicide: a chemical or agent that kills microorganisms, such as bacteria, and prevents them from causing disease. Such compounds must be registered as pesticides with the EPA.

Geyser: from an Icelandic word meaning "roarer." A natural fountain of ground water, forced to the surface by steam at fairly regular intervals. Geysers occur when water deep in the earth is converted to steam by hot volcanic rock. Steam pressure then builds up against the water on top of it; this produces a spectacular eruption, after which the geyser subsides until steam pressure builds up enough to set it off again.

Glacial cycle: the term glacial cycle is here reserved for the ideal case of so long a continuance of glaciation under fixed climatic conditions, except for changes of climate with change of altitude because of degradation, that glacial erosion would be carried to its completion, truncating all the higher mountains at the snowline, and therefore causing snowfall to replace rainfall, and normal erosion to replace glacial erosion.

Glacial epoch: the Pleistocene epoch, the earlier of the two epochs comprised in the Quaternary period; characterized by the extensive glaciation of regions now free from ice.

Glacial lake: (1) A sheet of water owing its existence to the effects of the glacial period. They are of two classes, those excavated in the rock, and those produced by the irregular deposit of heaps of drift. (2) Lake fed by glacial meltwater. (3) Lake lying against or on a glacier.

Glacial retreat: a glacier is said to retreat when its front recedes. The ice may be actually moving forward, but the rate of backward melting at the front, if it exceeds the rate of forward movement, will cause the position of the front to recede.

Glaciation: alteration of the earth's solid surface through erosion and deposition by glacier ice.

Glacier ice: a body of ice developed from snow which becomes large enough to move from its place of accumulation.

Globigerina ooze: a calcareous marine deposit formed in deep water (but less than 3660 m) consisting chiefly of calcareous shells of Foraminifera, especially *Globigerina* spp. These are surface-dwelling forms, which reach the bottom only after death. Globigerina ooze contains more than 30% $CaCO_3$, of which the greater part consists of pelagic Foraminifera. The carbonate content averages 64% but ranges from 30 to 97%.

Globular: having a round or spherical shape.

Grab sample: (1) sample of rock or sediment taken more or less indiscriminately at any place. (2) Subaqueous sample of bottom sediment obtained by an instrument with movable jaws that close after being dropped to the bottom.

Graded sediment: a sediment consisting chiefly of grains of the same size range; a sediment having a uniform or equable distribution of particles from coarse to fine.

Gradient: (1) Slope, particularly of a stream or a land surface. Measurements are expressed in percent, feet per mile, or degrees. (2) Change in value of one variable with respect to another variable, especially vertical or horizontal distance, e.g., gravity, temperature, magnetic intensity, electric potential.

Gradient analysis: an alternative to community classification that views communities as continuous or nearly continuous along spatial and temporal gradients (such as elevation, moisture, stand, age). Instead of assigning communities to fixed types, gradient analysis views and quantifies community characteristics and species composition along these gradients. *Direct gradient analysis* arranges samples along predefined direct gradients, such as elevation or stand age. *Indirect gradient analysis,* or *ordination,* uses various mathematical techniques to arrange samples based on their compositional similarities; thus, on an ordination, samples with similar species composition lie close together on the axis, dissimilar samples are placed further apart on the axis.

Grain: a unit of weight equivalent to 65 mg or 2/1000 of an ounce.

Grain loading: the rate of emission of particulate matter from a polluting source. Measurement is made in grains of particulate matter per cubic foot of gas emitted.

Grain size: a term relating to the size of mineral particles that make up a rock or sediment.

Gram stain: a technique used to differentiate bacteria. Gram positive bacteria stain violet and gram negative bacteria stain red.

Granite: a course-grained igneous rock consisting mainly of quartz and alkali feldspar.

Graphite: black compacted crystalline carbon, used as neutron moderator and reflector in reactor cores.

Gravel: (1) Accumulation of rounded water-worn pebbles. The word gravel is generally applied when the size of the pebbles does not much exceed that of an ordinary hen's egg. The finer varieties are called sand and the coarser varieties are called shingle. (2) An accumulation of rounded rock or mineral pieces larger than 2 mm in diameter. Divided into granule, pebble, cobble, and boulder gravel. (3) Consists of rock grains or fragments with a diameter range of from 76 mm (3 in.) to 4.76 mm (rentention on a No. 4 sieve). The individual grains are usually more or less rounded. (4) Accumulation of uncemented pebbles; Pebble gravel. May or may not include interstitial sand ranging from 50 to 70% of total mass.

Gravimeter: a device used to measure differences in the earth's gravitational field.

Gravity: the force of attraction that causes objects on earth to fall toward the center of the earth. The universal law of gravity as first given by Newton states that every particle in the universe attracts every other particle with a force that is proportional to the product of their masses and inversely proportional to the square of the distances between the particles.

Grazing: the feeding by zooplankton upon phytoplankton.

Green belts: certain areas restricted from being used for buildings and houses; they often serve as separating buffers between pollution sources and concentrations of population.

Greenhouse effect: the heating effect of the atmosphere upon the earth. Light waves from the sun pass through the air and are absorbed by the earth. The earth then reradiates this energy as heat waves that are absorbed by the air, specifically by carbon dioxide. The air therefore behaves like glass in a greenhouse, allowing the passage of light but not of heat. Many scientists theorize that an increase in the atmospheric concentration of CO_2 can eventually cause an increase in the earth's surface temperature. The thermal effect resulting when comparatively short-wavelength solar radiation penetrates the atmosphere rather freely, only to be largely absorbed near and at the earth's surface, whereas the resulting long-wavelength terrestrial radiation thus formed passes upward with great difficulty. The effect is because the absorption bands of water vapor, ozone, and carbon dioxide are more prominent in the wavelengths occupied by terrestrial radiation than the short wavelengths of solar radiation. Hence the lower atmosphere is almost perfectly transparent to incoming radiation, but partially opaque to outgoing longwave radiation. The effect is increased when the atmosphere content of carbon dioxide, particulates, water vapor, or ozone is increased, and vice versa.

Grid: (1) A systematic array of points or lines (a) at or along which field observations are made or (b) for which computations are made. (2) The control electrode in thermionic tubes.

Grit: (1) Sand, especially coarse sand. (2) Coarse-grained sandstone. (3) Sandstone with angular grains. (4) Sandstone with grains of varying size, producing a rough surface. (5) Sandstone suitable for grindstones.

Groin; groyne: a shore protection and improvement structure (built usually to trap littoral drift or retard erosion of the shore).

Ground cover: grasses or other plants grown to keep soil from being blown or washed away.

Ground swell: a long, high ocean swell. Also, this swell as it rises to prominent height in shallow water. Not usually so high or dangerous as blind rollers.

Ground water: phreatic water (which see). That part of the subsurface water which is in the zone of saturation; the supply of fresh water under the earth's surface in an aquifer or soil that forms the natural reservoir. It can be found as deep as several miles.

Ground-water runoff: ground water that is discharged into a stream channel as spring or seepage water.

Ground-water surface: a level below which the rock and subsoil (down to unknown depths) are full of water; also known as the ground-water level or water table.

Group selection: elimination of a group of individuals with a detrimental genetic trait, caused by competition with other groups lacking the trait. Often called intergroup selection.

Gross production: the amount of organic matter photosynthesized by plants over a certain period and within a certain area or volume.

Gross productivity: rate at which energy or nutrients are assimilated by an organism, a population, or an entire community.

Growth: increase in size, weight, power, etc.

Growth curve: the curve representing the growth phases usually of bacteria during the life of a culture.

Guano: (1) Applied to deposits of the excrement of bats, birds, or other animals. (2) A substance found in great amounts on some coasts or islands frequented by marine birds; it is composed chiefly of their excrement, and is rich in phosphates and nitrogenous matter. It is extensively used for soil fertilizer.

Gut: (1) A narrow passage such as a strait or inlet. (2) A channel in otherwise shallower water, generally formed by water in motion.

Guyot: A flat-topped submarine seamount.

Habit: the mode of life of an organism, in contrast to its habitat.

Habitat: the sum total of environmental conditions of a specific place that is occupied by an organism, a population, or a community; the environment in which an organism lives; place where an animal or plant normally lives, often characterized by a dominant plant form or physical characteristic (i.e., the stream habitat, the forest habitat); the environment in which the life needs of a plant or animal are supplied.

Habitat breadth (*W*). Distribution of a population among types of habitats.

Habitat mosaic: describes the total ecosystem's community patterns. Natural variance in the environment produces an entire mosaic of communities that reflect topography, moisture, stand age, and other variables. It is this natural mosaic which plant ecologists attempt to describe and model using the methods of *classification, gradient analysis,* or other techniques.

Habitat selection: preference for certain habitats.

Hadal: that depth zone of the oceans below 6500 m.

Half-life: the time it takes certain materials, such as persistent pesticides or radioactive isotopes, to lose half their strength. For example, the half-life of DDT is 15 years; the half-life of radium is 1580 years. The time it takes for one-half the atoms of a radioactive isotope to decay into another isotope. The half-life is different for each radioactive element. Period of time within which half the nuclei in a sample of radioactive material undergo decay; characteristic constant for each particular species of nucleus. The time period in which half the initial number of atoms of a radioactive element disintegrate into atoms of the element into which they change directly.

Halocline: a zone, usually 50 to 100 m below the surface and extending to perhaps 1000 m, where the salinity changes rapidly. The salinity change is greater in the halocline than in the water above or below it.

Halogen: any one of the elements in Group VIIb of the periodic table (F, Cl, Br, I).

Halophile: a microorganism that lives at high salinity.

Hammermill: a broad category of high-speed equipment that uses pivoted or fixed hammers or cutters to crush, grind, chip, or shred solid wastes.

Haploid: referring to a cell or organism that contains one set of chromosomes (see **ploidy**).

Hardpan: (1) a hard impervious layer, composed chiefly of clay, cemented by relatively insoluble materials; does not become plastic when mixed with water and definitely limits the downward movement of water and roots. It can be shattered by explosives. (2) Placer mining: applied to layers of gravel occurring a few feet below the surface and cemented by limonite.

Hard water: Water containing dissolved minerals, such as calcium, iron and magnesium. The most notable characteristic of hard water is its inability to lather soap. Some pesticide chemicals will curdle or settle out when added to hard water.

Hazardous air pollutant: according to law, a pollutant to which no ambient air quality standard is applicable and that may cause or contribute to an increase in mortality or in serious illness. Asbestos, beryllium, and mercury, for example, have been declared hazardous air pollutants.

Headland: (1) Any projection of the land into the sea; generally applied to a cape or promontory of some boldness and elevation. (2) In soil conservation (a) the source of a stream, (b) the water upstream from a structure.

Headwater: the beginning of a stream or river, its source or its upstream portion.

Heat: energy in the form of motion of molecules; a form into which all other types of energy may be converted.

Heat budget: amount of heat seasonally absorbed and given off by a body of water.

Heat budget, annual: (of a lake) the amount of heat necessary to raise its water from the minimum temperature of winter to the maximum summer temperature.

Heat capacity: a ratio of the amount of heat absorbed or released by an object to the change in temperature of the object; in other words, the amount of heat necessary to raise the temperature of the object.

Heat exchanger: boiler, in which hot coolant from reactor core raises steam to drive turbogenerator; also intermediate heat exchanger.

Heat engine: system drawing its energy from a concentration of heat (at a higher temperature than the surroundings).

Heat island effect: an air circulation problem peculiar to cities. Tall buildings, heat from pavements, and concentrations of pollutants create a haze dome that prevents rising hot air from being cooled at its normal rate. A self-contained circulation system is put in motion that can be broken by relatively strong winds. If such winds are absent, the heat island can trap high concentrations of pollutants and present a serious health problem.

Heating season: the coldest months of the year, when pollution emissions are higher in some areas because of increased fossil-fuel consumption.

Heavy hydrogen, heavy water: see **deuterium, deuterium oxide.**

Heavy metals: metallic elements with high molecular weights, generally toxic in low concentrations to plant and animal life. Such metals are often residual in the environment and exhibit biologic accumulation. Examples are mercury (Hg), chromium (Cr), cadmium (Cd), arsenic, lead (Pb), zinc (Zn), and copper (Cu).

Hectare: metric unit of land measure; 10,000 square meters, 0.4047 acres.

Helium: light, chemically inert gas used as coolant in high-temperature reactors.

Hellgrammites (Corydalidae): Dobsonfly larvae. Full-grown, larvae are 2 to 3 in. in length; they have a dark brown rough-looking skin, large jaws, and posterior hooks. The aquatic larval stage lasts 2 to 3 years. They are secretive and predaceous, living under rocks and debris in flowing water. These larvae are considered one of the finest live baits by fishermen. Pupation occurs on shore, under rocks and debris near the stream edge. The terrestrial adults are short lived.

Hepatic vein: the vein leading from the liver.

Herbicide: a pesticide chemical used to destroy or control the growth of weeds, brush, and other undesirable plants. See **pesticide.** Substances or a mixture of substances intended to control or destroy any vegetation.

Herbivore: an organism that feeds on vegetation.

Heredity: genetic transmission of traits from parents to offspring.

Hermaphrodite: an organism that has the reproductive organs of both sexes.

Heterocyst: a specialized vegetative cell in certain filamentous blue-green algae; larger, clearer, and thicker walled than the regular vegetative cells.

Heterogeneity: the variety of qualities found in an environment (habitat patches) or a population (genotypic variation).

Heterosis: situation in which the heterozygous genotype is more fit than either homozygote. Also called overdominance.

Heterostyly: variation in the spatial arrangement of sexual organs in flowers within a population that promotes outcrossing.

Heterotroph: a microorganism that utilizes organic compounds for its energy and carbon requirements.

Heterotrophic bacteria: bacteria that use organic material, produced by other organisms, for their food.

Heterotrophic organism: organism dependent on organic matter for food.

Heterotrophy: use of organic materials for food, as by herbivorous, predatory, and detritus-feeding organisms (see **Autotrophy**).

Heterozygous: containing two forms (alleles) of a gene, one derived from each parent.

Hibernation: reduction of biologic activity by organisms during winter or, more generally, during cold periods.

High-density polyethylene: a material often used in the manufacture of plastic bottles that produces toxic fumes if incinerated.

High tide: maximum height reached by each rising tide.

High-water line: in a strict interpretation the intersection of the plane of mean high tide with the shore. The shoreline delineated on the nautical charts of the United States Coast and Geodetic Survey is an approximation of the high tide line.

Hi-volume sampler: a device used in the measurement and analysis of suspended particulate pollution. Also called a Hi-Vol.

Holism: a doctrine that life in all forms and including the inorganic environment form an interacting, integrated system.

Holistic ecology: the study of populations and communities as a whole in consideration of relationships between organisms and the total environment.

Holocene: see **Pleistocene.**

Holomictic lakes: lakes where bottom and surface waters are mixed periodically.

Holoplankton: animals that live their complete life cycle in the floating state. Synonym: permanent plankton.

Holotype: the single specimen chosen by the original author of a species; it is the name bearer for that species and any concept of the species must include it. A specimen so selected by a later author is no longer called a holotype but is termed the lectotype or neotype.

Home range: an area, from which intruders may or may not be excluded, to which an individual restricts most of its normal activities (see **territory**).

Homeostasis: the buffering capacity of an ecosystem that allows it to resist perturbations; the tendency of a biologic system to remain at or return to normal. after or during an outside stress; resistance to change and a tendency toward equilibrium; maintenance of constant level conditions in the face of a varying external environment.

Homeothermy: ability to maintain constant body temperature in the face of fluctuating environmental temperatures; warm bloodedness.

Homologous chromosomes: corresponding chromosomes in the male and female gametes, which pair during meiosis.

Homology: the condition of having similar evolutionary or developmental origin.

Homothermous: having the same temperature throughout.

Homozygous: containing two identical alleles at a gene locus.

Horizon: (1) A particular level, with or without thickness; e.g., fossils can be collected from a horizon, therefore that horizon has thickness. (2) The subdivision of the soil solum or profile.

Hot: a colloquial term meaning highly radioactive.

Humic: derived from plants, carbonacious. Cf. bituminous.

Humic acid: a degradation-resistant material found during the decomposition of vegetation. (1) Any of the various complex organic acids supposedly formed by the partial decay of organic matter; an indefinite term, but widely used. (2) Black, acidic, organic matter extracted from soils, low-rank coals, and other decayed plant substances by alkalis.

Humus: decomposed organic material. Substance of organic origin that is fairly but not entirely resistant to further bacterial decay. It is black and it has a higher carbon content, commonly 52–58%, and lower nitrogen content that the original material. It may accumulate subaerially in soil or subaqueously in sediment.

Humus layer: the top portion of the soil, which owes its characteristic features to its content of humus. The humus may be incorporated or unincorporated in the mineral soil.

Hybridization: crossing of individuals from genetically different strains, populations, or sometimes species.

Hydration: the chemical combination of water with another substance.

Hydraulic: (1) Of or pertaining to fluids in motion; conveying, or acting, by water; operated or moved by means of water, as hydraulic mining. (2) Hardening or setting under water, as hydraulic cement.

Hydraulic dredging: removal of underwater deposits by pumping them as a slurry through a pipe.

Hydraulic gradient: (1) Pressure gradient; as applied to an aquifer it is the rate of change of pressure head per unit of distance of flow at a given point and in a given direction. (2) As applied to streams, the slope of the energy grade line, or slope of line representing the sum of kinetic and potential energy along the channel length. It is equal to the slope of the water surface in steady, uniform flow. (3) A vector point function equal to the increase in hydraulic head per unit distance in direction of greatest decrease in rate.

Hydric: an extremely wet, often submerged environment. Plants that characterize such an environment are called *hydrophytic*.

Hydrocarbons: a vast family of compounds containing carbon and hydrogen in various combinations, found especially in fossil fuels. Some hydrocarbons are major air pollutants, some may be carcinogenic, and others may contribute to photochemical smog.

Hydrogen sulfide (H_2S): A malodorous gas made up of hydrogen and sulfur with the characteristic odor of rotten eggs. It is emitted in the natural decomposition of organic matter and is also the natural accompaniment of advanced stages of eutrophication. Hydrogen sulfide is also a byproduct of refinery activity and the combustion of oil during power plant operations. In heavy concentrations it can cause illness.

Hydrograph: a graph showing stage, flow, velocity, or other properties of water with respect to time.

Hydrologic cycle: the complete cycle of phenomena through which water passes, commencing as atmospheric water vapor, passing into liquid and solid forms as precipitation, thence along or into the ground surface, and finally returning to the form of atmospheric water vapor by means of evaporation and transpiration. Synonym: water cycle.

Hydrology: the scientific study of the water found on the earth's surface, in its subsurface, and in the atmosphere; the science dealing with the properties, distribution and circulation of water and snow.

Hydrolysis: the process by which a compound reacts chemically with water to form new substances. For example, ferric chloride, when placed in water, forms some molecules of both hydrochloric acid and ferric hydroxide.

Hydrometer test: the standard test measures the settling time for an individual sediment particle to fall a distance of 14 in. The test requires a 1000-ml cylinder for the settling chamber, using distilled water containing a flocculant as the settling medium. The test applies Stoke's law to the equivalent diameter of an individual particle.

Hydrosphere: (1) The water portion of the earth, as distinguished from the solid part which is called the lithosphere (which see). (2) In a more inclusive sense, the water vapor in the atmosphere, the sea, the rivers, and the ground waters. (3) The liquid and solid water that rests on the lithosphere, including the solid, liquid, and gaseous materials that are suspended or dissolved in the water.

Hydrostatic head: the height of a vertical column of water, the weight of which, if of unit cross-section, is equal to the hydrostatic pressure at a point.

Hygroscopic water: water that is so tightly held by the attraction of soil particles that it cannot be removed except as a gas, by raising the temperature above the boiling point of water.

Hyperosmotic: having a salt concentration greater than that of the surrounding medium.

Hypolimnion: the region of a body of water that extends from the thermocline to the bottom of the lake and is removed from surface influence. The cold, lower level of a large body of water. The cold, oxygen-poor part of a lake (see **epilimnion**).

Hypoosmotic: having a salt concentration less than that of the surrounding medium.

Hysteresis: (1) a lag in the return of an elastically deformed specimen to its original shape after the load has been released. (2) An effect, involving energy loss, found to varying degrees in magnetic, electric, and elastic media when they are subjected to variation by a cyclical applied force. In such media the polarization or stress is not a single-valued function of the applied force, or, stated in another way, the state of the medium depends on its previous history as well as the instantaneous value of the applied force.

Ice age: a glacial period or part of a glacial period; most frequently refers to the last glacial period.

Ice rafting: transport of rock particles and other materials by floating ice.

Igneous rocks: rocks formed by the solidification of molten magma. Magma is composed of numerous minerals (mainly silicates) and gases derived from the earth's crust and mantle and is in a melted state.

Imhoff tank: an anaerobic sewage treatment tank in which the solids are withdrawn from the bottom of the tank.

Immature soil: (1) a soil in which erosion exceeds the rate at which the soil develops downward. (2) Soil has not reached maturity as shown by lack of zonation, e.g., young alluvial soils.

Impedance: the rate at which a substance can absorb and transmit sound.

Implementation plan: a document of the steps to be taken to insure attainment of environmental quality standards within a specified time period. Implementation plans are required by various laws.

Impoundment: a body of water, such as a pond, confined by a dam, dike, floodgate, or other barrier.

Incineration: the controlled process by which solid, liquid, or gaseous combustible substances are burned and changed into gases; the residue produced contains little or no combustible material.

Incinerator: an engineered apparatus used to burn waste substances and in which all the combustion factors—temperature, retention time, turbulence, and combustion air—can be controlled.

Incubation: growth of a microbial culture under specific environmental conditions.

Independent assortment: the separate inheritances of genes occurring on different chromosomes. Whether a particular trait is inherited from the male or female parent is not influenced by the parental origin of other traits (see **linkage groups**).

Index of refraction: a characterizing number that expresses the ratio of the velocity of light *in vacuo* to the velocity of light in the substance. The conventional symbol is *n*.

Indirect competition: exploitation of a resource by one individual that reduces the availability of that resource to others (see **direct competition**).

Individual distance: distance within which an individual will neither tolerate intruders nor will itself approach another individual.

Inert gas: a gas that does not react with other substances under ordinary conditions.

Inertial separator: an air pollution control device that uses the principle of inertia to remove particulate matter from a stream of air or gas.

Infection: disease caused by growth of a microorganism.

Infiltration: (1) the flow of a fluid into a substance through pores or small openings; it connotes flow into a substance, in contradistinction to the word percolation, which connotes flow through a porous substance. (2) The flow or movement of water through the soil surface into the ground. (3) The deposition of mineral matter among the grains or pores of a rock by the permeation or percolation of water carrying it in a solution. (4) The material filling a vein as though deposited from a solution in water.

Infiltration/inflow: total quantity of water entering a sewer system. Infiltration means entry through such sources as defective pipes, pipe joints, connections, or manhole walls. Inflow signifies discharge into the sewer system through service connections from such sources as area or foundation drainage, springs and swamps, storm waters, street wash waters, or sewers.

Infiltration rate: maximum rate at which soil can absorb rain or shallow impounded water.

Influent: (1) A tributary stream or river; effluent. (2) A stream or stretch of a stream is influent with respect to ground water if it contributes water to the zone of saturation. The upper surface of such a stream stands higher than the water table or other piezometric surface of the aquifer to which it contributes.

Infrared: that portion of the electromagnetic spectrum with wave-lengths of from 0.7 to about 1.0 μm just beyond the red end of the visible spectrum.

Inheritance: genetic transmission of traits from parent to offspring.

Inhibition: prevention of microbial growth.

Innate capacity for increase (r_0): the intrinsic growth of a population under ideal conditions without the restraining effects of competition.

Inoculation: introduction of microorganisms into an environment or a culture medium.

Inoculum: material, such as bacteria, placed in compost or other medium to initiate biologic action; the microorganisms inoculated into a medium.

Inshore: in beach terminology, the zone of variable width extending from the shoreface through the breaker zone.

Instar: an insect or other arthropod in any of the forms between successive molts. Insects with complete metamorphosis usually have several larval instars, and both the pupa and adult (or imago) are also considered instars.

Integrated pest control: a system of managing pests by using biologic, cultural, and chemical means.

Intercale: (v.) to insert among others as a bed or stratum of lava between layers of other material; to interstratify.

Interceptor sewers: sewers used to collect the flows from main and trunk sewers and carry them to a central point for treatment and discharge. In a combined sewer system, where street runoff from rains is allowed to enter the system along with sewage, interceptor sewers allow some of the sewage to flow untreated directly into the receiving stream, to prevent the plant from being overloaded.

Interface: the boundaries or surfaces between two different materials; the major interfaces in the ocean are the water–atmosphere, water–biosphere, and water–sediment.

Interference: direct antagonism between individuals whether by behavioral or chemical means.

Interspecific: referring to interactions that occur between individuals of different species.

Interstate carrier water supply: a water supply that can be used for drinking or cooking purposes aboard common carriers (planes, trains, buses, and ships) operating interstate. Interstate carrier water supplies are regulated by the Federal government.

Interstate waters: according to law, waters defined as (1) rivers, lakes, and other waters that flow across or form a part of state or international boundaries; (2) waters of the Great Lakes: (3) coastal waters—whose scope has been defined to include ocean waters seaward to the territorial limits and waters along the coastline (including inland streams) influenced by the tide.

Intertidal: that section of the seashore which lies between tide marks.

Intracellular: inside the cell.

Intraspecific: referring to interactions that occur between individuals of the same species.

Intrinsic rate of increase (r_m): exponential growth rate of a population with a stable age distribution.

Introgression: incorporation of genes of one species into the genes pool of another species.

Inversion: an atmospheric condition where a layer of cool air is trapped by a layer of warm air so that it cannot rise. Inversions spread polluted air horizontally rather than vertically so that contaminating substances cannot be widely dispersed. An inversion of several days can cause an air pollution episode. Chromosomal: reversal of the order of gene loci within a segment of a chromosome.

Invertebrate: (1) Without a backbone or spinal column. (2) Of or pertaining to all the phyla of animals exclusive of the Chordata or animals with notochord or backbones.

Investment ratio: purchased feedback energy divided by free natural energy where both are expressed in fossil fuel equivalents.

Iodine: (as iodine-131) biologically hazardous fission product of short half-life (8 days), which tends to accumulate in the thyroid gland.

Ion: an atom or group of atoms having an electrical charge. Most of the atoms in sea water are in the ionic form. An ion having a positive charge is called a cation; one having a negative charge is called an anion. An atom shorn of one or more electrons and therefore electrically charged.

Ion exchange: the use of chemicals, usually in columns, to remove an ion from water and replace it with another ion.

Ionic substitution: the partial or complete proxying of one or more types of ions for one or more other types of ions in a given structural size in a crystal lattice.

Ionization chamber: a device roughly similar to a Geiger counter that reveals the presence of ionizing radiation.

Ionizing radiation: radiation that can deliver energy in a form capable of knocking electrons off atoms, turning them into ions.

Ionosphere: the highest layer of the earth's atmosphere in which ionization takes place. It lies above the stratosphere; its lower limit is about 56.3-km high in daytime and about 96.6-km during nights. The ionosphere reflects radio signals.

Iron bacteria: usually refers to bacteria that causes the precipitation of iron oxide through their metabolic processes. However, in certain acid environments with relatively low oxidation potential there are bacteria that reduce iron, also.

Irradiated: of reactor fuel, having been involved in a chain reaction and having thereby accumulated fission products; in any application, exposed to radiation.

Island: (1) a tract of land, usually of moderate extent, surrounded by water; distinguished from a continent or the mainland, as an island in the sea, an island in a river. (2) A body of land extending above and completely surrounded by water at the mean high-water stage. An area of dry land entirely surrounded by water or a swamp; an area of swamp entirely surrounded by open water.

Isobath: (1) Line on a marine map or chart joining points of equal depth usually in fathoms below mean sea level. (2) A line on a land surface all points of which are the same vertical distance above the upper or lower surface of an aquifer may be called an isobath of the specified surface, or merely a line of equal depth to the surface.

Isolation (ecologic): avoidance of competition between two species by differences in food, habitat, activity period, or geographic range.

Isolation (reproductive): prevention of successful interbreeding between individuals of opposite sex. By definition such individuals belong to different species.

Isopleth: (1) a line on a map or chart drawn through points of equal size or abundance (2) a line of constant composition, as in a binary temperature versus composition plot.

Isotherm: any line connecting points of equal temperature.

Isotope: form of an element with the same number of protons in its nucleus as all other varieties of element but a different number of neutrons from other varieties of the element; elements having an identical number of protons in their nuclei but differing in the number of their neutrons. Isotopes have the same atomic number, differing atomic weights, and almost but not quite the same chemical properties. Different isotopes of the same element have different radioactive behavior.

Iteration: (1) repeated occurrence of similar evolutionary trends in successive offshoots of a group. (2) A series of logical steps followed in an analysis of a complex problem.

Jetty: (1) In the United States on open seacoasts, a structure extending into a body of water and designed to prevent shoaling of a channel by littoral materials and to direct and confine the stream or tidal flow. Jetties are built at the mouth of a river or tidal inlet to help deepen and stabilize a channel. (2) In Great Britain jetty is synonymous with wharf or pier.

Juvenile water: water that has been trapped for ages far below the earth's surface, so that it cannot take part in the hydrologic cycle.

Kame terrace: (1) A terracelike body of stratified drift deposited between a glacier and an adjacent valley wall. (2) A terrace of glacial sand and gravel, deposited between a valley ice lobe (generally stagnant) and the bounding rock slope of the valley. (3) Remnant of a depositional valley surface built in contact with glacial ice.

Karyotype: characteristic chromosomes of a particular species.

Katabatic wind: a wind that flows down slopes that are cooled by radiation, the direction of flow being controlled orographically. Such winds are the result of downward convection of cooled air. Synonyms: mountain wind, canyon wind, gravity wind.

Kelp bed: an ecologic system dominated by any of various large, coarse, brown seaweeds belonging to the brown algae.

Kilometer: a length of 1000 m, equal to 3280.8 f or 0.621 mile; the chief unit for long distances in the metric system.

Kilowatt: one thousand watts.

Kinetic energy: that energy of a body which is in its motion.

Kjeldahl nitrogen, total: includes ammonia and organic nitrogen but does not include nitrate or nitrite nitrogen.

Knot: a unit of velocity equal to one nautical mile (6080 ft) per hour. It is approximately equal to 50 cm/s or 1.69 ft/s.

Krypton: a chemically inert gas; the isotope krypton-858 is a troublesome fission product at present released to the atmosphere from reprocessing plants.

Lacustrine: refers to a lake.

Lag phase: the phase of a microbial growth curve, directly after inoculation, when the cells are adapting to the new medium.

Lagoon: in waste water treatment, a shallow pond, usually manmade, where sunlight, bacterial action, and oxygen interact to restore waste water to a reasonable state of purity; a shallow pond or lake separated from the sea by a shallow bar or bank. (1) Body of shallow water, particularly one possessing a restricted connection with the sea. (2) Water body within an atoll or behind barrier reefs or islands.

Landfill: See **sanitary landfill.**

Land-use planning: the sum of human activities that plan for the use of the land; it may be good, bad, or indifferent.

Larva: the wormlike form of an insect on issuing from the egg.

Latent heat: the amount of heat absorbed or released by a substance during a change of state, under conditions of constant temperature and pressure. The latent heat of evaporation relates to the amount of heat necessary to go from the liquid to the gaseous state.

Lateral sewers: pipes running underneath city streets that collect sewage from homes or businesses.

Lava: liquid rock that comes to the surface, usually from a volcano or a fissure.

LC$_{50}$: median lethal concentration, a standard measure of toxicity. LC$_{50}$ indicates the concentration of a substance that will kill 50% of a group of experimental insects or animals.

Leach: to wash or drain by percolation. To dissolve minerals or metals out of the ore, as by the use of cyanide or chlorine solutions, acids, or water.

Leachate: liquid that has percolated through solid waste or other mediums and has extracted dissolved or suspended materials from it.

Leaching: the process by which soluble materials in the soil, such as nutrients, pesticide chemicals, or contaminants, are washed into a lower layer of soil or are dissolved and carried away by water. Removal of soluble compounds from leaf litter or soil by water.

Lead: a heavy metal that may be hazardous to human health if breathed or ingested.

Lee: (1) shelter, or the part or side sheltered or turned away from the wind. (2) Chiefly nautical: the quarter or region toward which the wind blows.

Leeward: the direction toward which the wind is blowing.

Lek: a communal courtship area on which several males hold courtship territories to attract and mate with females; sometimes called an arena.

Lens: a body of ore, rock, sand, or water thick in the middle and thin at the edges; similar to a double convex lens.

Lentic: standing water, lakes, ponds, swamps, etc.

Levee: an artificial or a natural bank confining a stream to its channel or, if artificial, limiting the area of flooding; also a landing place, pier, or quay.

Lichen: a symbiotic association between an alga and a fungus.

Life cycle: the phases, changes, or stages an organism passes through during its lifetime. See **ontogeny.**

Life form: characteristic structure of a plant or animal.

Lift: in a sanitary landfill, a compacted layer of solid waste and the top layer of cover material.

Light water: ordinary water—to distinguish it from heavy water.

Limiting resource: a resource that is scarce relative to demand for it (see **resource**).

Limnetic, limnic: pertaining to fresh water.

Limnology: the study of the physical, chemical, meteorologic and biologic aspects of fresh waters. The scientific study of fresh water, especially that of ponds and lakes. In its broadest sense it deals with all physical, chemical, meteorologic, and biologic conditions pertaining to such a body of water.

Linkage group: genes on a single chromosome.

Lithosphere: (1) In plate tectonics: a layer of strength relative to the underlying asthenosphere for deformations at geologic rates, which includes the crust and part of the upper mantle and is on the order of 100 km in thickness. (2) In geochemistry and general geology: the silicate shell of the earth; includes mantle and crust, part of the classification lithosphere, hydrosphere, atmosphere, biosphere (which see).

Littoral: the benthic zone between high tides out to a depth of about 200 m; living in the shallow waters of lakes or the sea. (1) Belonging to, inhabiting, or taking place on or near the shore. (2) The benthonic environment between the limits of high and low tides.

Littoral current: longshore current. Generated by waves breaking at an angle to the shoreline, which move usually parallel to and adjacent to the shoreline within the surf zone.

Littoral drift: (1) applied to the movement along the coast of gravel, sand, and other material composing the bars and beaches. (2) The material moved in the littoral zone under the influence of waves and currents.

Littoral zone: (1) strictly, zone bounded by high- and low-tide levels. (2) Loosely, zone related to the shore, extending to some arbitrary shallow depth of water.

Load: (1) in erosion and corrosion the material that is transported may be called the "load." The load is transported by two methods: one portion floats with the water and another portion is driven along the bottom. (2) The sediment moved by a stream, whether in suspension or at the bottom, is its load. (3) The quantity of material actually transported by a current. This is usually somewhat less than the actual capacity of the current.

Loam: a soil composed of a mixture of clay, silt, sand, and organic matter.

Loess: a homogeneous, nonstratified, unindurate deposit consisting predominantly of silt, with subordinate amounts of very fine sand and/or clay; a rude vertical parting is common at many places.

Longshore currents: currents in the near-shore region that run essentially parallel to the coast.

Lotic: running water; creek, stream, river.

Macro- (Gr. Makro-): prefix meaning large, long, visibly large.

Macroorganisms: Plant, animal, or fungal organisms visible to the unaided eye; organisms that may be seen without the aid of magnification.

Magma: molten rock material derived from the earth's crust and mantle. When it is extruded and flows on the earth's surface (above or below water) it is called lava.

Magnetic anomaly: a departure of the regular pattern of the earth's magnetic pattern. The departure may be caused by the concentration of magnetic minerals or by a buried (or exposed) igneous rock.

Magnetometer: an instrument used to measure the direction and intensity of the earth's magnetic field.

Map scale: the ratio of the distance between two points shown on a map and the actual distance between the points on the earth's surface. Scale is commonly expressed as a representative fraction (RF) as $1/1000$.

Marsh: a low-lying tract of soft, wet land that provides an important ecosystem for a variety of plant and animal life but often is destroyed by dredging and filling. Marshes proper are shallow lakes, the waters of which are either stagnant or activated by a very feeble current; they are, at least in the temperate zone, filled with rushes, reeds, and sedge and are often bordered by trees that thrive on plunging their roots into the muddy soil. In the tropical zone a large number of marshes are completely hidden by a multitude of plants or forests of trees, between the crowded trunks of which the black and stagnant water can only here and there be seen. A tract of low, wet ground, usually mirey and covered with rank vegetation. It may, at times, be sufficiently dry to permit tillage or hay cutting but requires drainage to make it permanently arable. It may be very small and situated high on a mountain or of great extent and adjacent to the sea. In biologic usage: a herbaceous plant-dominated ecosystem in which the rooting medium is inundated for long periods if not continually.

Masking: covering over of one sound or element by another. Quantitatively, masking is the amount the audibility threshold of one sound is raised by the presence of a second masking sound. Also used in regard to odors.

Mass spectrometer: an instrument for separation and measurement of isotopic species by their mass.

Mass transport: the net transfer of water by wave action in the direction of wave travel. See **orbit.**

Mating system: pattern of matings between individuals in a population, including number of simultaneous mates, permanence of pair bond, degree of inbreeding, and so on.

Matrix: the natural material in which any metal, fossil, pebble, crystal, etc., is embedded.

Mature soil: a soil with well-developed characteristics produced by the natural processes of soil formation, and in equilibrium with its environment. Generally consists of several differently characterized zones. See **soil profile.**

Maximum-power principle: systems with more energy flow displace those with less flow.

Meander: (1) One of a series of somewhat regular and looplike bends in the course of a stream, developed when the stream is flowing at grade by lateral shifting of its course toward the convex sides of the original curves. (2) a land survey traverse along the bank of a permanent natural body of water.

Meandering stream: a characteristic habit of mature rivers; may be defined as winding freely on a broad flood plain, in rather regular river-developed curves.

Megafauna: animals, living or fossil, that are large enough to be seen and studied with the naked eye.

Megaflora: plants, living or fossil, that are large enough to be seen and studied with the naked eye.

Megawatt: one million watts.

Mechanical turbulence: the erratic movement of air caused by local obstructions, such as buildings.

Meiosis: a series of two divisions by cells destined to produce gametes, involving pairing and segregation of homologous chromosomes, and reducing chromosome number from diploid to haploid. Recombination occurs during meiosis (see **mitosis**).

Meiotic drive: preferential inclusion of chromosomes with one type of allele, or of one of the sex chromosomes, in gametes.

Melanism: occurrence of black pigment, usually melanin.

Meltdown: of reactor core, consequence of overheating that allows part or all of the solid fuel in a reactor to reach the temperature at which cladding and possibly fuel and support structure liquefy and collapse.

Melt water; meltwater: water resulting from the melting of snow or glacier ice.

Mercury: a heavy metal, highly toxic if breathed or ingested. Mercury is residual in the environment, showing biologic accumulation in all aquatic organisms, especially fish and shellfish. Chronic exposure to airborne mercury can have serious effects on the central nervous system.

Meroplankton: animals that spend only a portion of their life as plankton, ultimately becoming nekton or benthos.

Mesic: a moist environment, but drier than a hydric environment, and seldom submerged. Characteristic plants are called *mesophytic*.

Metabolism: the process that includes the formation of protoplasm by organisms from food or photosynthesis, the eventual breakdown of the protoplasm, and the release of waste products and energy; the biochemical processes in a cell.

Metamorphic rocks: rocks that have undergone large changes in temperature, pressure, or chemical environment. These metamorphic rocks were already rocks prior to these large changes.

Metamorphosis: an abrupt change in form during development that fundamentally alters the function of the organism, often called complete metamorphosis. Incomplete metamorphosis refers to more gradual change.

Meter, metre: a unit of length equivalent in the United States to 39.37 in. exactly.

Methane: colorless, nonpoisonous, and flammable gaseous hydrocarbon. Methane (CH_4) is emitted by marshes and by dumps undergoing anaerobic decomposition.

MGD: millions of gallons per day.

mg/1: milligrams per liter; equivalent to 1 part per million.

mg/m: milligrams per cubic meter.

Micheles–Menten equation: an equation used to describe the kinetics of enzymatic activity.

μg/l: micrograms per liter, equivalent to 1 part per billion or 0.001 milligrams per liter.

Microcosm: a miniature habitat.

Microenvironment: see **microhabitat,** which is the preferred term.

Microfauna: animals, living or fossil, that are normally so small that their details must be studied under a microscope.

Microflora: living or fossil animals too small to be seen with the naked eye. The term is sometimes used synonymously with paleoflora but is more frequently applied to living microscopic algae and fungi. A very localized or small group of plants; plants occupying a very small habitat.

Microhabitat: the particular parts of the habitat that an individual encounters in the course of its activities.

Micromillimeter: $^1/_{1,000,000}$ mm; abbreviated mμm.

Microorganism: an organism that can be seen only through a microscope.

Microsere: a series of successional ecologic stage that occur within a microhabitat, such as a tree stump.

Microtemperature structure: small-scale, in terms of vertical or horizontal dimensions, changes in temperatures.

Microbes: minute plant or animal life. Some disease-causing microbes exit in sewage.

Micron: a unit of length equal to the 1 one-millionth of a meter and usually denoted by the symbol μm.

Microphotograph: greatly enlarged photograph made through a microscope.

Microplankton: plankton ranging in size from 60 μm to 1 mm, including most phytoplankton.

Migration: (1) The movement of oil, gas, or water through porous and permeable rock. Parellel (longitudinal) migration is movement parallel to the bedding plane; transverse migration is movement across the bedding planes. (2) Plotting of seismic reflection time information to allow for the fact that reflecting points do not lie vertically beneath the point of observation. Involves both reflector dip and variation of velocity. Often neglects effects perpendicular to the seismic line.

Millipore Filters: a series of membrane filters commonly used to remove bacteria. A typical bacterial filter has a pore size of 0.45 μm.

Mimicry: resemblance of an organism to some other organism or object in the environment; evolved to deceive predators or prey into confusing the organism with what it mimics (see **Batesian mimicry, Mullerian mimicry, aggressive mimicry**).

Mineralization: (1) The process of replacing the organic constituents of a body by inorganic fossilization. (2) The addition of inorganic substances to a body. (3) The act or process of mineralizing. (4) The process of converting or being converted into a mineral, as a metal into an oxide, sulfide, etc.

Mist: Liquid particles in air formed by condensation of vaporized liquids. Mist particles vary from 500 to 40 μm in size. By comparison, fog particles are smaller than 40 μm. in size.

Mitosis: division of somatic or germ line cells to yield two daughter cells that are genetically identical to the parent cell (see **meiosis**).

Mixed liquor: a mixture of activated sludge and water containing organic matter undergoing activated sludge treatment in the aeration tank.

Mobile source: a moving source of air pollution such as an automobile.

Moderator: material whose nuclei are predominantly of low atomic weight (e.g., light water, heavy water, graphite) used in reactor core to slow down fast neutrons and increase the probability of their being absorbed in uranium-235 or plutonium-239 to cause fission.

Mohorovic discontinuity: the sharp change in seismic velocity, occurring at about 11 km depth in the ocean and 35 km depth under land, that defines the top of the earth's mantle. This discontinuity, commonly called the Moho, may represent either a chemical or a phase change in the layering of the earth.

Monitoring: periodic or continuous determination of the amount of pollutants or radioactive contamination present in the environment.

Monoecy: in plants, the occurrence of reproductive organs of both sexes on the same individual, either in different flowers (hermaphrodite) or in the same flowers (perfect flowers; see **dioecy**).

Monomictic lake: a lake with only one overturn per year. There are two types: (1) Warm, with water never below 4°C and overturn in the winter, and (2) cold, with water never above 4°C and overturn in the summer.

Monsoon: a term for seasonal winds, usually applied to the changing wind patterns in the Indian Ocean.

Morph: a specific form, shape, or structure.

Morph ratio cline: a gradual geographic change in the frequency of morphs in a population, usually associated with a gradual change in ecologic conditions.

Morphometry: the physical shape and form of a water body.

Mortality (m_x): ratio of the number of deaths of individuals to the population, often described as a function of age (x) (see **death rate**).

MPN: most probable number. This abbreviation denotes the most probable number of bacteria per 100 ml of sample.

Muck: (1) a dark colored soil, commonly in wet places, that has a high percentage of decomposed or finely comminuted organic matter. See **Peat.** (2) Earth, including dirt, gravel, hardpan, and rock, to be or being excavated; overburden. (3) A layer of earth, sand, or sediment lying immediately above the sand or gravel containing, or supposed to contain, gold in placer mining districts. It may itself contain some traces of gold. (4) to excavate or remove muck from (v.).

Muck soils: soils made from decaying plant materials.

Mud: (1) A slimy and sticky mixture of water and finely divided particles of solid or earthy material with a consistency varying from that of a semifluid to that of a soft plastic sediment. (2) A wet mixture of silt and clay.

Mud flat: a muddy, low-lying strip of ground by the shore, or an island, usually submerged more or less completely by the rise of the tide.

Mulch: a layer of wood chips, dry leaves, straw, hay, plastic strips, or other material placed on the soil around plants to retain moisture, to prevent weeds from growing, and to enrich soil.

Mullerian mimicry: mutual resemblance of two or more conspicuously marked, distasteful species to enhance predator avoidance.

Multiple gene inheritance: determination of a quantitatively varying trait by the additive effects of more than one allele.

Mutant: an organism with a changed characteristic resulting from a genetic change.

Mutation: the effect of a chemical change in the DNA of a chromosome; some are visible, most are not visible, many are deleterious. Mutations are the raw material of evolution. Any change in the genotype of an organism occurring at the gene, chromosome, or genome level; usually applied to changes in genes to new allelic forms.

Mutation rate (μ): probability that an allele will be altered to a different form in a generation or other time period.

Mutualism: relationship between two or more species that benefits all parties to the relationship; a benevolent association between two microorganisms that benefits both; a symbiotic relationship in which both members benefit. Neither can survive without the other.

Mycoorhizae: close association of fungi and tree roots in the soil which facilitates the uptake of minerals by trees.

Nannoplankton: plankton in the size range 5 to 60 μm defined as uncatchable in standard plankton nets.

Nansen Bottle: a special bottle used for sampling water at ocean depths. It is open at both ends when lowered to the desired depth, but the ends close when the bottle is flipped over, trapping the water sample.

Natural gas: fuel gases obtained from decomposing organic matter in the earth; a fuel gas occurring naturally in certain geologic formations. Natural gas is usually a combustible mixture of methane and hydrocarbons.

Natural selection: the natural process by which the organisms best adapted to their environment survive and those less well adapted are eliminated; elimination of an inferior genetic trait from a population through differential survival and reproduction of individuals bearing that trait.

Neap tide: low tides which occur about every 2 weeks when the moon is in its quarter positions. Tide occurring near the time of quadrature of moon. Neap tidal range is usually 10 to 30% less than the mean tidal range.

Near-shore current system: current system caused primarily by wave action in and near the breaker zone, which consists of four parts: the shoreward mass transport of water; longshore currents; seaward return flow, including rip currents; and longshore movement of expanding heads of rip currents.

Necrosis: death of plant cells resulting in a discolored, sunken area or death of the entire plant.

Negative feedback: tendency of a system of counteract externally imposed change and return to a stable state.

Nekton: swimming organism, including fish; any animal or group of animals that lead or leads an active swimming life; swimming animals that can direct their own movements against the action of marine currents.

Neritic: the part of the pelagic environment that extends from the near-shore zone out to a depth of about 200 m; the water overlying the continental shelf; related to shallow water on the margins of the sea, generally that overlying the continental shelf. When combined with oceanic, e.g., neritooceanic, refers to the shallow water above 100 fathoms (200 m) in the open oceans.

Neritic zone: (1) that part of the sea floor extending from the low-tide line up to a depth of 200 m. (2) A part of the pelagic division of the oceans with water depths of less than 200 m.

Net aboveground productivity (NAP): accumulation of biomass in aboveground parts of plants (trunks, branches, leaves, flowers, and fruits) over a specified period; usually expressed on an annual basis.

Net energy: high-quality energy produced in a process in excess of high-quality energy used in the process.

Net production: more production than the consumption necessary for a process.

Net production efficiency: percentage of assimilated energy that is incorporated into growth and reproduction.

Net reproductive rate (R): number of offspring a female can be expected to bear during her lifetime.

Neutral alleles: alleles of the same gene having identical fitness.

Neutron: uncharged particles; consitutent of nucleus ejected at high energy during fission, capable of being absorbed in another nucleus and bringing about further fission or radioactive behavior.

Niche: the place of an organism is an ecosystem. All the components of the environment with which the organism or population interacts.

Nidicolous: referring to young birds that hatch naked with their eyes closed and that are unable to move about (also called altricial).

Nidifugous: referring to young birds that hatch covered with down with their eyes open and that are able to move about (also called precocial).

Nitrate: biologically it represents the final form of nitrogen from the oxidation of organic nitrogen compounds. A salt of nitric acid; a compound containing the radical $(NO_3)^-$.

Nitric oxide (NO): a gas formed in great part from atmospheric nitrogen and oxygen when combustion takes place under high temperature and high pressure, as in internal combustion engines. NO is not itself a pollutant; however, in the ambient air, it converts to nitrogen dioxide, a major contributor to photochemical smog.

Nitrification: the biologic oxidation of ammonia to nitrate.

Nitrite: a salt of nitrous acid; a compound containing the radical NO_2^-.

Nitrogen cycle: the numerous biogeochemical processes that result in the cycling of nitrogen in the many chemical forms required by various organisms.

Nitrogen fixation: the reduction of atmospheric nitrogen to cell nitrogen by microorganisms.

Nitrogenous wastes: wastes of animal or plant origin that contain a significant concentration of nitrogen.

NO: a notation meaning oxides of nitrogen. See **nitric oxide.**

Noise: any undesired audible signal. In acoustics, for example, noise is any undesired sound.

Nomenclature: (1) The naming of divisions in any scientific taxonomic scheme. (2) The names used in systematic classification, as distinguished from other technical terms, as Linnaean nomenclature.

Nonevolutionary responses: adaptive changes made by the organism in response to changes in the environment (see **regulatory, acclimatory,** and **developmental responses**).

Nonpoint waste source: a general, unconfined waste discharge; not a point source waste.

Nonrenewable resource: see **resource.**

Nonstabilizing factors: influences on population growth that are independent of the size of the population.

Normal atmospheric pressure: standard pressure, usually taken to be equal to that of a column of mercury 760 mm in height; approximately 14.7 pounds per square inch.

NTA: Nitrilotriacetic acid, a compound once used to replace phosphates in detergents.

Nuclear event: any nuclear explosion, premediated or accidental.

Nuclear energy: energy released by reactions involving the nucleus of atoms. The two types of reactions are fission, splitting the nuclei of heavy elements, and fusion, combining the nuclei of light elements.

Nuclear power plant: any device, machine, or assembly that converts nuclear energy into some form of useful power, such as mechanical or electric power. In a nuclear electric power plant, heat produced by a reactor is generally used to make steam to drive a turbine that in turn drives an electric generator.

Nucleic acid: complex organic substance that is the genetic material in all known organisms. See **DNA; RNA.**

Nucleus: the portion of a cell that contains DNA. (1) A kernel; a central mass or point about which other matter is gathered or to which an accretion is made. (2) A usually spherical or ovoid protoplasmic body found in most cells and considered as a directive center of many protoplasmic activities, including the transmission of hereditary characteristics. (3) In radiolarians, a rounded mass of protoplasm enclosed in a delicate membrane. (4) Embryonic gastropod shell, commonly consisting of one to four whorls. (5) The central portion of an atom; the chief constituents are protons and neutrons. Synonym: protoconch.

Nuclide: Nucleus of isotope; nuclear species. A species of atom characterized by the constitution of its nucleus; thus abundances of isotopes are described as abundances of various nuclides.

Numerical response: change in the population size of a predatory species as a result of a change in the density of its prey (see **functional response**).

Nutrients: elements or compounds essential as raw materials for organism growth and development, for example, carbon, oxygen, nitrogen, and phosphorus; compounds or ions that plants need for the production of organic matter; a substrate used to stimulate growth; mineral raw materials necessary for growth.

Nutrient budget: a quantitative determination of the major nutrients flowing to, retained within, and discharged from a system.

Obligate: essential for growth, as in obligate anaerobe.

Ocean: the great body of salt water that occupies two-thirds of the surface of the earth, or one of its major subdivisions; the sea as opposed to the land.

Ocean basin: that portion of the ocean seaward of the continental margin which includes the deep sea floor; that part of the floor of the ocean that is more than about 200 m below sea level.

Ocean current: (1) the name current is usually restricted to the faster movements of the ocean, whereas those in which the movement amounts to only a few miles a day are termed drifts. (2) A nontidal current constituting a part of the great oceanic circulation. Examples: Gulf Stream, Kuroshio, Equatorial currents.

Ocean tides: the twice daily rise and fall of sea level caused mainly by the attraction of the moon and earth.

Oceanic: related to the open ocean beyond the edge of the continental shelf, in contrast to shallow, near-shore waters.

Oceanic islands: islands that rise from deep water far from any continent, although they may occur in a close group, such as the Hawaiian Islands, the Azores, and the Galapagos.

Oceanographic station: oceanographic observations made while the research vessel is stopped.

Oceanography: embraces all studies pertaining to the sea and integrates the knowledge gained in the marine sciences that deal with such subjects as the ocean boundaries and bottom topography, the physics and chemistry of sea water, the types of currents, and the many phases of marine biology. Synonyms: oceanology, thalassography.

Octane: a liquid hydrocarbon of the paraffin series, formula C_8H_{18}.

Off gas: radioactive gas from within a reactor that is released to the atmosphere, usually after a delay to reduce its radioactivity.

Offshore bar: barrier beach (which see); an accumulation of sand in the form of a ridge, built at some distance from the shore and under water. It results chiefly from wave action. Not to be confused with barrier island (which see).

Offshore current: (1) Any current in the offshore zone. (2) Any current flowing away from shore.

Oil and grease: the kinds and quantities of oil and grease contained in river waters or sediments; originate from various sources up and down the river but mostly from accidental spills. The oil and grease is gradually incorporated into the bottom sediments by combining with silt particles and settling to the bottom.

Oil sand: (1) A general term for any rock containing oil. (2) Porous sandstone from which petroleum is obtained by drilled wells.

Oil shale: a fine-grained, laminated sedimentary rock that contains an oil-yielding organic material called kerogen. In the United States, oil shale is found primarily in Colorado; much larger deposits exist in several other countries. Upon heating, oil shales yield from 12 to 60 gal of oil per ton of rock, although the best United States oil shale produces less than 30 gal per ton.

Oil spill: the accidental discharge of oil into oceans, bays, or inland waterways.

Methods of oil spill control include chemical dispersion, combustion, mechanical containment, and absorption.

Oil–water interface: a surface that forms the boundary between a body of ground water and an overlying body of petroleum that saturates the rock.

Old age: that state in the development of streams and land forms when the processes of erosion are decreasing in vigor and efficiency or the forms are tending toward simplicity and subdued relief.

Oligotrophic: term used for a lake, pond, etc., low in plant nutrients, minerals, and organisms; low in productivity; but with oxygen at all depths. Referring to a body of water with a low nutrient content and low productivity, usually characterized by extremely clear water. Refers to lakes with considerable oxygen in the bottom waters and with limited nutrient matter.

Oligotrophic lakes: deep lakes that have a low supply of nutrients and thus contain little organic matter. Such lakes are characterized by high water transparency and high dissolved oxygen.

Oligotrophic waters: waters with a small supply of nutrients; hence, they support little organic production.

Omnivore: an organism whose diet is broad, including both plant and animal foods.

Ontogeny: the life history, or development, of an individual, as opposed to that of the race (phylogeny).

Ooze: a fine-grained pelagic deposit that contains more than 30% of material of organic origin.

Opacity: degree of obscuration of light. For example, a window has zero opacity; a wall is 100% opaque. The Ringelmann system of evaluating smoke density is based on opacity.

Open burning: uncontrolled burning of wastes in an open dump.

Open dump: see **dump.**

Opportunistic species: a species that takes advantage of temporary or local conditions. (Populations of opportunistic species usually fluctuate markedly.)

Optimum: the range in pH, temperature, or other parameter at which growth or enzymatic activity is greatest.

Orbit: the path described by a body in its revolution around another body.

Ordination: see **gradient analysis.**

Organic: derived from living organisms; in chemistry, any compound containing carbon; a molecule that is of biologic origin and contains carbon. The opposite is inorganic.

Organic deposits: rocks and other deposits formed by organisms or their remains.

Organic reef: a sedimentary rock aggregate composed of the remains of colonial-type organisms, mainly marine.

Organism: any living human, plant, or animal.

Organoleptic: affecting or employing one or more of the organs of special sense.

Organophosphates: a group of pesticide chemicals containing phosphorus, such as malathion and parathion, intended to control insects. These compounds are short lived and therefore do not normally contaminate the environment. However, some organophosphates, such as parathion, are extremely toxic when initially applied and exposure to them can interfere with the normal processes of the nervous system, causing convulsions and eventually death. Malathion, in contrast, is low in toxicity and relatively safe for humans and animals; it is a common ingredient in household insecticide products.

Orogeny: the process of forming mountains, particularly by folding and thrusting.

Oscillation: regular fluctuation, variation, movement back and forth, as in ocean waves; regular fluctuation through a fixed cycle above and below some mean value.

Osmoregulation: regulation of the salt concentration in cells and body fluids.

Osmosis: the movement of dissolved ions or molecules through a semipermeable membrane. As osmotic pressure results when a difference in concentration exists on either side of the membrane. The greater the difference the higher the osmotic pressure; the flow will be toward the more concentrated solution. Diffusion across a membrane.

Outcrop: an exposure of rocks at the earth's surface.

Outfall: the mouth of a sewer, drain, or conduit where an effluent is discharged into the receiving waters.

Outwash: drift deposited by meltwater streams beyond active glacier ice.

Outwash plain: (Obs.) Outwash apron; overwash apron; marginal plain; outwash gravel plain; washed gravel plain. A plain composed of material washed out from the ice.

Overburden: material of any nature, consolidated or unconsolidated, that overlies a deposit of useful materials, ores, or coal, especially those deposits that are mined from the surface by open cuts.

Overdominance: see **heterosis.**

Overfire air: air forced into the top of an incinerator to fan the flame.

Overlay: a record or map on a transparent medium which may be superimposed on another record.

Overstory species: species that grow in the top stratum of a community, such as a forest's canopy or the top shrub stratum in the shrubfield.

Overturn: vertical mixing of layers of large bodies of water caused by seasonal changes in temperature.

Oxbow: a curved lake, created when a bend is abanded by a river that has changed its course; a crescent-shaped lake formed in an abandoned river bend that has become separated from the main stream by a change in the course of the river.

Oxidant: any oxygen-containing substance that reacts chemically in the air to produce new substances. Oxidants are the primary contributors to photochemical smog.

Oxidation: a chemical reaction in which oxygen unites or combines with other elements. Organic matter is oxidized by the action of aerobic bacteria; therefore oxidation is used in waste water treatment to break down organic wastes. The process whereby an element or compound combines with oxygen, or whereby electrons are removed from an ion or atom.

Oxidation pond: a manmade lake or pond in which organic wastes are reduced by bacterial action. Often oxygen is bubbled through the pond to speed the process.

Ozone (O_3): a pungent, colorless, toxic gas. Ozone is one component of photochemical smog and is considered a major air pollutant.

Oxygen-minimum zone: a layer below the surface where the oxygen content is very low or zero.

Pacific-type coastline: trend of folded belts are parallel to the coast. Contrasts with Atlantic type of coast.

Package plant: a prefabricated or prebuilt waste water treatment plant.

Packed tower: an air pollution control device in which polluted air is forced upward through a tower packed with crushed rock or wood chips, while liquid is sprayed downward on the packing material. The pollutants in the air stream either dissolve or chemically react with the liquid.

Paleoecology: the study of ancient ecology as recorded in fossil remains.

Paleomagnetism: the study of variations in the earth's magnetic field as recorded in ancient rocks.

Paleontology, palaeontology: (1) the science that treats fossil remains, both animal and vegetable. (2) The science that deals with the life of past geologic ages. It is based on the study of the fossil remains of organisms.

Palisade: a picturesque, extended rock cliff rising precipitately from the margin of a stream or lake and of columnar structure.

Pan: a shallow, basinlike depression without vegetation or outlet for drainage.

PAN: peroxyacetyl nitrate, a pollutant created by the action of sunlight on hydrocarbons and nitrogen oxides in the air. PAN are an integral part of photochemical smog.

Parameter: a specific part of an ecosystem under study.

Parasite: an organism that consumes part of the blood or tissues of its host, usually without killing the host. Parasites may live entirely within the host (endoparasites) or on its surface (ectoparasites).

Parasitic: the ability to grow on a living host. May not necessarily be associated with disease.

Parent material (soils): (1) the horizon of weathered rock or partly weathered soil material from which the soil is formed; horizon C of the soil profile. (2) The unconsolidated materials from which a soil develops.

Parthenogenesis: reproduction without fertilization by male gametes, usually involving the formation of diploid eggs whose development is initiated spontaneously.

Particulates: (1) Finely divided solid or liquid particles in the air or in an emission. Particulates include dust, smoke, fumes, mist, spray, and fog. (2) (Adj.) Of or relating to particles or occurring as minute particles.

Particulate loading: the introduction of particulates into the ambient air.

Pasteurization: the treatment of milk, alcoholic beverages, or other materials by heat to destroy most of the microorganisms.

Patch reefs: usually small, more or less isolated, organic buildups or areas of limestone deposition supported by a framework of organisms.

Pathogenic: causing or capable of causing disease. Having the potential to cause disease.

Pathogenic bacteria: bacteria capable of causing disease.

PCBs: Polychlorinate biphenyls, a group of organic compounds used in the manufacture of plastics. In the environment, PCBs exhibit many of the same characteristics as DDT and may, therefore, be confused with that pesticide. PCBs are highly toxic to aquatic life, they persist in the environment for long periods of time, and they are biologically accumulative.

Peak flood: the particular flood that covers the flood plain; not to be confused with the flood of flood classification, as some flood plains are covered by the 25-year flood, whereas others are covered only by the 500-year flood. Not to be confused with flood peak.

Peat: (1) A dark brown or black residuum produced by the partial decomposition and disintegration of mosses, sedges, trees, and other plants that grow in marshes and similar wet places. (2) Fibrous, partly decayed fragments of vascular plants that retain enough structure so that the peat can be identified as originating from certain plants (e.g., sphagnum peat or sedge peat).

Pedology: the science that treats of soils and their origin, character, and utilization.

Pebbles: smooth rounded stones ranging in diameter from 2 to 64 mm.

Pelagic: (1) pertaining to communities of marine organisms that live free from direct dependence on bottom or shore; the two types are the free-swimming forms (nektonic) and the floating forms (planktonic). (2) Related to water of the sea as distinct from the sea bottom. (3) Related to sediment of the deep sea as distinct from that derived directly from the land.

Pelagic sediments: sediments deposited in the deep sea that have little or no coarse-grained terrigenous material.

Penstock: a sluice for regulating flow of water; a conduit for conducting water.

Percent dry weight: percent dry weight $\times 10^4$ is equivalent to 1 part per million or 1 mg/kg.

Perched ground water: ground water separated from an underlying body of ground water by unsaturated rock. Its water table is a perched water table.

Perched water table: water table above an impermeable bed underlain by unsaturated rocks of sufficient permeability to allow movement of ground water.

Periphyton: the association of aquatic organisms attached or clinging to stems and leaves of rooted plants or other surfaces projecting above the bottom.

Percolate: to pass through fine interstices; to filter; as water percolates through porous stones.

Percolation: downward flow or infiltration of water through the pores or spaces of a rock or soil.

Perennial stream: streams that flow throughout the year and form source to mouth.

Periodic or tidal current: a current, caused by the tide-producing forces of the moon and the sun, which is part of the same general movement of the sea manifested in the vertical rise and fall of the tides.

Permafrost: permanently frozen ground (subsoil). Permafrost areas are divided into more northern areas, in which permafrost is continuous, and those more southern areas in which patches of permafrost alternate with unfrozen ground.

Permanent current: a current that runs continuously independent of the tides and temporary causes. Permanent currents include the fresh water discharge of a river and the currents that form the general circulatory systems of the oceans.

Permeability: the capacity of a solid to allow the passage of a liquid. The permeability of dirt or rock is determined by the number of pores or openings, their size and shape, and the number of interconnections between them.

Permeability coefficient: coefficient of permeability. The rate of flow of water in gallons a day through a cross-section of 1 ft^2 under a unit hydraulic gradient. The standard coefficient is defined for water at a temperature of 60°F. The field coefficient requires no temperature adjustment and the units are stated in terms of the prevailing water temperature.

Permeable: pervious. Hydrology: having a texture that permits water to move through it preceptibly under the head differences ordinarily found in subsurface water. A permeable rock has communicating interstices of capillary or supercapillary size.

Persistent pesticides: pesticides that will be present in the environment for longer than one growing season or 1 year after application.

Pest pressure: detrimental effects of parasites and seed predators on populations of animals and plants. (Pest pressure is thought to be particularly great in dense host populations.)

Pesticide: a chemical agent used to control pests. This includes insecticides for use against harmful insects; herbicides for weed control; fungicides for control of plant diseases; rodenticides for killing rats, mice, etc.; germicides used in disinfectant products; algaecides; slimicides; etc. Some pesticides can contaminate water, air, or soil and accumulate in humans, animals, and the environment, particularly if they are misused. Certain of these chemicals have been shown to interfere with the reproductive processes of predatory birds and possibly other animals.

Pesticide tolerance: a scientifically and legally established limit for the amount of chemical residue that can be permitted to remain in or on a harvested food or feed crop as a result of the application of a chemical for pest control purposes. Such tolerances or safety levels, established federally by the EPA, are set well below the point at which residues may be harmful to consumers.

Petri dish: a glass or plastic dish with a cover used to grow microorganisms on solid media.

pH: a measure of acidity or alkalinity of a material, liquid, or solid. pH is represented on a scale of 0 to 14, with 7 representing a neutral state, 0 representing the most acid, and 14 the most alkaline. A measure of the alkalinity or salinity of a solution. pH is the logarithm of the reciprocal of the hydrogen ion concentration, or $pH = \log 1/H^+$. A scale of acidity or alkalinity.

Phenols: a group of organic compounds that in very low concentrations produce a taste and odor problem in water. In higher concentrations, they are toxic to aquatic life. Phenols are by-products of petroleum refining, tanning, and textile, dye, and resin manufacture. Hydroxy derivatives of benzene usually found in waste products from oil refineries, and chemical plants.

Phenotype: physical expression in the organism of the interaction between the genotype and the environment; outward appearance of the organism.

Pheromone: externally produced chemicals that control behavior in other individuals of the same species. Allomones control behavior in individuals belonging to different species. Chemical substances for communication between individuals.

Phosphates: the salt or ester of a phosphoric acid.

Phosphorus: an element that although essential to life, contributes to the eutrophication of lakes and other bodies of water.

Phosphorus, total: includes both soluble and insoluble orthophosphates and condensed phosphates as well as organic and inorganic species.

Photic zone: the surface waters that are penetrated by sunlight.

Photoautotroph: an organism capable of utilizing light energy for growth; an organism that utilizes sunlight as its primary energy source for the synthesis of organic compounds.

Photochemical oxidants: secondary pollutants formed by the action of sunlight on the oxides of nitrogen and hydrocarbons in the air; they are the primary contributors to photochemical smog.

Photochemical smog: air pollution associated with oxidants instead of with sulfer oxides, particulates, etc. Produces necrosis, chlorosis, and growth alterations in plants and is an eye and respiratory irritant in humans.

Photogrammetry: the science and art of obtaining reliable measurements from photographs.

Photomicrograph: an enlarged or macroscopic photograph of a microscopic object, taken by attaching a camera to a microscope.

Photoperiodism: seasonal response by organisms to change in the length of the daylight period. (Flowering, germination of seeds, reproduction, migration, and diapause are frequently under photoperiodic control.)

Photosynthesis: the series of biochemical reactions involving light energy, carbon dioxide, and water used by photosynthetic organisms for growth; the production of organic substances and oxygen from carbon dioxide and water occuring in green plant cells supplied with enough light to allow chlorophyll to assist the transformation of the radiant energy into a chemical form.

Photosynthetic: that part of the ocean where photosynthesis is possible, usually defined by the availability of light. In the open ocean this zone usually extends from the surface to about 100 m.

Photosynthetic efficiency: percentage of light energy assimilated by plants; based either on net production (net photosynthetic efficiency) or on gross production (gross photosynthetic efficiency).

Phycosphere: the zone of water influenced by algal exudates.

Phyletic evolution: genetic changes that occur within an evolutionary line.

Phylogeny: the line, or lines, of direct descent in a given group of organisms. Also the study or the history of such relationships.

Phylum: one of the primary divisions of the animal and plant kingdom; a group of closely related classes of animals or plants.

Physical–chemical environment: all the nonbiologic factors that form a part of the environment for an organism.

Phytoecology: the branch of ecology concerned with the relationship between plants and their environment.

Phytoplankton: the plant portion of plankton, the plantlike unicellular organisms in water. The animal-like unicellular organisms in water. The animal-like unicellular organisms are zooplankton. Plant microorganisms, such as certain algae, living unattached in the water. See **plankton.** All the floating plants, such as diatoms, dinoflatellates, coccolithophores, and sargassum weed.

Phytotoxic: injurious to plants.

Piedmont: lying or formed at the base of mountains, as a piedmont glacier.

Piezometric surface: the theoretical level to which water should rise under its own pressure if tapped by a well or spring; the water level of an artesian well is one point on such a surface. An imaginary surface that everywhere coincides with the static level of the water in the aquifer. The surface to which the water from a given aquifer will rise under its full head.

Pig: a container, usually made of lead, used to ship or store radioactive materials.

Pile: a nuclear reactor.

Pinger: an acoustic device used to determine distance of instruments above the bottom.

Plane table: a simple surveying instrument by means of which one can plot the lines of a survey directly from the observations. It consists of a drawing board on a tripod, with a ruler, the ruler being pointed at the object observed.

Planimeter: an instrument for measuring the area of any plane figure by passing a tracer around the bounding plane.

Planimetry: the determination of horizontal distances, angles, and areas by measurements on a map.

Plankton: floating or weakly swimming organisms that are carried by the currents in a body of water. The plankton range in size from microscopic plants to large jellyfish. Organisms of relatively small size, mostly microscopic, that either have relatively small powers of locomotion or drift in the water subject to the action of waves and currents. The usually microscopic animal and plant life found floating or drifting in the ocean or in bodies of fresh water, used as food by fish. Microscopic floating aquatic plants (phytoplankton) and animals (zooplankton). Holoplankton (which see); floating organisms of the floating or drifting marine life.

Plankton bloom: a large concentration of plankton within an area, caused by a rapid growth of the organisms. The large numbers of plankton can color the water, causing in some instances a red tide. A sudden rapid increase (usually geometric) to an enormous number of individual plankters under certain conditions. See also **bloom.**

Planktonic: floating

Plastids: a body in a plant cell that contains photosynthetic pigments.

Plate tectonics: a theory of global-scale dynamics involving the movement of many rigid plates of the earth's crust. Considerable tectonic activity occurs along the margins of the plates, where buckling and grinding occur as the plates are propelled by the forces of deep-seated mantle convection currents. This has resulted in continental drift and changes in shape and size of oceanic basins and continents.

Pleiotrophy: influence of one gene on the expression of more than one trait in the phenotype.

Pleistocene: a geologic epoch that ended about 10,000 years ago and lasted about 1 to 2 million years. This epoch has been subdivided into four glacial stages and three interglacial stages. The last of the Pleistocene glacial stages is called the Wisconsin stage. The period we are now in, the Holocene epoch, is not part of the Pleistocene and may be an interglacial stage.

Ploidy: number of chromsome complements, or sets, contained by a cell; one, haploid; two, diploid; three, triploid; four, tetraplid; and so on.

Plume: the visible emission from a flue or chimney.

Plutonium: heavy artificial metal made by neutron bombardment of uranium; fissile, highly reactive chemically, extremely toxic alpha emitter.

Pluvial: caused by the action of rain; pertaining to deposits by rain water or ephemeral streams.

Poikilothermy: "cold bloodedness", temperature conforming.

Point source: in air pollution, a stationary source of a large individual emission, generally of an industrial nature. This is a general definition; point source is legally and precisely defined in Federal regulations. See **area source.**

Point waste source: any discernible, confined, and discrete conveyance, such as any pipe, ditch, channel, tunnel, or conduit, from which pollutants are or may be discharged.

Poised stream: condition of a river that is neither eroding nor depositing sediment.

Pollen: a fine dust produced by plants; a natural or background air pollutant.

Pollutant: any natural or artificial substance that enters the ecosystem in such quantities that it does harm to the ecosystem in any way; any introduced gas, liquid, or solid that makes a resource unfit for a specific purpose.

Pollution: the presence of matter or energy whose nature, location, or quantity produces undesired environmental effects: the natural or unnatural addition of anything to an ecosystem to the extent that it harms all or any part of that ecosystem.

Polyandry: see **polygamy.**

Polychaetes: Segmented worms; mainly marine, living in muds, in both deep and shallow waters.

Polyelectrolytes: synthetic chemicals used to speed flocculation of solids in sewage.

Polygamy: a mating system in which a male pairs with more than one female at one time (polygyny) or a female pairs with more than one male (polyandry).

Polygyny: see **polygamy.**

Polymictic lake: a lake that is continually mixing or that has very short stagnation periods.

Polymorphism: organisms that can occur in several different forms, independent of sexual differences.

Polypeptide: a polymer formed either by synthesis from amino acids or by protein degradation.

Polysaccharide: a polymeric carbohydrate formed by synthesis from monosaccharides or disaccharides (sugars). Starch, cellulose, and glycogen are polysaccharides.

Polyvinyl chloride (PVC): a common plastic material that releases hydrochloric acid when burned.

Population equivalent: a means of expressing the strength of a waste by equating it to the strength of waste from an equivalent number of persons.

Population trajectory: movement of a point on a population graph representing simultaneous changes in two populations.

Pore space: the open space between particles in a rock or soil.

Porosity: the ability of rock and other earth materials to hold water in open spaces or pores; the percentage of such open space in relation to total volume.

Porosity: the ratio of the aggregate volume of interstices in a rock or soil to its total volume. It is usually stated as a percentage.

Potable: drinkable. Said of water and beverages.

Potable water: water suitable for drinking or cooking purposes by both health and aesthetic standards.

Potential energy: energy in an inactive form that is the result of relative position or structure instead of motion, as in a coiled spring or stored chemicals.

Potential evapotranspiration: see evapotranspiration.

ppb: parts per billion; equivalent to $1\mu g$/liter.

ppm: parts per million; the unit commonly used to represent the degree of pollutant concentration where the concentrations are small. Larger concentrations are given in percentages. For example, BOD is represented in ppm whereas suspended solids in water are expressed in percentages. In air, ppm is usually a volume/volume ratio; in water, it is a weight/volume ratio. Equivalent to 1 mg/l.

Precipitate: a solid that separates from a solution because of some chemical or physical change or the formation of such a solid.

Precipitation: the discharge of water, in liquid or solid state, out of the atmosphere, generally upon a land or water surface. The quantity of water that has been precipitated (as rain, snow, hail, sleet) measured as a liquid. The process of separating mineral constituents from a solution by evaporation (halite, anhydrite) or from magma to form igneous rocks.

Precipitators: in pollution control work, any of a number of air pollution control devices usually using mechanical/electric means to collect particulates from an emission.

Precocial: see **nidifugous.**

Predator: an organism that lives by eating other organisms, its prey.

Pressure: the force per unit area upon an object.

Pressure vessel: large container of welded steel or prestressed concrete within which are the reactor core and other reactor internals.

Pretreatment: in waste water treatment, any process used to reduce pollution load before the waste water is introduced into a main sewer system or delivered to a treatment plant for substantial reduction of the pollution load.

Primary production: assimilation (gross) or accumulation (net) of energy and nutrients by green plants and other autotrophs.

Primary productivity: production of photosynthetic organisms, the first step in the food chain.

Primary shorelines: shorelines where the coastal region has been mainly formed by terrestrial agents, such as rivers, glaciers, deltas, volcanos, folding, and faulting.

Primary treatment: the first stage in waste water treatment in which substantially all floating or settleable solids are mechanically removed by screening and sedimentation.

Primordial: pertaining to the beginning or initial times of the earth's history.

Procaryotic: a protist with a simple structure; its nucleus has no nuclear membrane.

Process weight: the total weight of all materials, including fuels, introduced into a manufacturing process. The process weight is used to calculate the allowable rate of emission of pollutant matter from the process.

Producer: organism, human being, or industry that generates high-quality energy by transforming and combining low grades of sunlight and other energy sources and raw materials in excess of its own use.

Productivity: the production of organic material. see **gross productivity; net above-ground productivity, primary production.**

Profile section: a diagram showing the shape of a surface as it would appear in vertical section. The placement of the profile section is indicated by a line drawn on an accompanying map.

Profundal zone: the deep and bottom water area beyond the depth of effective light penetration; all of the lake floor beneath the hypolimnion; in lakes, the bottom and deep water areas beneath the light compensation level.

Protandry: course of development of an individual during which its sex changes from male to female.

Protein: the basic organic material of living matter, consisting of chains of amino acids; about 6% is nitrogen. Protein occurs in all animal and vegetable matter and is essential to the diet of animals.

Proteins: organic polymers composed of amino acids.

Proteolytic enzymes: enzymes that degrade proteins.

Protist: a one-celled organism belonging to the kingdom Protista.

Proton: positively charged particle; constituent of nucleus.

Protoplanets: the early planets which preceded and developed into the present planets (according to the condensation theory of the origin of the planets).

Protoplasm: the material inside a living cell.

Province: a large area or region unified in some way and considered as a whole.

Proximate factors: aspects of the environment that organisms use as cues for behavior; for example, day length. (Proximate factors are often not directly important to the organism's wellbeing.)

Psammon: interstitial organisms found between sand grains.

Pulverization: the crushing or grinding of material into small pieces.

Pumping station: a station at which sewage is pumped to a higher level. In most sewer systems pumping is unnecessary; waste water flows by gravity to the treatment plant.

Pupa: an intermediate, usually quiescent, form assumed by insects after the larval stage and maintained until the beginning of the adult stage.

Pure culture: a culture or medium containing one organism.

Putrescible: capable of being decomposed by microorganisms with sufficient rapidity to cause nuisances from odors, gases, etc. For example, kitchen wastes or dead animals.

Pycnocline: a zone where the water density rapidly increases. The increase is greater than that in the water above or below it. The density change, or pycnocline, results from changes in temperature and salinity.

Pyrolysis: chemical decomposition by extreme heat.

Q_{10}: see **Temperature coefficient.**

Quaywall: a heavy gravity or platform structure fronting on navigable water used as a retaining structure at the inshore ends of slips between piers.

Quench tank: a water-filled tank used to cool incinerator residues.

Quicksilver: mercury where it occurs as a native mineral.

Rad: radiation absorbed dose; measure of exposure to radiation.

r **and** K **selection:** alternative expressions of selection on traits that determine fecundity and survivorship to favor rapid population growth at low population density (r) or competitive ability at densities near the carrying capacity (K).

Race: (1) if a shallow or narrow passage, or "strait," intervenes in the course of the currents, the tide may there produce a rapid rush of water or a race. (2) Local group of organisms differing from other groups; less significant than subspecies.

Rad: a unit of measurement of any kind of radiation absorbed by humans.

Radiation: the emission of fast atomic particles or rays by the nucleus of an atom. Some elements are naturally radioactive, whereas others become radioactive after bombardment with neutrons or other particles. The three major forms of radiation are alpha, beta, and gamma.

Radiation standards: regulations that include exposure standards, permissible concentrations, and regulations for transportation.

Radioactive decay: spontaneous, radioactive transformation of one nuclide to another. Synonyms: decay, radioactive disintegration, disintegration.

Radioactive element: an element capable of changing spontaneously into another element by the emission of charged particles from the nuclei of its atoms. For some elements, e.g., uranium, all known isotopes are radioactive; for others, e.g., potassium, only one of several isotopes is radioactive. Radioactive isotopes of most elements can be prepared artificially, but only a few elements are naturally radioactive.

Radioactive isotope: a form of an element that emits radiation. Also called radionuclide.

Radioactive pollutants: either artificial or natural radionuclides that produce, or that potentially produce, radiation that is injurious to part or all of the ecosystem.

Radioactivity: the property shown by some elements of changing into other elements by the emission of charged particles from their nuclei.

Radioactivity: behavior of substance in which nuclei are undergoing transformation and emitting radiation; note that radioactivity produces radiation—the two terms are not equivalent.

Radiobiology: the study of the principles, mechanisms, and effects of radiation on living matter.

Radiocarbon dating: the determination of the age of a material by measuring the proportion of the isotope ^{14}C (radiocarbon) in the carbon it contains. The method is suitable for the determination of ages up to a maximum of about 30,000 years.

Radioecology: the study of the effects of radiation on species of plants and animals in natural communities.

Radioisotopes: radioactive isotopes. Radioisotopes, such as cobalt-60, are used in the treatment of disease.

Radiolarian ooze: deposits of siliceous ooze that are made up largely of radiolarian skeletons and that are at depths of 12,000 to 25,000 ft (3700 to 7200 m) but that in the geologic past may have been deposited at much shallower depths.

Radionuclide: radioactive nuclide.

Radium: intensely radioactive alpha-emitting heavy element.

Rafting: transportation of material be means of attachment to ice, plants or other floating material.

Rare gases: those gases (such as krypton, xenon, and argon) that are present in the earth's atmosphere in very small quantities.

Rasp: a device used to grate solid waste into a more manageable material, ridding it of much of its odor.

Raw sewage: untreated domestic or commercial waste water.

Reach: (of a river) an extended portion of water, as in a straight portion of a stream or river; a level stretch, as between locks in a canal; an arm of the sea extending into the land. A promontory, tongue, or extended portion of land.

Reactivity: measure of ability of assembly of fissile material to support sustained chain reaction. The coefficient of reactivity is a measure of the way the reactivity of an assembly changes in response to any other change, as for instance of temperature.

Reactor: an assembly of nuclear fuel capable of sustaining a fission chain reaction.

Recalcitrant chemical: one that is not easily degraded by microorganisms.

Recapitulation: the theory (if applied to entire organisms it is now rejected in large part) that the individual development of an organism passes through stages resembling the adult conditions of its ancestors. It is still possible that if the individual is not recapitulated, individual characters may be recapitulated. Cf. Haeckel's law.

Receiving waters: rivers, lakes, oceans, or other bodies that receive treated or untreated waste waters.

Recessiveness: failure of an allele to influence the phenotype when present in heterozygous form.

Recharge: intake. The processes by which water is absorbed and is added to the zone of saturation, either directly into a formation or indirectly by way of another formation. Also, the quantity of water that is added to the zone of saturation.

Recombination: exchange of genetic information between paired homologous chromosomes during meiosis.

Recruitment: addition by reproduction of new individuals to a population.

Recycling: the process by which waste materials are transformed into new products in such a manner that the original products may lose their identity.

Red tide: a proliferation or bloom of a certain type of plankton with red to orange coloration, which often causes massive fish kills. Though they are a natural phenomenon, blooms are believed to be stimulated by phosphorus and other nutrients discharged into waterways by humans.

Reduction: a chemical reaction in which oxygen is removed or electrons are accepted by an electron acceptor.

Reef: (1) a range or ridge of rocks lying at or near the surface of water, especially one of coral. (2) A rock structure, either moundlike or layered, built by such sedentary organisms as corals and usually enclosed in rock of a differing lithology.

Reef patch: a term for all coral growths that have grown up independently in lagoons of barriers and atolls.

Reflected wave: the wave that is returned seaward when a wave impinges upon a very steep beach, barrier, or other reflecting surface.

Refraction: the deflection of the direction of wave propagation when waves pass obliquely from one region of velocity to another.

Refractometer: an instrument for determining the index of refraction of a mineral or of a liquid, etc.

Refuse: see **solid waste.**

Refuse reclamation: the process of converting solid waste to saleable products. For example, the composting of organic solid waste yields a saleable soil conditioner.

Regime: the dynamic activities and cycles of particular physical element of an ecosystem, e.g., water regime.

Regulatory response: a rapid, reversible physiologic or behavioral response by an organism to change in its environment.

Relative abundance: proportional representation of a species in a sample or a community.

Relic geomorph: a landform that has survived the forces of erosion and decay, such as an erosional remnant.

Relic sediments: sediments whose character represent not present-day conditions, but a past environment.

Rem: a measurement of radiation dose to the internal tissue of humans. (Acronym for roentgen equivalent man.)

Remote sensing: the measurement of acquisition of information on some property of an object or phenomenon by a recording device that is not in physical or intimate contact with the object or phenomenon under study. The technique employs such devices as cameras, lasers, infrared and ultraviolet detectors, microwave and radio frequency receivers, and radar systems. The practice of data collection in the wavelengths from ultraviolet to radio regions. This restricted sense is the practical outgrowth from airborne photography. It is sometimes called rapid reconnaissance.

Renewable resource: see **resource.**

Rep: a unit of measurement of any kind of radiation absorbed by humans.

Repose, angle of: the slope at which any given deposited material will come to rest under a given set of physical conditions. Eolian sands generally have higher angles of repose than aqueous sands of the same size grade.

Reservoir: a pond, lake, tank, or basin, natural or manmade, used for storage, regulation, and control of water.

Residence time: the residence time of an element in sea water is defined as the total amount of the element in the ocean divided by the rate of introduction of the element or the rate of its precipitation to the sediments.

Resistivity: the resistance to electric current of a three-dimensional, unbounded medium, as opposed to resistance that refers to the electric impedance of confined conductors. It is the reciprocal of conductivity and is usually expressed in ohm-meters. Synonym: specific resistance.

Resolution: (1) the separation of a vector into its components. (2) The sharpness with which the images of two closely adjacent spectrum lines, etc., may be distinguished. (3) Gravity or magnetic prospecting: the indication in some measured quantity, such as the vertical component of gravity, of the presence of two or more close but separate disturbing bodies. (4) Seismologic prospecting: the ability to indicate separately two closely adjacent interfaces.

Resource: a substance or object required by an organism for normal maintenance, growth, and reproduction. If the resource is scarce relative to demand, it is referred to as a limiting resource. Nonrenewable resources (such as space) occur in fixed amounts and can be fully utilized; renewable resources (such as food) are produced at a fixed rate, with which the rate of exploitation attains an equilibrium.

Resource recovery: the process of obtaining materials or energy, particularly from solid waste.

Respiration: an oxidation process whereby organic matter is used by plants and animals and converted to energy. Oxygen is used in this process and carbon dioxide and water are liberated. Aerobic metabolism in a cell. The energy for growth in respiratory processes is obtained by transfer of electrons from a substrate to oxygen as the ultimate electron acceptor.

Reverberation: the persistence of sound in an enclosed space after the sound source has stopped.

Reverse osmosis: an advanced method of waste treatment relying on a semipermeable membrane to separate waters from pollutants. Desalination by filtration of water through a semipermeable membrane. Ultrafiltration is used to filter larger molecules.

Reversing thermometer: a thermometer attached to a Nansen bottle (which see) that registers the temperature of sea water at the depth where a sample of the sea water is obtained.

Rhizobium: the bacterium associated with legumes in nitrogen fixation.

Rhizosphere: the zone of soil influenced by plant root exudates.

Richter scale: the range of numerical values of earthquake magnitude, devised in 1935 by the seismologist C. F. Richter. Very small earthquakes, or microearthquakes, can have negative magnitude values. In theory there is no upper limit to the magnitude of an earthquake. However, the strength of earth materials produces an actual upper limit of slightly less than 9.

Riffle: a rock or gravel bar in a stream over which backed up water runs at greater than normal speed during periods of low flow. Riffles tend to occur at relatively uniform intervals along most streams. A shallow extending across the bed of a stream; a rapid of comparatively little fall.

Rift valley: a deep fracture or break about 25 to 50 km wide, extending along the crest of the midocean ridge. Magma upwellings occur here as the midocean ridge spreads apart along the rifts, adding volcanic rock to the sea floor.

Rill: a very small trickling stream of water; a very small brook; a minute stream that flows away from a beach as a wave subsides; a small channel made by circulating water in the wall, floor, or ceiling of a cave. One of several relatively long (up to several hundred kilometers), narrow (1 to 2 km), trench- or cracklike valleys commonly occurring on the moon's surface. Rilles may be extremely irregular with meandering courses (sinuous rilles); they have relatively steep walls and usually flat bottoms. Rills are essentially youthful features and apparently represent fracture systems originating in brittle material. Synonym: Rille; rima.

Ringelmann chart: a series of illustration ranging from light grey to black used to measure the opacity of smoke emitted from stacks and other sources. The shades of grey simulate various smoke densities and are assigned numbers ranging from 1 to 5. Ringelmann No. 1 is equivalent to 20% dense; No. 5 is 100% dense. Ringelmann charts are used in the setting and enforcement of emission standards.

Riparian rights: rights of a landowner to the water on or bordering his or her property, including the right to prevent diversion or misuse of upstream water.

Rip current: a narrow, seaward flowing current that results from breaking of waves and subsequent accumulation of water in the near-shore zone.

Riprap: (1) broken rock used for revetment; the protection for bluffs or structures exposed to wave action, foundations, etc. (2) Foundation or wall of broken rock thrown together irregularly.

Rip tide: an improper name for rip current. A rip current has no relation to the tide; hence rip tide in this connection is a misnomer.

Rise: (1) (Oceanic) an elevation that ascends gradually with an angle of only a few minutes of arc, irrespective of whether it is wide or narrow or of its vertical development. On account of its flatness, the rise apparently plays only a subordinate part in suboceanic topography; however, rises carry the chief features of suboceanic relief, so that if an ocean floor were changed into dry land they would act as the main watersheds (Ger., Schwellen; Fr., *seuil*). (2) A long and broad elevation of the deep sea floor that ascends gently and smoothly. (3) Spring rising from fractures in limestone. (4) Point at which an underground stream comes to the surface. Synonym: resurgence; emergence.

River basin: the total area drained by a river and its tributaries.

River bed: the bed of a river is the channel that contains its waters. By the term bed is understood all the space ordinarily covered by water and lying between the lands on each side of the stream.

River valley: the depression made by the stream and by its various processes that precede and accompany the development of the stream.

Rock weathering: the chemical and physical processes that cause rocks to decay and eventually form soil.

Rodenticide: a chemical or agent used to destroy or prevent damage by rats or other rodent pests. See **pesticide.**

Rough fish: those fish species considered to be of poor fighting quality when taken on tackle or of poor eating quality; for example, gar, suckers. Most rough fish are more tolerant of widely changing environmental conditions than are game fish.

Rubbish: a general term for solid waste—excluding food waste and ashes—taken from residences, commercial establishments, and institutions.

Rubble: all accumulations of loose angular fragments not water worn or rounded like gravel. Loose angular water-worn stones along a beach.

Runoff: the portion of rainfall, melted snow, or irrigation water that flows across ground surface and eventually is returned to streams. Runoff can pick up pollutants from the air or the land and carry them to the receiving waters. The discharge of water through surface streams; the quantity of water discharged through surface streams, expressed usually in units of volume such as gallons, cubic feet, or acre-feet.

Rural: having to do with farming, agriculture, or country life; not urban.

S: standard deviation.

Safe yield: that rate of which water can be withdrawn from an aquifer without depleting the supply to such an extent that withdrawal at this rate is harmful to the aquifer itself, or to the quality of the water, or is no longer economically feasible.

Salinity: measure of the quantity of total dissolved solids in water. A measure of the total concentration of dissolved solids in a saline water. Specifically, for sea water, the total of dissolved solids, expressed in grams per kilogram, when all carbonate has been converted to oxide, Br and I replaced by Cl, and all organic matter completely oxidized.

Salt lakes: lakes that contain a predominating amount of sodium chloride in solution, and usually magnesium chloride as well as magnesium and calcium sulfate.

Salinity: the degree of salt in water. The total amount of dissolved material in sea water. It is measured in parts per thousand by weight in 1 kg of sea water. The salt content of fresh water is usually reported 0.5 parts per thousand, whereas the average concentration of seawater is 35 parts per thousand.

Salmonella: the genus *Salmonella* includes several species pathogenic to humans. Primary diseases are typhoid fever and food poisoning.

Salt marsh: a marsh periodically flooded by salt water.

Salt pan: a shallow lake of brackish water. Also written panne. A large pan for making salt by evaporation; a salt works.

Salt-water encroachment or intrusion: the phenomenon occurring when a body of salt water, because of its greater density, invades a body of fresh water. It can occur either in surface- or ground-water bodies. The balance between the two, in static situations, is expressed by the Ghyben–Herzberg formula. When this invasion is caused by oceanic waters, it is called sea water intrusion.

Salts: a group of soluble compounds, including sodium chloride (common table salt) dissolved from earth and rocks by the water that flows through them (see **leaching**).

Salvage: the utilization of waste materials.

Sandy gravel: gravel containing 50 to 75% of sand.

Sanitation: the control of all the factors in the physical environment of humans that exercise or can exercise a deleterious effect on our physical development, health, and survival.

Sanitary landfill: a site for solid waste disposal using sanitary landfilling techniques.

Sanitary landfilling: an engineered method of solid waste disposal on land in a manner that protects the environment; waste is spread in thin layers, compacted to the smallest practical volume, and covered with soil at the end of each working day.

Sanitary sewers: sewers that carry only domestic or commercial sewage. Storm water runoff is carried in a separate system. See **sewer.**

Saprophyte: a nonpathogenic microorganism.

Saturated hydrocarbon: a hydrocarbon in which there are no double or triple bond linkages. Therefore, the hydrocarbon is more stable. Contrast unsaturated hydrocarbon.

Saturated zone: the zone of saturation; a subsurface zone below which all rock pore space is filled with water.

Savanna, savannah: a tract of level land having a wet soil except during periods of dry weather, and supporting grass and other low vegetation, with but a scattered growth of pine or other trees and bushes. Sometimes applied to tracts of open prairie land.

Scrap: discarded or rejected materials that result from manufacturing or fabricating operations and are suitable for reprocessing.

Screening: the removal of relatively coarse floating and suspended solids by straining through racks or screens.

Scrubber: an air pollution control device that uses a liquid spray to remove pollutants from a gas stream by absorption or chemical reaction. Scrubbers also reduce the temperature of the emission.

SCUBA: self-contained underwater breathing apparatus.

Sea: (1) an ocean or a large body of (usually) salt water smaller than an ocean. (2) Waves caused by wind at the place and time of observation; waves within their area of generation; usually they are irregular without a definite pattern. (3) State of the ocean or lake surface in regard to waves.

Sea slide: submarine slide similar to a landslide.

Seamounts: isolated elevations, usually higher than 1000 m, on the sea floor. Usually they resemble an inverted cone in shape.

Search image: a behavioral selection mechanism that enables predators to increase searching efficiency for prey that are abundant and worth capturing.

Secchi disk: a circular metal plate, 20 cm in diameter, the upper surface of which is divided into four equal quadrants and so painted that two quandrants directly opposite each other are black and the intervening ones white. Relative turbidity of water is measured by submerging this disk and lowering it until it disappears from view.

Secondary shorelines: shorelines where the coastal region has been mainly formed by marine or biologic agents, such as coral reefs, barrier beaches, and marshes.

Secondary treatment: waste water treatment, beyond the primary stage, in which bacteria consume the organic parts of the wastes. This biochemical action is accomplished by use of trickling filters or the activiated sludge process. Effective secondary treatment removes virtually all floating and settleable solids and approximately 90% of both BOD_5 and suspended solids. Customarily, disinfection by chlorination is the final stage of the secondary treatment process. A process used to oxidize organic matter in wastes by conversion to bacteria under high aeration.

Sedentary: attached, as an oyster, barnacle, or similar shelled invertebrate. In sedimentation, formed in place without transportation by the underlying rock or by the accumulation of organic material; said of some soils, etc.

Sedgwick-rafter counting cell: a plankton-counting cell consisting of a brass or glass receptacle, 50 by 20 by 1 mm, sealed to a 1 by 3 in. glass microscope slide. A rectangular cover glass large enough to cover the whole cell is required. The cell has a capacity of exactly 1 ml.

Sediment: (1) Solid material settled from suspension in a liquid. (2) Solid material, both mineral and organic, that is in suspension, is being transported, or has been moved from its site of origin by air, water, or ice and has come to rest on the earth's surface either above or below sea level.

Sediment load: the solid material transported by a stream, including bed material load and wash load.

Sedimentary rocks: rocks that have formed from the accumulation of particles (sediment) in water or from the air.

Sedimentation: in waste water treatment, the settling out of solids by gravity.

Sedimentation tanks: in waste water treatment, tanks where the solids are allowed to settle or to float as scum. Scum is skimmed off and settled solids are pumped to incinerators, digesters, filters, or other means of disposal.

Seepage: water that flows through the soil.

Segment: a portion of a river basin the surface waters of which have common hydrologic characteristic (or flow regulation patters); common natural, physical, chemical, and biologic processes; and common reactions to external stresses, such as discharging of pollutants.

Segregation: separation of alleles derived from the male and female parent into different gametes during meiosis.

Seiche: an oscillation of a body of water whose period is determined by the resonant characteristics of the containing basin as controlled by the physical dimensions. These periods generally range from a few minutes to an hour or more. (Originally the term was applied only to lakes but now is also applied to harbors, bays, oceans, etc.)

Seismicity: relates to earth movements or earthquakes; an area of high seismicity has numerous earthquakes.

Selection: see **artificial selection, group selection, natural selection, sexual selection, sibling selection, social selection.**

Selective death: a death attributed to the deleterious expression of a genotype; a death that would have been avoided had the individual had the optimum genotype.

Selective herbicide: a pesticide intended to kill only certain types of plants, especially broad-leafed weeds, and not harm other plants, such as farm crops or lawn grasses. The leading herbicide in the United States is 2,4-D. A related but stronger chemical used mostly for brush control range, pasture, and forest lands and on utility or highway rights of way is 2,4,5-T. Use of the latter chemical has been somewhat restricted because of laboratory evidence that it or a dioxin contaminant in 2,4,5-T can cause birth defects in test animals.

Self-purification: the ability of a body of water to rid itself of pollutants.

Senescence: the process of growing old. Sometimes used to refer to lakes nearing extinction. Gradual deterioration of function in an organism leading to an increased probability of death; aging.

Septic tank: an underground tank used for the deposition of domestic wastes. Bacteria in the wastes decompose the organic matter, and the sludge settles to the bottom. The effluent flows through drains into the ground. Sludge is pumped out at regular intervals. A tank used in rural area for disposal of domestic wastes.

Seral: a successional community that has not yet reached maturity or climax. Successional communities usually exhibit an increase in standing crop as gross productivity exceeds respiration, and thus net productivity is greater than zero.

Sere: a sequence of ecologic communities from pioneer stage to climax community. A series of stages of community change in a particular area, involving replacement of populations and leading toward a stable state (see **succession**).

Serial samples: samples collected according to some predetermined plan, such as along the intersections of grid lines or at stated distances or times. The method is used to insure random sampling.

Sessile: applied to organisms that are closely attached to other objects or to the substrate; an organism that is permanently attached to a surface.

Seston: the living and nonliving bodies of plants or animals that float or swim in the water.

Settleable solids: suspended solids that will settle in quiescent water or sewage in a reasonable period. Such period is commonly, though arbitrarily, taken as 2 hr. Bits of debris and fine matter heavy enough to settle out of waste water.

Settling chamber: in air pollution control, a low-cost device used to reduce the velocity of flue gases, usually by means of baffles, promoting the settling of fly ash.

Settling tank: in waste water treatment, a tank or basin in which settleable solids are removed by gravity.

Sewage: the total of organic waste and waste water generated by residential and commercial establishments.

Sewage lagoon: see **lagoon.**

Sewer: any pipe or conduit used to collect and carry away sewage or storm water runoff from the generating source to treatment plants or receiving streams. A sewer that conveys household and commercial sewage is called a sanitary sewer. If it transports runoff from rain or snow, it is called a storm sewer. Often storm water runoff and sewage are transported in the same system, called a combined sewer.

Sewerage: the entire system of sewage collection, treatment, and disposal. Also applies to all effluent carried by sewers whether it is sanitary sewage, industrial wastes, or storm water runoff.

Sewered population: the population served by waste water collecting sewers.

Sex chromosomes: a pair of nonhomologous chromosomes, called X and Y in mammals, which determine the sex of the organism (XX is female and XY is male in mammals).

Sexual environment: interactions of an organism with mates or potential mates.

Sexual selection: selection by one sex for specific characteristics in individuals of the opposite sex, usually exercised through courtship behavior.

Shale oil: a crude oil obtained from bituminous shales, especially in Scotland, by submitting them to destructive distillation in special retorts.

Shelf break: the sharp break in slope which marks the edge of the continental shelf and the beginning of the continental slope.

Shelf ice: the extension of glacial ice from land into coastal waters. Shelf ice, which may be several hundred feet thick, is in contact with the bottom near shore but not at its seaward terminus. An ice shelf.

Shelf sea: the water that rests upon a sea shelf (continental shelf).

Shellfish: any aquatic invertebrate with a hard external covering; more commonly mollusks and crustaceans.

Shield: (1) A continental block of the earth's crust that has been relatively stable over a long period of time and has undergone only gentle warping, in contrast to the strong folding of bordering geosynclinal belts. Mostly composed of Precambrian rocks. (2) A disk-shaped formation standing edgewise at a high angle in a cave. (3) In animals, a protective structure likened to a shield, as a large scale, carapace, or lorica. (4) A wall that protects workers from harmful radiation released by radioactive materials.

Shielding: wall of material (concrete, lead, water) surrounding a source of radiation to reduce its intensity.

Shingle: loosely and commonly, any beach gravel that is coarser than ordinary gravel, especially if consisting of flat or flattish pebbles and cobbles.

Shingle beach: a beach whose surface is covered with shingle rock. see **shingle.**

Shoal: A part of the area covered by water in the sea, lake, or river where the depth is little; a bank always covered, though not deeply. (v.) To become shallow gradually. A detached elevation of the sea bottom comprised of any material except rock or coral, and which may endanger surface navigation.

Shoran: a high-frequency radiowave location system using microwave pulses used for offshore and airborne prospecting operations. Two stations are located at fixed points; the third is on the mobile station whose location is desired. The fixed stations broadcast pulses, the mobile station rebroadcasts them, and the roundtrip time is measured by means of cathode-ray screens to an accuracy of $\pm$ 7.6 m.

Shoreline: the place where land and water meet.

Sibling selection: selection within a brood of offspring for interaction, both competitive and cooperative, that increases their evolutionary fitness.

Sibling species: species with similar appearance that are incapable of breeding with each other.

Sieve analysis: determination of the percentage distribution of particle size by passing a measured sample of soil or sediment through standard sieves of various sizes.

Sill: a ridge separating one partially closed ocean basin from the ocean or another basin.

Silt: (1) a clastic sediment, most of the particles of which are between 1/16 and 1/256 mm in diameter. (2) Soil consisting of 80% of more silt (0.05 to 0.002 mm.) and less than 12% clay; finely divided particles of soil or rock. Often carried in cloudy suspension in water and eventually deposited as sediment.

Sinking: a method of controlling oil spills that employs an agent to entrap oil droplets and sink them to the bottom of the body of water. The oil and sinking agent are eventually biologically degraded.

Skimming: the mechanical removal of oil or scum from the surface of water.

Slack water: essentially currentless water, such as that occurring in flooded area beyond a stream channel or in an estuary at high or low tide.

Slope stability: the resistance of a natural or artificial slope to landslide failure.

Sludge: the construction of solids removed from sewage during waste water treatment. Sludge disposal is then handled by incineration, dumping, or burial. The microbial cells that settle out during aerobic sewage treatment.

Slumping: the sliding or moving of sediments down a submarine slope.

Smog: Generally used as an equivalent of air pollution, particularly associated with oxidants.

Smoke: solid particles generated as a result of the incomplete combustion of materials containing carbon.

Snail: an organism that typically possesses a coiled shell and crawls on a single muscular foot. Air breathing snails, called pulmonates, do not have gills but typically obtain oxygen through a "lung" or pulmonary cavity. At variable intervals most pulmonate snails come to the surface of the water for a fresh supply of air. Gill breathing snails possess an interval gill through which dissolved oxygen is removed from the surrounding water.

Social adaptation: any adaptation that facilitates behavioral interactions among individuals of the same species; usually does not include reproductive activities.

Social behavior: any direct interaction among distantly related individuals of the same species; usually does not include courtship, mating, or parent–offspring and sibling interactions.

Social environment: all the interactions that an individual has with other individuals of the species, other than reproductive activities.

Social facilitation: enhancement of any behavior by association with other individuals engaged in similar behavior.

Social feedback: direct interaction by which some individuals exercise control over the activities of other individuals so as to regulate population processes.

Social group: a group of individuals of the same species formed by mutual attraction of individuals to each other and within which individuals are interdependent to some degree for their wellbeing.

Social pathology: a syndrome of physiologic and behavioral disturbances, caused by crowding, that leads to reduced fecundity and increased mortality.

Social selection: selection for genetic traits by social behavior (which see).

Society: a group of organisms of the same species characterized by specialization of individual roles, divisions of labor, and mutual dependence.

Soft detergents: biodegradable detergents.

Soil: that earth material which has been so modified and acted upon by physical, chemical, and biologic agents that it can support rooted plants. The term as used by engineers includes, in addition to the above, all regolith.

Soil conditioner: a biologically stable organic or inorganic material, such as humus compost, vermiculite, or perlite, that makes soil more amenable to the passage of water and to the distribution of fertilizing material, providing a better medium for necessary soil bacteria growth.

Soil horizon: a layer of soil that is distinguished from adjacent layers by characteristic physical properties, such as structure, color, or texture. The letters A, B, and C are used to designate soil horizons. The A horizon is the upper part. It consists of mineral layers of maximum organic accumulation or layers from which clay materials, iron, and aluminum have been lost, or both. The B horizon lies under the A. It consists of weathered material with an accumulation of clay, iron, or aluminum or with a more or less blocky or prismatic structure, or both. The C horizon, under the B, is the layer of unconsolidated, weathered parent material. Not all these horizons are present in all soils. Sometimes O or H is used for the unaltered organic debris at the surface.

Soil profile: succession zones or horizons (which see), beginning at the surface, that have been altered by normal soil-forming processes of which leaching, oxidation, and accretion are particularly important.

Soil surveys: dynamic methods used in engineering to test the site of buildings, dams, bridges and similar structures for possibilities of compaction or of earthquake or other vibration damage.

Solid waste: useless, unwanted, or discarded material with insufficient liquid content to be free flowing. Also see **waste.** (1) Agriculture: solid waste that results from the raising and slaughtering of animals and the processing of animal products and orchard crops. (2) Commercial: waste generated by stores, offices, and other activities that do not actually turn out a product. (3) Industrial: waste that results from industrial processes and manufacturing. (4) Institutional: waste originating from educational, health care, and research facilities. (5) municipal: residential and commercial solid waste generated within a community. (6) Pesticide: the residue from the manufacturing, handling, or use of chemicals intended for killing plant and animal pests. (7) residential: waste that normally originates in solid waste (continued) of a residential environment; sometimes called domestic solid waste.

Solid waste management: the purposeful, systematic control of the generation, storage, collection, transport, separation, processing, recycling, recovery, and disposal of solid wastes.

Solubility: the degree which a substance mixes with another substance.

Sonic boom: the tremendous booming sound produced as a vehicle, usually a supersonic jet airplane, exceeds the speed of sound and the shock wave reaches the ground.

Sorting coefficient: this term applies to the degree of sorting of sediment particles and is a measure of the spread of the distribution. It is defined statistically as the standard deviation of grain size spread. Higher values indicate less homogeneous, poorly sorted mixtures, whereas lower values indicate more homogeneous, better sorted mixtures.

Soot: Agglomerations of tar-impregnated carbon particles that form when carbonaceous material does not undergo complete combustion.

Sorption: A term including both adsorption and absorption. Sorption is basic to many processes used to remove gaseous and particulate pollutants from an emission and to clean up oil spills.

Sound channel: a sound channel can occur where the sound velocity reaches a minimum value. Sound in this zone is refracted upward or downward back into the zone, with little energy loss. Thus, the sound traveling in this channel can be transmitted over distances of many thousands of kilometers.

Sounding: a measured depth of water. On hydrographic charts the soundings are adjusted to a specific plane of reference (sounding datum). The determination of the depth of the ocean. It can be done by lowering a line to the bottom or, electronically, by noting how long sound takes to travel to the bottom and return.

Sounding line: a line, wire, or cord used in sounding. It is weighted at one end with a plummet (sounding lead). Synonym: lead line.

SO_x: a symbol meaning oxides of sulfur.

Spartina: the dominant plant in a salt marsh community.

Specialization: restriction of an organism's or a population's activities to a portion of the environment; a trait that enables an organism (or one of its organs) to modify (or differentiate) in order to adapt to a particular function or environment.

Speciation: separation of one population into two or more reproductively isolated, independent evolutionary units.

Species: (both singular and plural) an organism or organisms forming a natural population or group of populations that transmit specific characteristics from parent of offspring. They are reproductively isolated from other populations with which they might breed. Populations usually exhibit a loss of fertility when hybridizing. A very closely related group of organisms within a genus.

Specific heat: amount of energy that must be added or removed to change the temperature of a substance by a specific amount. By definition, 1 cal of energy is required to raise the temperature of 1 g of water by 1°C.

Spectrophotometer: optical instrument for comparing the intensities of the corresponding colors of two spectra.

Spermatophyte: seed-forming plants.

Spoil: dirt or rock that has been removed from its original location; specifically, materials that have been dredged from the bottoms of waterways.

Spontaneous liquefaction: when water-saturated sediments are subjected to a sudden shock, shear, or increase in pore water pressure, the internal grain to grain contacts within the sediment may change. If this happens, the entire sediment mass may move or flow, similar to an avalanche.

Spore: a resistant form of a microorganism that can survive adverse conditions for very long periods of time.

Spring: an opening in the surface of the earth from which ground water flows.

Spring tide: a tide that occurs at or near the time of new and full moon and that rises highest and falls lowest from the mean level. High tides that occur about every 2 weeks, when the moon is full or new.

Stability: the ability to resist perturbations. Inherent capacity of any system to resist change.

Stabilization: the process of converting active organic matter in sewage sludge or solid wastes into inert, harmless material.

Stabilization ponds: See **lagoon, oxidation pond.**

Stabilizing factors: factors that tend to restore a system to its equilibrium state; specifically, the class of density-dependent factors that act to restore populations to equilibrium size.

Stable age distribution: proportion of individuals in various age classes in a population that has been growing at a constant rate.

Stable air: an air mass that remains in the same position instead of moving in its normal horizontal and vertical directions. Stable air does not disperse pollutants and can lead to high buildups of air pollution.

Stack: a smokestack; a vertical pipe or flue designed to exhaust gases and suspended particulate matter.

Stadia rod: a graduated rod used with an instrument of the stadia class to measure the distance from the observation point to the place where the rod is positioned.

Stagnation: lack of wind in an air mass or lack of motion in water. Both cases tend to entrap and concentrate pollutants.

Stand of tide: an interval at high or low water when there is no sensible change in the height of the tide. The water level is stationary at high and low water for only an instant, but the change in level near these times is so slow that it is not usually perceptible.

Standard grain size test: a sieve analysis is used to separate the finest fraction from the coarse fraction. The standard hydrometer test is then applied to the fine fraction.

Standing crop: the biota present in an environment at a selected point in time. See **biomass.**

Station: the position at which instruments are set up or specimens are collected.

Stationary phase: the phase in a growth curve in which the death rate equals the growth rate.

Stationary source: a pollution emitter that is fixed rather than moving, as an automobile.

Substrate: the molecule that reacts with an enzyme or that is used for growth.

Succession: sequence of stages and changes that occurs in an ecologic system as it goes from some starting condition to a steady state; the continuous replacement in time of one community by another. Ultimately leads to a climax stable community. Replacement of populations in a habitat through a regular progression to a stable state.

Sulfur dioxide (SO_2): A heavy, pungent, colorless gas formed primarily by the combustion of fossil fuels. Sulfur dioxide damages the respiratory tract as well as vegetation and materials and is considered a major air pollutant.

Sump: a depression or tank that serves as a drain or receptacle for liquids for salvage or disposal.

Supergene: a series of genes, often with related functions, so closely placed on a chromosome that virtually no recombination occurs between them.

Supplement: something added, especially to make up for a lack or deficiency.

Surface runoff: that part of the precipitation that passes over the surface of the soil to the nearest surface stream without first passing beneath the surface.

Surf: surf is a term for breaking waves in a coastal area.

Surf zone: the area between the outermost breaker and the limit of wave up rush.

Surface water: water on the earth's surface exposed to the atmosphere as rivers, lakes, streams, the oceans. See **ground water.**

Surfactant: an agent used in detergents to cause lathering. Composed of several phosphate compounds, surfactants are a source of external enrichment thought to speed the eutrophication of our lakes.

Surge: the name applied to wave motion with a period intermediate between that of the ordinary wind wave and that of the tide, say from ½ to 60 min. It is of low height, usually less than 10 cm.

Surge zone: the region between the breaker zone and the 15 to 18 m depth contour, where the effect of sea waves and swell produces oscillatory surges causing sediment transport and abrasive erosion.

Survivorship (1_x): proportion of newborn individuals that will be alive at a given age.

Surveillance system: a monitoring system to determine environmental quality. Surveillance systems should be established to monitor all aspects of progress toward attainment of environmental standards and to identify potential episodes of high pollutant concentrations in time to take preventive action.

Suspended solids (SS): small particles of solid pollutants in sewage that contribute to turbidity and that resist separation by conventional means. The examination of suspended solids and the BOD test constitute the two main determinations for water quality performed at waste water treatment facilities.

Swale: a slight, marshy depression in generally level land.

Swamp: A marsh dominated by woody vegetation including trees.

Swell: waves that have traveled away from their generating area. These waves have a more regular pattern than sea waves.

Swimmer's itch: a rash produced on bathers by a parasitic flatworm in the cercarial stage of its life cycle. The organism is killed by the human body as soon as it penetrates the skin; however, the rash may persist for a period of about 2 weeks.

Symbiosis: the intimate living together of two kinds of organisms where such association is of mutual advantage. Intimate, and often obligatory, association of two species, usually involving coevolution. Symbiotic relationships can be parasitic or mutualistic. Two organisms living closely together for their mutual benefit.

Sympatric: occurring in the same place, usually referring to areas of overlap in species distributions.

Sympatric speciation: speciation without geographic isolation; reproductive isolation occurring between segments of a single population.

Synergism: the cooperative action of separate substances so that the total effect is greater than the sum of the effects of the substances acting independently.

Synoptic measurements: numerous measurements taken simultaneously over a large area.

System: a combination of parts organized into a unified whole, usually processing a flow of energy.

Systemic pesticide: a pesticide chemical that is carried to other parts of a plant or animal after it is injected or taken up from the soil or body surface.

Taiga: an open, usually coniferous forest, near the arctic.

Tailings: second grade or waste material derived when raw material is screened or processed.

Taxon cycle: cycle of expansion and contraction of the geographic range and population density of a species or higher taxonomic category.

Taxonomic classification: a system method of classifying animals and plants.

Tectonics: study of the broader structural features of the earth and their causes.

Telemetry: a method, usually electronic, of measuring something and then transmitting the measurements to a receiving station.

Temperature coefficient: rate of increase in an activity or process over a 10°C increase in temperature. Also referred to as the Q_{10}.

Terrace: a plateau of the side of a valley, representing an old floodplain no longer reached by the river below.

Terrigenous sediments: sediments composed of material derived from the land. Usually these deposits are found close to land.

Territory: any area defended by one or more individuals against intrusion by others of the same or different species (see **home range**).

Tertiary treatment: waste water treatment beyond the secondary, or biologic, stage that includes removal of such nutrients as phosphorus and nitrogen and a high percentage of suspended solids. Tertiary treatment, also known as advance waste treatment, produces a high-quality effluent.

Tetraploid: referring to a cell or organism containing four sets of chromosomes (see **ploidy**).

Thalassic: of or pertaining to the sea.

Thermal conductance: rate at which heat flows through a substance.

Thermal pollution: degradation of water quality by the introduction of a heated effluent. Primarily a result of the discharge of cooling waters from industrial processes, particularly from electric power generation. Even small deviations from normal water temperatures can affect aquatic life. Thermal pollution usually can be controlled by cooling towers.

Thermal waste: waste heat from industrial plants that is discharged into the atmosphere or water.

Thermistor: a heat-sensitive device that can be used to measure temperature; a heat-sensitive device used in bolometric measurements of power in high-frequency electric circuits.

Thermocline: the sharp zone in a stratified body of water that separates the epilimnion from the hypolimnion. The layer in a body of water in which the drop in temperature equals or exceed 1°C for each meter or approximately 3 ft of water depth. A zone where the water temperature decreases more rapidly than the water above or below it. This zone usually starts from 10, 50, 500 m below the surface and can extend to cover 1500 m in depth. The layer of water in which rapid change in temperature takes place, usually between the upper warm water layer and lower cold water layer in large bodies of water.

Thermocouple: a union of two conductors, as, bars or wires of dissimilar metals joined at their extremities for producing a thermoelectric current. Synonym: thermojunction, thermoelectric couple.

Thermophile: an organism that has a high temperature optimum. Compare mesophile and psychrophile.

Threshold dose: the minimum dose of a given substance necessary to produce a measurable physiologic or psychologic effect.

Tidal bore: a large wave of tidal origin that will travel up some rivers and estuaries.

Tidal correction: a correction applied to gravitational observations to remove the effect of earth tides on gravimetric observations.

Tidal currents: Currents caused by tides.

Tidal day: the time of the rotation of the earth with respect to the moon, or the interval between two successive upper transits of the moon over the meridian of a place, about 24.84 solar hours (24 hr and 50 min) in length or 1.035 times as great as the mean solar day.

Tidal flat: a marsh or muddy land area which is covered and uncovered by the rise and fall of the tide.

Tidal friction: the frictional effect of the tides, especially in shallow waters, lengthening the tidal epoch and tending to retard the rotational speed of the earth and so increase very slowly the length of the day.

Tidal pool: a pool of water remaining on a beach or reef after recession of the tide.

Tidal prism: the total amount of water that flows into the harbor or out again with movement of the tide, excluding any fresh-water flow.

Tidal range: the difference between the level of water at high tide and low tide.

Tidal wave: In astronomic usage, restricted to the periodic variations of sea level produced by the gravitational attractions of the sun and the moon. Commonly and incorrectly used for a large sea wave caused by a submarine earthquake or volcanic eruption, which is properly called a tsunami (which see). Sometimes used for a hurricane wind or a severe gale, properly called a storm wave.

Tide: the periodic rise and fall of oceans and bodies of water connecting them, caused chiefly by the attraction of the sun and moon.

Time lag: delay in response to a change.

Tolerance: the relative capability of an organism to endure an unfavorable environmental factor; the amount of a chemical considered safe on any food to be eaten by man or animals. Also see **pesticide tolerance.**

Topography: the configuration of a surface area including its relief, or relative elevations, and the position of its natural and manmade features. The study or description of the physical features of the earth's surface. The physical features of a district or region, such as are represented on maps, taken collectively; especially, the relief and contour of the land.

Topsoil: the fertile, dark colored surface soil, or A horizon.

Toxicant: a substance that kills or injures an organism through its chemical or physical action or by altering its environment; for example, cyanides, phenols, pesticides, or heavy metals. Especially used for insect control.

Toxicity: the quality or degree of being poisonous or harmful to plant or animal life.

Toxic pollutants: a combination of pollutants, including disease-carrying agents, that after discharge and upon exposure, ingestion, inhalation, or assimilation into any organism can cause death or malfunctions in such organisms or their offspring.

Toxin: a poison produced by an organism.

Trace element: a chemical element required in extremely small quantities by a microorganism.

Trace metals: metals found in small quantities or traces, usually because of their insolubility.

Transection sampling: samples collected at defined intervals along an imaginary line connecting two points.

Transit time: in nutrient cycles, average time that a substance remains in a particular form; ratio of biomass to productivity.

Translocation: switching of a segment of a chromosome to another chromosome.

Transpiration: the exhalation of water vapor by the leaves of plants.

Transpiration efficiency: grams of net primary production per 1000 g of water transpired by plants.

Trickling filter: a device for the biologic or secondary treatment of waste water consisting of a bed of rocks or stones that support bacterial growth. Sewage is trickled over the bed enabling the bacteria to break down organic wastes.

Triploid: referring to a cell or organism containing three sets of chromosomes (see **ploidy**).

Tritium: hydrogen-3; nucleus contains one proton plus two neutrons; radioactive.

Trophic: pertaining to food or nutrition.

Trophic level: the level of an organism in a community based on its feeding requirements. Phytoplankton are the first level; zooplankton are at the second. Position in the food chain determined by the number of energy transfer steps to that level: 1, producer; 2, herbivore; 3, 4, 5, carnivore.

Trophic structure: organization of the community based on feeding relationships among its species.

Trophogenic region: the superficial layer of a lake in which organic production from mineral substances takes place on the basis of light energy.

Tropholytic region: the deep layer of the lake where organic dissimilation predominates because of light deficiency.

Troposphere: the layer of the atmosphere extending 7 to 10 miles above the earth. Vital to life on earth, it contains clouds and moisture that reach earth as rain or snow.

Tsunami: this Japanese word, meaning "storm wave," refers to a giant ocean wave produced by a seismic disturbance beneath the ocean floor. The wave is sometimes mistakenly called a tidal wave, which actually results from the pull of the sun and moon. A long-period wave caused by a submarine earthquake, slumping, or volcanic eruptions. Tsunamis have heights of only a few centimeters in deep water but can reach several tens of meters before they break on the beach.

Tundra: one of the level or undulating treeless plains characteristic of artic regions, having a black muck soil and permanently frozen subsoil; vast ecologic system of the treeless plains of the Arctic regions or high mountains above the timber line.

Turbidite: a generally coarse-grained sediment that is deposited from a turbidity current. These sediments are generally found imbedded with the fine-grained muds typical of the deep sea.

Turbidity current: a turbid relatively dense current composed of water and sediments that flows downslope through less dense sea water. The sediment eventually settles out, forming a turbidite.

Turbidimeter: a device used to measure the amount of suspended solids in a liquid.

Turbidity: cloudiness caused by sediment suspended in water. Rivers are generally at their most turbid following a rainstorm, when extraordinary amounts of sediment carried into the stream by the runoff have not yet settled. A thick, hazy condition of air caused by the presence of particulates or other pollutants, or the similar cloudy condition in water caused by the suspension of silt or finely divided organic matter.

Turbulence: a flow of water in which the motion of individual particles appears irregular and confused.

Turificidae: aquatic segmented worms that exhibit marked population increases in aquatic environments containing organic decomposable wastes.

UTM: Universal Transverse Mercator coordinate system, a metric system that covers the entire earth. The coordinates for a unit of land are expressed in kilometers north and east from a fixed reference point in one of 15 UTM zones. Thus, a square of land enclosed by incremental UTM coordinates (between 5355 and 5356 North and between 243 and 244 East) is 1 km on a side and contains an area of 100 hectares.

Ultimate factors: aspects of environment that are directly important to the wellbeing of an organism (for example, food).

Underflow: the downstream movement of ground water through permeable rock beneath a riverbed.

Understory species: species that grow under the top stratum of a community.

Undertow: a supposed undersurface flow return of surface wave water taking place after the wave has broken on the beach. These returning waters are now thought to be concentrated into definite surface currents and are called rip current (which see).

Upwelling: the movement of water from depth to the surface; vertical movement of water currents, usually near coasts and driven by onshore winds, that brings nutrients from the depths of the ocean to surface layers; light surface water transported away from a coast (by action of winds parallel to it) and replaced near the coast by heavier subsurface water.

Uranium: heaviest natural element, dark grey metal; isotopes 233 and 235 are fissile, 238 is fertile; an alpha emitter.

Urban: pertaining to city areas.

Urban runoff: storm water from city streets and gutters that usually contains a great deal of litter and organic and bacterial wastes.

Vaccine: a culture of a pathogenic microorganisms in which the ability to cause disease has been eradicated by killing or attenuation of the cells. Inoculation of the treated cells provides immunity from the disease.

Vadose water: suspended water. A term used to designate subsurface water above the zone of saturation in the zone of aeration.

Vagrant animal: an animal that customarily moves about by its own volition, either continuously or intermittently.

Vagrant benthos: bottom-dwelling organisms that are capable of movement on, in, or above the substratum.

van der Waals' forces: the weak attraction exerted by all atoms on one another, resulting from the mutual interaction of the electrons and nuclei of the atoms. Sometimes called the van der Waals' bond.

Vapor: the gaseous phase of substances that normally are either liquids or solids at atmospheric temperature and pressure; for example, steam and phenolic compounds.

Vapor plume: the stack effluent consisting of flue gas made visible by condensed water droplets or mist.

Vaporization: the change of a substance from the liquid to the gaseous state. One of three basic contributing factors to air pollution; the others are attrition and combustion.

Variance: sanction granted by a governing body for delay or exception in the application of a given law, ordinance, or regulation.

Vector: disease vector: a carrier, usually an arthropod, that is capable of transmitting a pathogen from one organism to another.

Velocity: the rate of motion in a given time, e.g., 50 km/hr.

Ventifact: a general term for any stone or pebble shaped, worn, faceted, cut, or polished by the abrasive or sandblast action of wind-blown sand, generally under desert conditions; e.g., a dreikanter.

Versatile: adaptable to many uses or functions.

Vinyl chloride (VC): a gaseous chemical suspected of causing angiosarcoma, a rare form of cancer of the liver.

Virion: a viral cell.

Volatile: evaporating readily at a relatively low temperature.

Volatile solids: percent weight loss on ignition of the volatile organics contained in a sample.

Volcanic rocks: rocks that have resulted from volcanic eruptions.

Wandering: the compound movement of sweeping meanders in a swinging meander belt.

Waste: Also see **solid waste.** (1) bulky waste: items whose large size precludes or complicates their handling by normal collection, processing, or disposal methods. (2) Construction and demolition waste: building materials and rubble resulting from construction, remodeling, repair, and demolition operations. (3) Hazardous waste: wastes that require special handling to avoid illness or injury to persons or damage to property. (4) Special waste: waste that requires extraordinary management. (5) Wood pulp waste: wood or paper fiber residue resulting from a manufacturing process. (6) Yard waste: plant clippings, prunings, and other discarded material from yards and gardens. Also known as yard rubbish.

Waste water: water carrying wastes from homes, businesses, and industries that is a mixture of water and dissolved or suspended solids.

Wasting disease: a disease of eel gross (*Zostera marina*) often producing epidemics caused by the slime mold *Labyrinthula* sp.

Water content: water contained in porous sediment or sedimentary rock, generally expressed as a ratio of water weight to dry sediment weight.

Watercourse: a stream of water; a river or brook. A natural channel for water; also a canal for the conveyance of water, especially in draining lands.

Water cycle: hydrologic cycle (which see).

Water, film: water held tenaciously by the soil particles, not free to move in the interstices.

Water, interstitial: water that exists in the interstices or voids in a rock, or other porous medium.

Water, juvenile: water from the interior of the earth that is new or that has never been a part of the general system of ground-water circulation, e.g., magmatic water.

Water pollution: the addition of sewage, industrial wastes, or other harmful or objectionable material to water in concentrations or quantities sufficient to result in measurable degradation of water quality.

Water quality criteria: the levels of pollutants that affect the suitability of water for a given use. Generally, water use classification includes public water supply, recreational use, propagation of fish and other aquatic life, agricultural use; and industrial use. A scientific requirement on which a decision or judgment may be based concerning the suitability of water quality to support a designated use.

Water quality standard: a plan for water quality management containing four major elements: the use (recreation, drinking water, fish and wildlife propagation, industrial, or agricultural) to be made of the water; criteria to protect those uses; implementation plans (for needed industrial–municipal waste treatment improvements) and enforcement plans, and an antidegradation statement to protect existing high-quality waters. A plan that is established by governmental authority as a program for water pollution prevention and abatement. Usually it includes a delineation of designated water uses, criteria, and pollution abatement, implementation, and surveillance programs.

Watershed: the area contained within a drainage divide above a specified point on a stream; the area drained by a given stream.

Water supply system: the system for the collection, treatment, storage, and distribution of potable water from the sources of supply to the consumer.

Water table: the level to which ground water rises or the surface of the zone of saturation. The upper surface of a zone of saturation except where that surface is formed by an impermeable body. Locus of points in soil water at which the pressure is equal to atmospheric pressure.

Wave attenuation: the decrease in the wave form or height with distance from its origin.

Wave crest: the highest part of a wave.

Wavelength: the horizontal distance between two wave crests (or similar points on the wave form) measured parallel to the direction of travel of the wave.

Wave period: the time required for successive wave crests to pass by a fixed point.

Wave refraction: the change in direction of waves that occurs when one portion of the wave reaches shallow water and is slowed down while the other portion is in deep water and moving relatively fast.

Wave trough: the lowest part of a wave between two successive crests.

Wave velocity: the velocity or speed at which the wave form proceeds. It is equal to the wavelength divided by the wave period.

Weather meteor: (1) the state of the atmosphere, defined by measurement of the six meteorologic elements, viz., air temperature, barometric pressure, wind velocity, humidity, clouds, and precipitation. (2) Geology: to undergo or endure the action of the atmosphere; to suffer meteorologic influences.

Weir: an obstruction placed in a stream, diverting the water through a prepared opening for measuring or controlling the rate of flow.

Wetlands: swamps or marshes, especially as areas preserved for wildlife.

Withdrawal use: the use of surface or underground water that is later returned to the hydrologic cycle, although not necessarily to the same place. An example is most water piped into homes and industrial plants.

Within-habitat specialization: restriction of the ways in which individuals use a habitat.

Woodlot: a small area of land occupied by trees.

Work: physical or mental effort exerted to do or make something; a force exerted for a distance.

$\bar{X}$: mean; average.

Xeric: a dry environment with low available moisture. Characteristic species are called xerophytic.

Xylophagous: refers to organisms that consume wood.

Yeasts: a group of fungi that produce characteristic budding cells rather than hyphae.

Yield: (lake) (1) the amount of water that can be taken continuously from a lake for any economic purpose. (2) The amount of organic matter (plant and animal) produced by a lake, either naturally or under management.

Yield ratio: ratio of energy yielded to high-quality energy invested, where both are expressed in fossil fuel equivalents.

Zeolite: a particular type of authigenic mineral.

Zonation: stratigraphy: the condition of being arranged or distributed in bands or zones, generally more or less parallel to the bedding.

Zone of aeration: a subsurface zone containing water under pressure less than that of the atmosphere, including water held by capillarity, and containing air or gases generally under atmospheric pressure. This zone is limited above by the land surface and below by the surface of the zone of saturation, i.e., the water table.

Zone of capillarity: an area that overlies the zone of saturation and contains capillary voids, some or all of which are filled with water that is held above the zone of saturation by molecular attraction acting against gravity.

Zone of saturation: a subsurface zone in which all the interstices are filled with water under pressure greater than that of the atmosphere. Although the zone may contain gas-filled interstices or interstices filled with fluids other than water, it is still considered saturated. This zone is separated from the zone of aeration (above) by the water table. Synonyms: saturated zone, phreatic zone.

Zooecology: the branch of ecology concerned with the relationships between animals and their environment.

Zooplankton: the animal forms of plankton, e.g., jellyfish. Planktonic animals that supply food for fish; those plankton that are of the animal kingdom; animal microorganisms living unattached in water. They include small crustacea, such as daphnia and cyclops, and single-celled animals, such as protozoa. See **plankton.**

"Zero population growth" (ZPG): a theory advocating that there be no increase in population, that each person only replace her- or himself, and that birth control be practiced in all countries.

Index

Page numbers in *italic* type refer to illustrations.